Nucleic Acids and Molecular Biology

Volume 37

Series Editor

Allen W. Nicholson, Department of Biology, Temple University, Philadelphia, PA, USA

Since its inception in 1987, Nucleic Acids and Molecular Biology has emerged as a high-visibility series that brings to the scientific community in-depth analyses of cutting-edge topics in the field. Each volume offers a focused set of reviews, written by leading experts, that provides a comprehensive, multi-perspective analysis of the chosen topic. As such, Nucleic Acids and Molecular Biology serves as an enduring research resource for scientists, academicians, and students alike.

Tessa M. Burch-Smith
Editor

Chloroplast Gene Expression: Regulation, Stress Signaling and Biotechnology

Springer

Editor
Tessa M. Burch-Smith
Donald Danforth Plant Science Center
Saint Louis, MO, USA

ISSN 0933-1891 ISSN 1869-2486 (electronic)
Nucleic Acids and Molecular Biology
ISBN 978-3-031-70097-2 ISBN 978-3-031-70098-9 (eBook)
https://doi.org/10.1007/978-3-031-70098-9

This Springer imprint is published by the registered company Springer Nature Switzerland AG
The registered company address is: Gewerbestrasse 11, 6330 Cham, Switzerland

To Cassie and Savannah Plummer

Preface

Plastids are the descendants of cyanobacterial endosymbionts of ancient eukaryotes. They are well known for their contributions to plant growth and development through their primary metabolism that includes carbon fixation through photosynthesis and the production of biomolecules like amino acids. There is emerging and continued interest in the specialized metabolites of chloroplasts that are important for plant defense and adaptation to environmental niches, and which are the source of pharmaceuticals and other compounds that are critical for human consumption and industry. Understanding plastid function is therefore important for harnessing these organelles for advancing biotechnological applications that could address many challenges.

Chloroplasts, specialized plastids, are semi-autonomous organelles, encoding remnants of their prokaryotic genomes while most genes have been transferred to the eukaryotic nucleus of their hosts. Coordinated expression of the chloroplast genome and the nuclear genes encoding chloroplast proteins is therefore essential for proper chloroplast biogenesis and function. Over the last four decades, how the genetic information encoded by the chloroplast genome is converted to the chloroplast proteome has been intensely studied. It is now clear that the chloroplast uses prokaryotic mechanisms for transcription and translation but that nuclear proteins play major roles in the chloroplast gene expression.

This book discusses how the chloroplast genome is expressed, describing transcription and subsequent processing of the resulting mRNA by numerous nucleus-encoded RNA binding proteins. The translation of mature mRNA is also described, and how transcription and translation respond to environmental stimuli is also considered. The chloroplast is an important initiator of signaling to direct plant responses to the environment, and a description of how chloroplasts are involved in responses to stress is also presented. Also presented is a discussion of the biotechnology of chloroplasts and the potential for engineering these organelles for the production of high-value compounds or novel growth capabilities.

This book is intended as a useful reference for our colleagues, and we especially hope that it will be helpful to students who are just beginning their study of these fascinating organelles.

Saint Louis, MO, USA Tessa M. Burch-Smith

Contents

About the Editor and Contributors

About the Editor

Tessa M. Burch-Smith is Associate Member and Principal Investigator at the Donald Danforth Plant Science Center, MO, USA. She completed her undergraduate studies at the University of the West Indies and then received her Ph.D. from Yale University under the mentorship of Dr. S. P. Dinesh-Kumar. Her thesis work in plant innate immunity concerned plant response to Tobacco mosaic virus, leading her to investigate plasmodesmata and intercellular communication with Dr. Patricia Zambryski at the University of California, Berkeley, for her postdoctoral work. Current research in her lab examines the connection between chloroplast signaling and intercellular communication with a focus on how signals from the chloroplast vary in response to changing chloroplast gene expression.

Contributors

Mohammad F. Azim Department of Biochemistry & Cellular and Molecular Biology, University of Tennessee, Knoxville, TN, USA

Donald Danforth Plant Science Center, St. Louis, MO, USA

Meng Chen Department of Botany and Plant Sciences, Institute for Integrative Genome Biology, University of California, Riverside, CA, USA

Xuemei Chen State Key Laboratory for Protein and Plant Gene Research, Peking-Tsinghua Joint Center for Life Sciences, School of Life Sciences and Beijing Advanced Center of RNA Biology (BEACON) Peking University, Beijing, China

Tessa M. Burch-Smith Donald Danforth Plant Science Center, St. Louis, MO, USA

Nora Flynn Department of Botany and Plant Sciences, University of California, Riverside, USA

Kamel Hammani Institut de Biologie Moléculaire des Plantes, Centre National de la Recherche Scientifique (CNRS), Université de Strasbourg France, Strasbourg, France

Matthew D. Lemke The School of Plant Sciences, University of Arizona, Tucson, AZ, USA

Ansul Lokdarshi Department of Biology, Valdosta State University, Valdosta, GA, USA

Tyra N. McCray 10x Genomics, Pleasanton, CA, USA

Alessandro Occhialini Department of Plant Sciences and Center for Agricultural Synthetic Biology (CASB), University of Tennessee, Knoxville, TN, USA

Snigdha Rai The School of Plant Sciences, University of Arizona, Tucson, AZ, USA

David W. Tano The School of Plant Sciences, University of Arizona, Tucson, AZ, USA

Jesse D. Woodson The School of Plant Sciences, University of Arizona, Tucson, AZ, USA

Andrea A. Zanini Donald Danforth Plant Science Center, St. Louis, MO, USA

Abbreviations

ABA	abscisic acid
aSD	anti-Shine-Dalgarno
GUS	β-glucuronidase
HR	homologous recombination
IR	inverted repeats
JA	jasmonic acid
LSC	large single copy region
MORF	multiple organellar RNA editing factor
mTERF	mitochondrial transcription termination factor
NEP	nuclear-encoded RNA polymerase
NTP	nucleoside triphosphate
ORRM	organelle RNA recognition motif protein
PCD	programmed cell death
PEP	plastid-encoded RNA polymerase
PS	photosystem
PNPase	polynucleotide phosphorylase
PPR	pentatricopeptide repeat
RBP	RNA-binding protein
PSRP	plastid-specific ribosomal proteins
RH	RNA helicase
RIP	RNA-editing factor interacting protein
RNase	ribonuclease
ROS	reactive oxygen species
SD	Shine-Dalgarno
SSC	small single copy region
Ub	ubiquitin
UTR	untranslated region

Chapter 1
Plastid Transcription: A Major Regulatory Point in Chloroplast Biogenesis

Nora Flynn, Xuemei Chen, and Meng Chen

Abstract Plastids are endosymbionts that retain their own, minimal genome that must be transcribed. This genome is known as the plastome. The host cell nucleus exerts tight control over the plastome to ensure the accurate timing of key developmental events, such as chloroplast biogenesis. Therefore, communication from the nucleus to the plastid, or anterograde signaling, is a vital step in plastome expression. Regulation of plastid transcription by the nucleus is tied to two DNA-dependent RNA polymerases that transcribe the plastome: the nuclear-encoded polymerase (NEP) and the plastid-encoded polymerase (PEP), which contains numerous nuclear-encoded PEP-associated proteins (PAPs). This chapter will cover the fundamentals and regulation of plastome transcription, with a focus on the role of plastid transcription in chloroplast biogenesis. In particular, the chapter will highlight the regulatory strategy of the nucleus to comprehensively control plastome gene expression through the assembly and activation of the PEP supercomplex, which is coordinated by both environmental and developmental factors.

Keywords Chloroplast · Organelle gene expression · Plastid-encoded polymerase (PEP) · PEP-associated proteins (PAP) · Sigma factors · Nuclear-encoded polymerase (NEP) · Anterograde signaling · Phytochrome · Phytochrome interacting factors (PIF)

N. Flynn · M. Chen (✉)
Department of Botany and Plant Sciences, Institute for Integrative Genome Biology, University of California, Riverside, CA, USA
e-mail: meng.chen@ucr.edu

X. Chen (✉)
National Key Laboratory for Protein and Plant Gene Research, Peking-Tsinghua Joint Center for Life Sciences, School of Life Sciences, Peking University, Beijing, China

Beijing Advanced Center of RNA Biology (BEACON), Peking University, Beijing, China
e-mail: xuemei.chen@pku.edu.cn

T. M. Burch-Smith (ed.), *Chloroplast Gene Expression: Regulation, Stress Signaling and Biotechnology*, Nucleic Acids and Molecular Biology 37, https://doi.org/10.1007/978-3-031-70098-9_1

1.1 Becoming Green: Not as Easy as it Seems

Emerging from its seed, a plant activates a sophisticated molecular network to develop green leaves. The greening of plants is due to the differentiation of a special plastid, the chloroplast. As a photosynthetic powerhouse, a factory for diverse chemical compounds, and an intricate signaling hub, the chloroplast is a critical organelle with a unique origin. The chloroplast arose from an endosymbiotic event that occurred over a billion years ago, where a photosynthetic, cyanobacterial-like ancestor was taken up by a eukaryotic cell (Mereschkowsky 1905; Sagan 1967; Martin and Kowallik 1999; Gould et al. 2008). Over evolutionary time, the cyanobacterial ancestor endosymbiont evolved into the plastids we see today, of which the chloroplast is only one of multiple specialized types (Kirk and Tilney-Bassett 1978).

As descendants of cyanobacteria, plastids maintain their own genomes (Ris and Plaut 1962; Sager and Ishida 1963), complete with transcriptional and translational activity (Kirk 1964; Zoschke and Bock 2018). However, this genome is subjected to nuclear control by the host cell, accomplished partially through the large-scale the biosynthesis of the photoprotection information to the nucleus (Arabidopsis Genome Initiative 2000; Kleine et al. 2009). Thus, a free-roaming bacterium transitioned to the modern-day chloroplast, with a minimal genome, known as the plastome, that contains less than 10% of the cyanobacterial genome (Martin and Herrmann 1998; Martin et al. 2002). With so many of its original genes transferred to the nucleus, and newly-evolved proteins targeted to the plastid, the chloroplast depends on protein import to obtain a majority of plastid-localized proteins (Abdallah et al. 2000; Kessler and Schnell 2009). Therefore, the chloroplast is genetically semi-autonomous, with all basic functions, such as transcription, translation, and photosynthesis, dependent on nuclear gene expression.

Similarly, chloroplast biogenesis during greening requires the interplay of both photosynthesis-associated nuclear genes (PhANGs) and plastome-encoded photosynthesis-associated genes (PhAPGs). For the purposes of this chapter, PhANGs are defined as a set of ~150 genes that were classified by Hwang et al. 2022, which directly relate to light harvesting complexes, chlorophyll biosynthetic enzymes, photosystem I and II complexes, electron transport, ATP synthase, and the Calvin cycle. During chloroplast biogenesis, PhANGs and PhAPGs are closely coordinated through signaling between the nucleus and the plastid, leading to the frequent synchronization of their expression. For this reason, PhANGs and PhAPGs are typically considered as coupled (Nott et al. 2006). The mirroring of the two genomes can be observed by applying chemical treatments to the chloroplast, such as lincomycin or norflurazon, to inhibit plastid translation or block the biosynthesis of the photoprotection pigment, carotenoid, respectively (Oelmüller 1989; Susek et al. 1993; Koussevitzky et al. 2007). By greatly disturbing chloroplast function and PhAPG transcription, PhANG transcription decreases in turn (Nott et al. 2006; Ruckle et al. 2012), demonstrating the dynamic communication between the plastid and the nucleus.

To achieve this tight coordination between the nucleus and the plastid, there must be an array of regulatory signals. From the plastid, retrograde signals update the nucleus on chloroplast functioning (Nott et al. 2006; de Souza et al. 2017). Meanwhile, from the nucleus, anterograde signals exert extensive influence on chloroplast activity, especially by controlling plastome transcription (Yoo et al. 2020). Compared to retrograde signals, determining the identity of the anterograde signals has proved difficult considering the sheer number of nuclear-encoded proteins essential for chloroplast functions. Nonetheless, in recent years, promising advances in this challenging field have begun to reveal some possible nucleus-to-plastid anterograde mechanisms to control plastid transcription.

This chapter will consider both a plastid and a nuclear perspective to review the current theories surrounding the regulation of plastome transcription for accurate timing of chloroplast biogenesis. We will begin by exploring the plastome and the genes that remain there, then will pivot to discuss the DNA-dependent RNA polymerases involved in transcribing these genes. Finally, we will delve into the regulation of chloroplast transcription and biogenesis with an emphasis on how these processes are influenced by light via photoreceptor signaling. In particular, we will highlight an important contrast between the strategies of transcriptional control in the nucleus and plastid: while the nucleus employs a gene-by-gene-based regulatory system, the plastid instead utilizes polymerase activity itself as a key point of control in plastome expression.

As a fascinating chimeric system, the chloroplast retains both its original bacterial machinery as well as learned eukaryotic methods to achieve regulated gene expression, creating a complex transcriptional landscape despite its overwhelmingly simple genome. This chapter will only focus on the importance of transcriptional regulation in the plastid, particularly during chloroplast biogenesis. However, it is worthwhile to note that post-transcriptional and post-translational processes are also critically important regulatory steps in plastid gene expression. For a review of these topics, we suggest a number of reviews (Stern et al. 2010; Barkan and Small 2014; Nawaz and Kang 2017; Zhao et al. 2020; Zeng et al. 2022; Zhang et al. 2023) and Chaps. 2 and 3 of this volume.

1.2 The Plastome: A Brief Review of Structure and Contents

Reflecting its small size, the plastome was one of the first genomes to be fully sequenced using first-generation sequencing approaches. The *Marchantia polymorpha* (liverwort) and *Nicotiana tabacum* (tobacco) plastomes were fully sequenced in the 1980s (Ohyama et al. 1986; Shinozaki et al. 1986) and were followed by the Arabidopsis plastome roughly a decade later (Sato et al. 1999). Generally, the plastome contains 100–200 genes, including rRNA, tRNA, and protein-coding genes. The plastid DNA displays a circular, quadripartite structure that is shared among both plant and algal lineages, with two single-copy regions and two inverted repeats, though there are some exceptions to this rule (Turmel et al. 2002; Wicke et al. 2011). The inverted repeats provide duplicated regions to increase the transcription of highly abundant genes, including rRNA. Meanwhile, the two single-copy regions encode genes associated with photosynthesis, carbon assimilation, and housekeeping. These two single-copy regions are distinguished as the large single-copy region, which houses most of the photosynthesis-associated protein-coding genes, and the small single-copy region, which, at least for most land plants, is associated with *ndh* genes for the NAD(P)H dehydrogenase complex, a mediator of electron transport (Peltier et al. 2016; Shikanai 2016). However, *ndh* genes have been lost multiple times, such as in *Chlamydomonas* or in the highly-reduced plastomes of parasitic plants such as *Epifagus* (dePamphilis and Palmer 1990; Martín and Sabater 2010; Wicke et al. 2011). Finally, both regions encode numerous tRNAs to support plastid translation, including around 35 tRNA genes that represent all 20 amino acids (Sugiura et al. 1989; Sato et al. 1999; Mohanta et al. 2023).

Similar to bacterial genomes, the plastome is organized in defined operons, with the vast majority of plastid genes transcribed as polycistronic units (Sugita and Sugiura 1996; Shahar et al. 2019; Castandet et al. 2019). For example, the plastome encodes core genes for the two photosystems, photosystem I (designated as *psa* genes) and photosystem II (designated as *psb* genes), with each having unique light absorption spectra during photosynthesis (Wollman 2001). Many photosystem-related genes are transcribed as polycistronic units, such as *psbD-psbC*, *psaA-psaB*, and *psbE-psbF-psbL-psbJ* (Sugita and Sugiura 1996). Similarly, many *ndh* genes locate to an operon with *psaC*, (*ndhD-psaC-ndhE-ndhG-ndhI-ndhA-ndhH*) (Kanno and Hirai 1993) and rRNA genes form an operon with some tRNAs (*rrn16-trnI-trnA-rrn23-rrn4.5-rrn5*) (Strittmatter et al. 1985; Sugita and Sugiura 1996; Shahar et al. 2019). Only a minority of genes, such as the large subunit of Rubisco (*rbcL*), the critical reaction center D1 protein of photosystem II (*psbA*), and many tRNAs, are considered monocistronic (Sugita and Sugiura 1996; Shahar et al. 2019).

1.3 What Transcribes the Plastome?

On its own, the plastome cannot fully achieve transcription. The chloroplast relies on the nucleus to provide many necessary factors for plastome transcription. For this reason, plastid protein import and protein complex assembly mark critical points of plastid transcriptional regulation.

What enzymes are involved in the transcription of the plastome? There are two DNA-dependent RNA polymerases (RNAPs) that transcribe the plastome—a bacterial-type, multi-subunit, plastid-encoded RNA polymerase (PEP) and a phage-type, single-subunit, nuclear-encoded RNA polymerase (NEP) (Liere and Börner 2007). The next sections will explore the discovery, structure, and activity of each of the polymerases.

1.3.1 PEP

1.3.1.1 Discovery of PEP

The endogenous PEP enzyme was first partially purified from maize (Bottomley et al. 1971; Smith and Bogorad 1974), wheat (Polya and Jagendorf 1971a, b), and the single-celled microorganism *Euglena gracilis* (Hallick et al. 1976). PEP was shown to be a stable, multisubunit complex (Bottomley et al. 1971; Smith and Bogorad 1974). When purified, the PEP transcription complex remained bound to chloroplast DNA and was able to transcribe this associated DNA. Therefore, the purified DNA-RNA-protein complexes containing the PEP were termed plastid Transcriptionally Active Chromosome (pTAC) (Hallick et al. 1976; Briat et al. 1979).

Biochemical analysis demonstrated that the purified PEP contains core bacterial subunits and therefore represents a bacterial-type plastid-encoded RNAP (Little and Hallick 1988; Hu and Bogorad 1990; Hu et al. 1991). Sequencing of the plastome revealed the genes homologous to the bacterial α, β, and β′ subunits: *rpoA, rpoB, and rpoC*, respectively. However, in plastids, the β′ subunit is split into β′ and β″, which are encoded by *rpoC1* and *rpoC2* (Ohyama et al. 1986; Shinozaki et al. 1986)—a feature shared by the ancestral cyanobacterial RNAP (Bergsland and Haselkorn 1991). In terms of organization, for most plant and algal lineages, *rpoB*, *rpoC1*, and *rpoC2* form an operon, while *rpoA* is part of a different operon that includes genes encoding ribosomal proteins (Liere and Börner 2007). Exceptions to this rule do exist, such as for *Chlamydomonas*, which displays fragmentation and rearrangement of these genes (Smith and Purton 2002). Despite these exceptions, overall, the core subunits of PEP are widely conserved among members of the green lineage, particularly the sequences of the catalytic domains of β, β′ and β″ (Sutherland and Murakami 2018; Ruedas et al. 2022). The α subunit is more involved in complex assembly and promoter binding and is less conserved than the catalytic domains.

The PEP core rarely exists or functions on its own. More commonly, PEP exists as part of a much larger supercomplex. Mass spectrometry (MS)-based proteomic analyses on the isolated pTAC complex from tobacco (*Nicotiana tabacum*) (Suzuki et al. 2004; Wu et al. 2024), Arabidopsis (Pfalz et al. 2006), and mustard (Steiner et al. 2011; Ruedas et al. 2022; Vergara-Cruces et al. 2024; do Prado et al. 2024) eventually identified fifteen nucleus-encoded accessory proteins—named PEP-associated proteins (PAPs) (Steiner et al. 2011). Many of the PAPs follow a dual naming convention due to their progressive identification. The Arabidopsis PAPs and their common identifiers are as follows: PAP1/pTAC3, PAP2/pTAC2, PAP3/pTAC10, PAP4/FSD3, PAP5/pTAC12/HEMERA, PAP6/FLN1, PAP7/pTAC14, PAP8/pTAC6, PAP9/FSD2, PAP10/TrxZ, PAP11/MurE-like, PAP12/pTAC7/ω, PAP13/FLN2, PAP14/pTAC18 and PAP15/PRIN2 (Table 1.1).

The PAPs are mostly eukaryotic additions that are not present in the polymerases of bacteria. In *Chlamydomonas*, many PAP orthologues are similarly not found and the PAPs that are present might not associate with PEP due to the lack of conservation

Table 1.1 Constituents of the PEP supercomplex

Component	Accession	Phenotype	Potential Interactors
Core			
rpoA/α	ATCG00740	Albino (De Santis-MacIossek et al. 1999)	**PAP5** (Ruedas et al. 2022; Wu et al. 2024) **PAP6** (Ruedas et al. 2022; Wu et al. 2024) **PAP13** (Wu et al. 2024)
rpoB/β	ATCG00190	Albino (Allison et al. 1996; De Santis-MacIossek et al. 1999)	**RpoC1** (Ruedas et al. 2022; Vergara-Cruces et al. 2024; Wu et al. 2024) **RpoC2** (Wu et al. 2024) **PAP3** (Wu et al. 2024) **PAP5** (Vergara-Cruces et al. 2024) **PAP1** (Wu et al. 2024)
rpoC1/β′	ATCG00180	Albino (De Santis-MacIossek et al. 1999)	**RpoB** (Ruedas et al. 2022; Vergara-Cruces et al. 2024; Wu et al. 2024) **RpoC2** (Wu et al. 2024) **PAP5** (Ruedas et al. 2022; Vergara-Cruces et al. 2024; Wu et al. 2024) **PAP14** (Vergara-Cruces et al. 2024) **PAP13** (Wu et al. 2024) **PAP1** (Vergara-Cruces et al. 2024; Wu et al. 2024)
rpoC2/β″	ATCG00170	Untested, assumed albino	**RpoB** (Wu et al. 2024) **RpoC1** (Wu et al. 2024) **PAP3** (Wu et al. 2024) **PAP5** (Wu et al. 2024) **PAP11** (Vergara-Cruces et al. 2024) **PAP9** (Vergara-Cruces et al. 2024; Wu et al. 2024) **PAP1** (Wu et al. 2024) **PAP2** (Wu et al. 2024)
PAP12/ω/pTAC7	AT5G24314	Albino/ivory (Pfalz and Pfannschmidt 2013; Yu et al. 2013)	**PAP3** (Yu et al. 2013; Chang et al. 2017) **PAP5** (Yu et al. 2013) **PAP7** (Yu et al. 2013) **PAP6** (Yu et al. 2013) **PAP1** (Vergara-Cruces et al. 2024)
Scaffold Module			
PAP3/pTAC10	AT3G48500	Albino/ivory (Steiner et al. 2011; Jeon et al. 2012; Chang et al. 2017)	**RpoB** (Wu et al. 2024) **RpoC2** (Wu et al. 2024) **PAP12** (Yu et al. 2013; Chang et al. 2017) **PAP5** (Wu et al. 2024) **PAP7** (Chang et al. 2017) **PAP4** (Vergara-Cruces et al. 2024; Wu et al. 2024) **PAP9** (Wu et al. 2024) **PAP6** (Wu et al. 2024) **PAP10** (Chang et al. 2017)
PAP5/pTAC12/HMR	AT2G34640	Albino/ivory (Pfalz et al. 2006a, 2015; Chen et al. 2010; Kendrick et al. 2022)	**RpoA** (Ruedas et al. 2022; Wu et al. 2024) **RpoB** (Vergara-Cruces et al. 2024) **RpoC1** (Ruedas et al. 2022; Vergara-Cruces et al. 2024; Wu et al. 2024) **RpoC2** (Wu et al. 2024) **PAP12** (Yu et al. 2013) **PAP3** (Wu et al. 2024)

in key interface residues (Pfalz and Pfannschmidt 2013; Díaz et al. 2018; Wu et al. 2024; do Prado et al. 2024). PAPs more frequently emerged in charophytes, the sister group of land plants, but the full PEP-PAP supercomplex evolved with land plants and most land plants have homologs of the PAPs (Wu et al. 2024). Therefore, PAPs evolved over time to become an important piece of the plastid transcriptional machinery in land plants, providing an additional layer of nuclear control over the plastid.

Tab. 1.1 (continued)

			PAP7 (Gao et al. 2011; Song et al. 2023) **PAP8** (Liebers et al. 2020; Vergara-Cruces et al. 2024) **PAP6** (Ruedas et al. 2022)
PAP7/ pTAC14	AT4G20130	Albino/ivory (Gao et al. 2011; Grübler et al. 2017)	**PAP12** (Yu et al. 2013) **PAP3** (Chang et al. 2017) **PAP5** (Gao et al. 2011; Song et al. 2023)
PAP8/pTAC6	AT1G21600	Albino/ivory (Pfalz et al. 2006a)	**PAP5** (Liebers et al. 2020; Vergara-Cruces et al. 2024) **PAP1** (Vergara-Cruces et al. 2024)
PAP11/ MurE-like	AT1G63680	Albino (Garcia et al. 2008; Kendrick et al. 2022)	**RpoC2** (Vergara-Cruces et al. 2024) **PAP13** (Vergara-Cruces et al. 2024) **PAP1** (Ruedas et al. 2022) **PAP2** (Wu et al. 2024)
PAP14/ pTAC18	AT2G32180	Untested, possible WT-like phenotype (Pfalz and Pfannschmidt 2013)	**RpoC1** (Vergara-Cruces et al. 2024) **PAP9** (Vergara-Cruces et al. 2024; Wu et al. 2024)
Protection Module			
PAP4/FSD3	AT5G23310	Pale, but *fsd2 fsd3* is albino (Myouga et al. 2008)	**PAP3** (Chang et al. 2017; Vergara-Cruces et al. 2024; Wu et al. 2024) **PAP9** (Myouga et al. 2008)
PAP9/FSD2	AT5G51100	Pale, but *fsd2 fsd3* is albino[1] (Myouga et al. 2008)	**RpoC2** (Vergara-Cruces et al. 2024; Wu et al. 2024) **PAP3** (Chang et al. 2017; Wu et al. 2024) **PAP14** (Vergara-Cruces et al. 2024; Wu et al. 2024) **PAP4** (Myouga et al. 2008)
Regulation Module			
PAP6/FLN1	AT3G54090	Albino/ivory (Arsova et al. 2010; Steiner et al. 2011; Gilkerson et al. 2012)	**RpoA** (Ruedas et al. 2022; Wu et al. 2024) **PAP12** (Yu et al. 2013) **PAP3** (Wu et al. 2024) **PAP5** (Ruedas et al. 2022) **PAP10** (Arsova et al. 2010; He et al. 2018; Wu et al. 2024) **PAP13** (Huang et al. 2013)
PAP10/TrxZ	AT3G06730	Albino/ivory (Arsova et al. 2010; Schröter et al. 2010)	**PAP3** (Chang et al. 2017) **PAP6** (He et al. 2018; Wu et al. 2024) **PAP13** (Arsova et al. 2010; Huang et al. 2013; He et al. 2018) **PAP15** (Díaz et al. 2018)
PAP13/FLN2	AT1G69200	Albino/ivory (Huang et al. 2013) or pale (Arsova et al. 2010; Gilkerson et al. 2012)	**RpoA** (Wu et al. 2024) **RpoC1** (Wu et al. 2024) **PAP6** (Huang et al. 2013) **PAP10** (Arsova et al. 2010; He et al. 2018) **PAP11** (Vergara-Cruces et al. 2024)
RNA/DNA Module			
PAP1/pTAC3	AT3G04260	Albino/ivory (Yagi et al. 2012)	**RpoB** (Wu et al. 2024) **RpoC1** (Vergara-Cruces et al. 2024; Wu et al. 2024) **RpoC2** (Wu et al. 2024) **PAP12** (Vergara-Cruces et al. 2024) **PAP8** (Vergara-Cruces et al. 2024) **PAP11** (Ruedas et al. 2022) **PAP2** (Ruedas et al. 2022; Wu et al. 2024)
PAP2/pTAC2	AT1G74850	Pale (Pfalz et al. 2006a)	**RpoC2** (Wu et al. 2024) **PAP11** (Wu et al. 2024) **PAP1** (Ruedas et al. 2022; Wu et al. 2024)
PAP15/ PRIN2	AT1G10522	Pale (Kindgren et al. 2012; Díaz et al. 2018)	**PAP10** (Díaz et al. 2018)

Components are colored by proposed functional module. Potential interactions only include interactions determined by crosslinking-MS or yeast two-hybrid assay. See Fig. 1.1 for additional interactions indicated by the cryo-EM structures

1.3.1.2 PEP Structure

Recent structural studies using cryo-electron microscopy (cryo-EM) revealed the complete structure of PEP complexes from white mustard and tobacco (Vergara-Cruces et al. 2024; Wu et al. 2024; do Prado et al. 2024). PEP is a 21-subunit

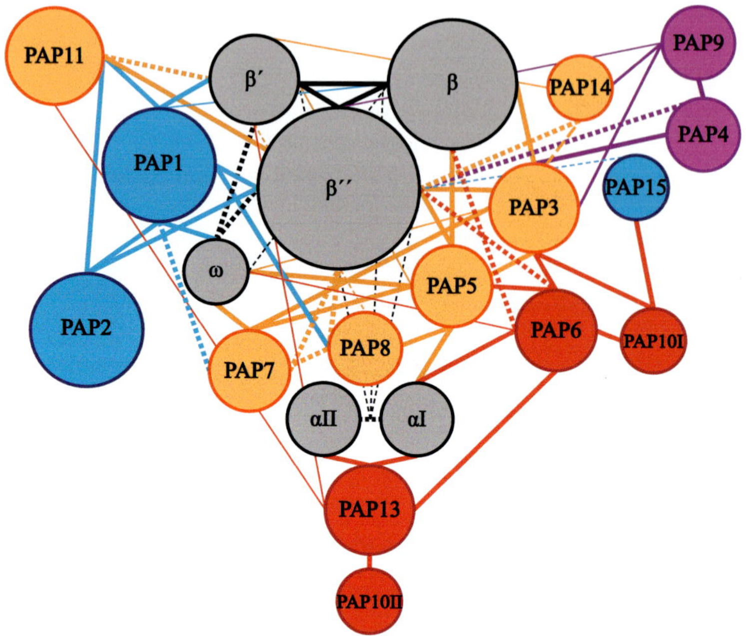

Fig. 1.1 The interaction network of the PEP supercomplex. PEP contains 21 subunits, 19 of which are unique. The colors correspond to the predicted functional modules in Table 1.1; the circle size is proportional to the size of the protein. Solid lines indicate that the interaction was confirmed by crosslinking-MS or yeast two-hybrid assay (Table 1.1). Dashed lines depict additional interactions that were found through cryo-EM (Vergara-Cruces et al. 2024; Wu et al. 2024; do Prado et al. 2024). Thinner lines pass behind at least one object, while thicker lines move directly between two objects

complex that contains six core subunits, $\alpha_2\beta\beta'\beta''$, which are similar in structure and sequence to bacterial RNAPs (Ruedas et al. 2022; Vergara-Cruces et al. 2024; Wu et al. 2024; do Prado et al. 2024). A notable difference between bacterial RNAPs and the PEP core is the ω subunit. The core bacterial RNAP is $\alpha_2\beta\beta'\omega$ (Zhang et al. 1999; Murakami et al. 2002; Vassylyev et al. 2002), where the ω subunit serves structural and functional roles (Patel et al. 2020). In the PEP, the nuclear-encoded PAP12 may be homologous to the ω subunit due to its similarity in folding and positioning within the PEP complex, as well as the conservation of key residues (Vergara-Cruces et al. 2024; Wu et al. 2024; do Prado et al. 2024). For the remainder of this chapter, PAP12 will be considered part of the PEP core. However, surprisingly, PAP12 is not present in all chlorophytes and charophytes, which raises some skepticism about whether it fully encompasses the roles of the ω subunit or if it has more limited roles to promote stability (Wu et al. 2024; do Prado et al. 2024).

The PEP core evolved to handle extensive interactions with PAPs. Besides PAP12, PEP includes fourteen other PAPs, with thirteen unique PAPs and two copies of the thioredoxin subunit, PAP10. These PAPs encircle the interconnected core subunits, with each core subunit in close association with multiple PAPs (Fig. 1.1 and Table 1.1). To support these numerous interactions with eukaryotic PAPs, the surface residues of the core are less conserved than the internal catalytic domains when compared with their bacterial counterparts (Ruedas et al. 2022). Further, PEP β″ contains a large lineage-specific region known as sequence insertion 3 (SI3) that may have roles in both protein interaction and polymerase activity (Lane and Darst 2010; Windgassen et al. 2014; Mazumder et al. 2020). In bacteria, SI3 is relatively short, with only a couple of sandwich-barrel hybrid-motif (SBHM) domains (Artsimovitch et al. 2003; Qayyum et al. 2024). Meanwhile, SI3 in cyanobacteria contains more SBHM domains (Qayyum et al. 2024), but is still approximately 100 residues shorter than the PEP SI3 (Vergara-Cruces et al. 2024; Wu et al. 2024; do Prado et al. 2024). The larger SI3 in PEP makes frequent contacts with PAPs, connecting the core with these essential eukaryotic proteins (Vergara-Cruces et al. 2024; Wu et al. 2024; do Prado et al. 2024). Overall, the PEP supercomplex forms a star-shape with five protruding arms created by the PAPs (Fig. 1.1) (Vergara-Cruces et al. 2024).

The positioning of PEP components is reflected in mutant phenotypes, with more peripheral PAPs demonstrating less severe phenotypes. The PEP core itself is critical for chloroplast development. Loss of any of the core subunits results in albino seedlings that require an external carbon source, such as sucrose, in order to survive (Allison et al. 1996; De Santis-Maclossek et al. 1999; Chateigner-Boutin et al. 2008; Pfannschmidt et al. 2015). Similarly, the loss of any of the PAPs results in a chloroplast-defective phenotype, though PAP14 remains untested (Table 1.1). Intermediate forms of the complex may be too unstable to achieve full PEP function, leading to the defects observed when even a single PAP is lost (Pfalz and Pfannschmidt 2013; Pfannschmidt et al. 2015). In particular, the PAPs that are most important for maintaining the integrity of the large complex demonstrate an albino phenotype upon their loss. These include PAP1, PAP3, PAP5, PAP6, PAP7, PAP8, PAP10, PAP11, PAP12, and, potentially, PAP13, many of which are also positioned close to the core (Pfalz et al. 2006; Garcia et al. 2008; Arsova et al. 2010; Schröter et al. 2010; Chen et al. 2010; Steiner et al. 2011; Gao et al. 2011; Jeon et al. 2012; Yagi et al. 2012; Gilkerson et al. 2012; Pfalz and Pfannschmidt 2013; Yu et al. 2013; Huang et al. 2013; Chang et al. 2017; Grübler et al. 2017; Vergara-Cruces et al. 2024; Wu et al. 2024; do Prado et al. 2024). Meanwhile, null mutants of *PAP2*, *PAP4*, *PAP9*, and *PAP15* are pale, implying that carotenoid biosynthesis remains intact. Following their slightly more mild phenotypes, these components are positioned towards the periphery of the complex, where they may function in transcriptional activity (Pfalz et al. 2006; Myouga et al. 2008; Steiner et al. 2011; Kindgren et al. 2012; Gilkerson et al. 2012; Pfalz and Pfannschmidt 2013; Díaz et al. 2018). For dark-grown plants, the *pap* mutants have no noticeable phenotype, suggesting that the addition of the PAPs serves as a switch for PEP transcription in light in some organisms (Ruedas et al. 2022).

The organization of the PAPs in the PEP complex can be divided into four modules based on their location and predicted functions (Steiner et al. 2011; Kindgren and Strand 2015; Vergara-Cruces et al. 2024; Wu et al. 2024; do Prado et al. 2024). First, the scaffold module (PAP3, PAP5, PAP7, PAP8, PAP11 and PAP14) contains the major PAPs for establishing PEP architecture. Second, the protection module

(PAP4 and PAP9) serves to shield the PEP from damaging reactive oxygen species. Third, the regulation module (PAP6, PAP10I, PAP10II, and PAP13) may be involved in redox-related processes. Finally, the RNA/DNA module (PAP1, PAP2, and PAP15) includes PAPs that may interact with nucleic acids during transcription. For a summary of the interactions of these modules and the core, see Fig. 1.1 and Table 1.1.

1.3.1.2.1 The Scaffold Module

The scaffold module serves as a docking site or bridge to connect the core with the other modules, with each PAP in the scaffold module fielding diverse interactions. Through its numerous interactions, the scaffold module is a major coordinator of PEP architecture, setting the correct positioning of PAPs around the core. However, the scaffolding PAPs could also have functions outside of PEP stability, as seen by their conserved domains. For example, PAP7 has a SET domain that is shared by enzymes with histone lysine methyltransferase function (Gao et al. 2011; do Prado et al. 2024). Meanwhile, PAP11 contains partially conserved residues for the substrate binding and catalytic activity of a Mur ligase (Garcia et al. 2008; do Prado et al. 2024). Whether these domains contribute to the activities of the PEP complex remains unknown and the specific roles of each of the PAPs requires further study.

1.3.1.2.2 The Protection Module

The protection module acts to defend the PEP complex from photosynthesis-related oxidation damage by converting precursors of reactive oxygen species to less-damaging forms. In support of this, both PAP4 and PAP9 belong to the iron superoxide dismutase family and loss of either protein leads to sensitivity to oxidative stress (Myouga et al. 2008; Steiner et al. 2011; Pfalz and Pfannschmidt 2013). In the PEP supercomplex, PAP4 and PAP9 form a heterodimer (Myouga et al. 2008) that is docked to the core through the scaffolding PAPs, PAP3 and PAP14 (Wu et al. 2024).

1.3.1.2.3 The Regulation Module

The regulation module may inform PEP activity through involvement in redox-related processes. PAP10 has a thioredoxin motif and disulfide-bond reductase activity, while PAP6 and PAP13 both have phosphofructokinase domains (Arsova et al. 2010; Chibani et al. 2011; Gilkerson et al. 2012; Wimmelbacher and Börnke 2014; Díaz et al. 2018). However, the mode of action of these components and the significance of these domains remain to be determined. For example, the albino phenotype of *pap10* could be rescued by a catalytically deficient PAP10, with slight chlorosis in the rescued plants (Wimmelbacher and Börnke 2014).

In the PEP supercomplex, two units of PAP10 (PAP10I and PAP10II) each form a heterodimer with PAP6 and PAP13, respectively (Arsova et al. 2010). The scaffolding PAP that is most involved in connecting the regulatory module with the core is PAP5. Specifically, PAP5 interacts with an α subunit to provide a bridge to attach the PAP6-PAP10I heterodimer to the PEP core, with the help of PAP3. On the other hand, the PAP13-PAP10II heterodimer is directly attached to the α subunits (Ruedas et al. 2022; Vergara-Cruces et al. 2024; Wu et al. 2024; do Prado et al. 2024). PAP6 might also make direct contact with an α subunit, which could have interesting mechanistic implications in the movement and controlled release of the α subunits during transcription (Vergara-Cruces et al. 2024; Wu et al. 2024).

Of note, PAP15, which is currently placed in the RNA/DNA module, plays a critical role in the redox regulation of PEP activity. The light-dependent reduction of an intermolecular disulfide bond converts PAP15 from a dimeric form to a monomeric form that is required for PEP activation (Kindgren et al. 2012; Díaz et al. 2018). Continued research on the specific functions of the PAPs within the PEP complex may result in assignment of PAPs to different or multiple modules.

1.3.1.2.4 The RNA/DNA Module

The RNA/DNA module is split among several arms of the star-shaped structure (Fig. 1.1). These PAPs are thought to interact with nucleic acids. PAP1 was shown to bind chloroplast DNA and contains both a SAP domain and pentatricopeptide repeat (PPR) motifs (Yagi et al. 2012; Wang et al. 2024). While the SAP domain might interact with entering DNA during recruitment of PEP, the PPR motifs may instead be more involved in the scaffolding of the complex itself (Vergara-Cruces et al. 2024; Wang et al. 2024). Similarly, PAP2 has an SMR domain and PPR motifs, but the SMR domain may be more involved in PAP-PAP interaction than nuclease activity. (Vergara-Cruces et al. 2024). However, the PPR motif of PAP2 is positioned to interact with nascent RNA (Vergara-Cruces et al. 2024; Wu et al. 2024). As for PAP15, it does not contain a known domain for binding nucleic acids, but it can bind to DNA *in vitro* and may assist the PEP complex with DNA binding (Kremnev and Strand 2014; Vergara-Cruces et al. 2024). As for the other modules, further research is required to determine the functions of these domains within the complex.

The scaffolding proteins PAP8 and PAP7 dock PAP1 and PAP2, creating a region of PEP that could specialize in interacting with nucleic acids (Vergara-Cruces et al. 2024; Wu et al. 2024; do Prado et al. 2024). PAP1 interacts with PAP2 to position PAP2's PPR domain near where the nascent RNA exits the complex. PAP2's PPR domain could have sequence-specific RNA-binding properties to bind a consensus sequence and direct polymerase pausing and downstream RNA processing events (Ruedas et al. 2022; Vergara-Cruces et al. 2024; Wu et al. 2024; do Prado et al. 2024). Therefore, together, the interaction between PAP1 and PAP2 endows the complex the ability to interact with entering DNA through PAP1's SAP domain and the ability to interact with nascent RNA through PAP2's PPR domain. As for PAP15, it binds directly to the β″-SI3 (Vergara-Cruces et al. 2024).

In addition, other PAPs outside of the RNA/DNA module could impact interactions with nucleic acids during transcription (Kindgren and Strand 2015; Pfannschmidt et al. 2015). For example, PAP3 contains an S1 domain with RNA-binding properties, but this domain might not be well-positioned in the PEP complex to interact with nucleic acids (Pfalz et al. 2006; Vergara-Cruces et al. 2024). Maize PAP5 has also demonstrated binding to single-stranded nucleic acids (Pfalz et al. 2015). Continued analysis of the activities of each of the PAPs is necessary to confirm their placement in each of these modules.

1.3.1.2.5 Sigma Factors and Promoter Recognition

The composition of PEP must accommodate the transient guidance by sigma (σ) factors. Similar to bacterial RNAPs, the PEP complex requires sigma factors for promoter binding and recognition (Burgess et al. 1969; Allison 2000; Schweer et al. 2010; Feklístov et al. 2014). The current models do not prohibit sigma factor binding and propose that sigma factors could associate with the complex by interacting with PAP11 (Vergara-Cruces et al. 2024).

Arabidopsis has six nucleus-encoded sigma factors (SIG1-SIG6) that recruit the PEP to similar consensus sequences as bacterial σ^{70}-type promoters, which have −35 (TGACA) and −10 (TATAAT) sequences (Gatenby et al. 1981; Liere and Börner 2007; Ortelt and Link 2021). The ancestors of plastid sigma factors, the bacterial σ^{70} family of proteins can be separated into three general categories in cyanobacteria: essential/nonessential proteins for viability, and alternative, stress-responsive proteins (Imamura and Asayama 2009; Yagi and Shiina 2016). Most, if not all, Arabidopsis sigma factors emerged from the essential sigma factor, SigA. The sigma factors that most likely branched off from SigA include SIG2, SIG3, SIG4 and SIG6, with the latter three emerging later from SIG2 itself (Shiina et al. 2009; Yagi and Shiina 2014, 2016; Pfannschmidt et al. 2015; Fu et al. 2021). While SIG3, SIG4 and SIG6 evolved in seed plants, SIG2 and SIG1 could have each emerged from a gene duplication event of SigA, or SIG1 might instead be derived from a different SigA homolog in the nonessential group of sigma factors (Pfannschmidt et al. 2015; Fu et al. 2021). Meanwhile, SIG5 is the most distinct among the six sigma factors. SIG5 may have evolved independently from the stress-responsive group in cyanobacteria, and appears orthologous to the *E. coli* sigma factor, RpoH, a stress-responsive protein (Fujiwara et al. 2000; Pfannschmidt et al. 2015; Yagi and Shiina 2016; Fu et al. 2021). Therefore, of the Arabidopsis sigma factors, SIG1, SIG2, and SIG5 are the most ancient, while SIG3, SIG4 and SIG6 evolved more recently.

Why does Arabidopsis contain so many sigma factors when the plastome itself is so small? One explanation is that sigma factors have been added over time in response to the high plastome mutation rate caused by photosynthetic stress. Point mutations, such as in promoter regions, could hinder recruitment of the PEP and negatively impact plastid functioning, but the nucleus could accommodate these changes through the evolution of additional sigma factors. This theory is endearingly known as the "spoiled kid hypothesis", where the buildup of mutations in the troublesome plastome is compensated for by the nucleus, creating surprising transcriptional complexity (Lerbs-Mache 2011; Lefebvre-Legendre et al. 2014). Perhaps this explains the frequent functional redundancy among the sigma factors, because, outside of the delayed greening of *sig2* and *sig6* mutants, there are few observable phenotypic differences upon the loss of a single sigma factor. However, more severe phenotypes emerge if two or more sigma factors are lost, and triple mutants can be lethal (Hanaoka et al. 2003; Ishizaki et al. 2005; Woodson et al. 2013; Bock et al. 2014; Chi et al. 2015).

Despite this functional overlap, the sigma factors do not only maintain transcriptional efficiency by merely mitigating undesirable mutations. Later sections of this chapter will discuss how sigma factors have evolved specific roles as transcriptional regulators to influence PEP activity.

1.3.2 NEP

1.3.2.1 Discovery of NEP

Even without PEP, the plastome is transcriptionally active, indicating that an alternative RNA polymerase must be able to function in the plastid. For example, deletion of *rpoB* in tobacco does not prevent plastid transcription (Allison et al. 1996) and the plastid-ribosome-deficient barley mutant, *albostrians*, similarly still displays plastid transcription (Hess et al. 1993). An even more extreme example is the parasitic plant *Epifagus virginiana*, which lacks genes encoding PEP subunits, yet still maintains plastid transcription (Morden et al. 1991). These observations indicated that a nuclear-encoded RNA polymerase must be operational in plant plastids to perform transcription even when PEP is not active.

The presence of NEP in land plants was finally confirmed with the identification of organelle-targeted, nucleus-encoded proteins with strong sequence similarities to the RNA polymerase of T3/T7 bacteriophages (Hedtke et al. 1997; Emanuel et al. 2004). NEP is likely the newer polymerase for plastid transcription because it is absent in algae (Liere et al. 2011; Yagi and Shiina 2016). However, in land plants, monocotyledonous species can contain two NEP genes, while dicotyledonous plants can have three NEP genes. Monocots have both an NEP targeted to the plastid (RPOTp) and an NEP targeted to the mitochondria (RPOTm). The additional NEP in dicots can be targeted to either organelle (RPOTmp) (Hedtke et al. 1997, 2000; Ortelt and Link 2014). Conservation of sequences between the different NEPs suggests that the additional NEP genes arose from gene duplication events of *RPOTm* (Hedtke et al. 1997).

1.3.2.2 NEP Promoter Recognition

Unlike the PEP complex, NEP functions as a single-subunit enzyme. RPOTmp and RPOTp likely serve redundant roles because the loss of one can be tolerated, though it appears that RPOTp is the predominant NEP in plastids, since its loss results in a more severe phenotype (Hricová et al. 2006). The loss of both RPOTp and RPOTmp is lethal, demonstrating the importance of NEP in land plants.

Due to its bacteriophage origin, NEP recognizes distinct promoters from PEP. In general, NEPs recognize two types of promoters (Weihe and Börner 1999; Liere and Börner 2007; Zhelyazkova et al. 2012; Ortelt and Link 2014; Börner et al. 2015). The most common promoter type is the Type I promoter, which has a YRT-element (pyrimidine-purine-thymidine). In dicots, a subclass of this promoter can be found, which contains both a YRT motif and a GAA-box. Type II promoters have a completely different sequence from the YRT-motif and are considered nonconsensus promoters. RPOTp and RPOTm have the intrinsic ability to recognize these promoter sequences *in vitro*, without assistance from a cofactor (Kühn et al. 2007). However, *in vivo*, currently unknown cofactors could be involved in directing NEP activity. On the other hand, RPOTmp, the newest of the three polymerases, seems to have lost the ability to recognize promoters as only a single-polypeptide enzyme (Kühn et al. 2007; Yagi and Shiina 2016). Nonetheless, RPOTmp is still involved in transcription of the plastome because it affects the accumulation of plastid transcripts (Baba et al. 2004).

RPOTp and RPOTmp may exhibit some preferences between these promoter types (Courtois et al. 2007; Swiatecka-Hagenbruch et al. 2008). RPOTp appears to prefer the Type I motif, characteristic of it being the principal NEP in plastids. Meanwhile, RPOTmp prefers the nonconsensus Type II motif, found in promoters of *clpP* and the *rrn16* operon. However, despite the limited specificity between the two plastid-targeted NEPs, they likely serve redundant functions, as RPOTmp can rescue transcription from most NEP promoters upon the loss of RPOTp (Swiatecka-Hagenbruch et al. 2008).

1.3.3 Shared or Divided Labor of the Plastid RNA Polymerases?

NEP and PEP share the responsibility of transcribing the plastome, but do not seem to collaborate. Despite the range of proteins isolated in pTACs, NEP was not co-isolated, suggesting that the two polymerases work independently (Pfannschmidt et al. 2015). During leaf development, NEP is activated first and is typically considered the major RNA polymerase of housekeeping genes, while PEP is the major RNA polymerase of photosynthesis-associated genes (Hajdukiewicz et al. 1997; Emanuel et al. 2004). This division of labor is canonically used to divide genes into three classes, where genes transcribed by PEP are Class I, genes transcribed by NEP are Class III, and genes transcribed by both enzymes are Class II (Hajdukiewicz et al. 1997).

However, it may be too simplistic to say that NEP and PEP have a strict division of labor for most genes. Most plastid genes have both PEP and NEP promoters (Zhelyazkova et al. 2012) and can be transcribed in the absence of PEP (Silhavy and Maliga 1998; Krause et al. 2000; Legen et al. 2002; Swiatecka-Hagenbruch et al. 2007; Ortelt and Link 2021). In fact, there are far more promoters than genes in the plastome (Zhelyazkova et al. 2012). There could be 400 transcription start sites in the Arabidopsis plastome of ~130 genes, averaging at a transcription start site every 600-nt or so (Castandet et al. 2019). These promoters can be found all over the plastome, with transcription start sites in regions not only preceding genes, but also within genes, antisense to genes, or in intergenic regions. The abundance of transcriptional events has led to speculation that the plastome can be entirely transcribed, though perhaps not equally (Barkan 2011; Shi et al. 2016; Sanitá Lima and Smith 2017; Palomar et al. 2022).

There are still some genes that are mostly or completely transcribed by only one of the polymerases. For example, NEP is essential for the transcription of the *rpoB-rpoC1-rpoC2* operon to initially generate the core subunits of PEP. NEP also largely transcribes other housekeeping genes like *accD*, part of the acetyl-CoA carboxylase for fatty acid biosynthesis, *clpP*, part of the Clp protease complex, and *ycf2*, an open reading frame that may be important in protein import (Hajdukiewicz et al. 1997; Silhavy and Maliga 1998; Börner et al. 2015; Ortelt and Link 2021; Puthiyaveetil et al. 2021; Palomar et al. 2022; Chang et al. 2023). On the other hand, PEP transcribes *psbA* and *rbcL*, the large subunit of ribulose bisphosphate carboxylase (Rubisco) (Ortelt and Link 2021). Considering the entirety of the plastome, PEP most frequently binds within the inverted repeats and large single-copy region (Palomar et al. 2022). However, PEP's presence is reduced in the small single-copy region, suggesting that NEP may be involved in transcribing these genes. Still, overall, plastid transcription by the polymerases is highly flexible and both can transcribe wide swaths of the plastome.

1.3.3.1 Regulation of the Plastid RNA Polymerases

NEP and PEP are regulated developmentally. Early in development, NEP is the predominant transcriber of the plastome (Emanuel et al. 2004) and NEP-associated transcripts, like *clpP*, tend to display the highest expression in younger tissues, like the base of a monocot leaf (Baumgartner et al. 1989, 1993; Sakai et al. 1992; Cahoon et al. 2004, 2008). Initial transcription of housekeeping genes prepares the plastid for chloroplast biogenesis because NEP is needed to transcribe the PEP core. PEP is canonically viewed as the predominant polymerase once it is produced and activated. However, the PEP core on its own has limited capabilities, and putting together the numerous components of the PEP supercomplex is complicated. Therefore, the assembly and activation of the PEP supercomplex itself is a major point of transcriptional control imposed on the plastid by the nucleus.

PEP assembly coordinates plastome transcription with light and developmental stage to accurately time chloroplast biogenesis. The remainder of this chapter will focus on the nuclear control of chloroplast development, which is accomplished through anterograde signaling. To explore plastid transcriptional control by the nucleus, we will discuss two models: chloroplast biogenesis during light-dependent seedling de-etiolation in dicots and chloroplast biogenesis during leaf development in monocots. Through these models, we will impart the importance of the PEP for collective control of the transcription of plastid photosynthesis genes.

1.4 Anterograde Signaling: Constant Dialogue Between the Nucleus and the Plastid

1.4.1 Anterograde Signaling for the Control of PEP by Light

Plastid gene expression is part of the developmental program of the host cell and, ultimately, is controlled by the host cell's nucleus (Taylor 1989; Pearce et al. 2017). However, the mechanism of nucleus-to-organelle communication, or anterograde signaling, has remained elusive for decades. The main challenge is the difficulty in teasing apart regulators from the battery of nuclear-encoded components required for organellar biogenesis and/or function (Taylor 1989). Finally, some anterograde signals may have been revealed by studying the role of photoreceptor signaling in the light-dependent regulation of PEP during de-etiolation of developing seedlings. This anterograde signaling not only controls the transcription of PhAPGs by PEP, but also synchronizes PhAPGs with PhANGs.

1.4.1.1 The Process of Greening: De-etiolation

The greening, or chloroplast biogenesis, of angiosperms (flowering plants) and some gymnosperms occurs only in the presence of light (Kirk and Tilney-Bassett 1978; Armstrong 1998). Light directly turns on chlorophyll biosynthesis in plastids by activating protochlorophyllide oxidoreductase to catalyze the final step of the biosynthesis pathway, the conversion of protochlorophyllide to chlorophyll a (Klein and Schiff 1972; Mapleston and Griffiths 1980; Santel and Apel 1981). In parallel, as an environmental signal, light reprograms hundreds of genes in the nucleus to initiate the developmental transition from heterotrophic growth, which is supported by seed-stored energy, to autotrophic growth, which relies on photosynthesis (Kirk and Tilney-Bassett 1978; Chen et al. 2004). In dicotyledonous plants, such as *Arabidopsis thaliana*, young seedlings that germinate underground adopt a dark-grown developmental program called skotomorphogenesis (or etiolation), which inhibits leaf development and promotes elongation of the embryonic stem (hypocotyl), allowing seedlings to emerge quickly from the soil into the light. Plastids in the leaves of dark-grown seedlings differentiate into photosynthetically inactive, non-green etioplasts. Upon illumination, seedlings undergo de-etiolation, a developmental transition to a light-grown developmental program called photomorphogenesis, which restricts hypocotyl elongation and promotes leaf development and chloroplast biogenesis.

1.4.1.2 Phytochromes and the Sensing of Light

Light is perceived by a suite of photoreceptors that play an essential role in initiating chloroplast biogenesis in both monocots and dicots (Takano et al. 2009; Strasser et al. 2010; Hu et al. 2013; Yoo et al. 2019). These include the red and far-red light-sensing phytochromes (phys) and the blue-light-absorbing cryptochromes (CRYs) (Chen et al. 2004). Here we will focus on phy signaling, because the mechanistic link between phy signaling and chloroplast biogenesis has been extensively studied. Phy is a bilin-containing pigment or a biliprotein that senses light using a covalently linked phytochromobilin as the chromophore (Rockwell et al. 2006; Burgie and Vierstra 2014). Light induces the isomerization of the C15-C16 carbon-carbon double bond of phytochromobilin between a C15-*Z* (*trans*) and a C15-*E* (*cis*) configuration, which in turn photoconverts phys between two relatively stable forms: a red light-absorbing, inactive phytochrome, Pr (C15-*Z*), and a far-red light-absorbing, active phytochrome, Pfr (C15-*E*) (Rockwell et al. 2006; Burgie and Vierstra 2014). In Arabidopsis, phys are encoded by five genes, *PHYA-E*, among which the gene products of *PHYA* and *PHYB* are the predominant sensors of continuous far-red and red light, respectively (Sharrock and Quail 1989; Parks and Quail 1993; Reed et al. 1993; Whitelam et al. 1993).

One of the earliest light responses at the cellular level is the translocation of photoactivated phys from the cytoplasm to the nucleus, where phys are further compartmentalized to photosensory subnuclear domains named photobodies (Sakamoto and Nagatani 1996; Yamaguchi et al. 1999; Kircher et al. 1999; Chen et al. 2003, 2005; Van Buskirk et al. 2012). Nuclear-localized phys initiate photomorphogenesis, including chloroplast biogenesis, by inducing transcription of nuclear photosynthesis genes, the PhANGs (Leivar et al. 2008a, b; Shin et al. 2009; Li et al. 2012; Luo et al. 2014). Light-based signaling by phys also synchronously induces the transcription of PhAPGs (Link 1982; Thompson et al. 1983; Bennett et al. 1984). As photoactivated phys and CRYs localize only to the nucleus and not to plastids (Fankhauser and Chen 2008; Wang et al. 2018), they must regulate plastid gene expression via anterograde signaling.

1.4.1.3 Light-regulated PEP Assembly and Transcription

Although it was observed that plastid mRNA levels were influenced by light, these light-induced changes could result from altered transcription of plastid genes or altered mRNA stability. Plastid run-on transcription assays were developed to distinguish between the two mechanisms (Mullet and Klein 1987; Deng et al. 1987). Using plastid run-on transcription assays, it was demonstrated that light-inducible transcription itself is a major control step during chloroplast biogenesis. However, intriguingly, the overall plastid transcription rates of individual plastid genes were similar. Therefore, differential accumulation of individual transcripts was mostly attributable to post-transcriptional regulation of RNA stability. These results raised the hypothesis that, unlike gene regulation in the nucleus, where cell signaling regulates individual genes via gene-specific transcription regulators, light regulates PhAPGs by controlling the general, collective transcriptional activity of plastids (Deng and Gruissem 1987; Klein and Mullet 1990).

Biochemical studies of the pTAC complex in mustard (*Sinapis alba* L.) first indicated that light regulates plastid transcription through the assembly of the PEP complex. While a 420-kDa PEP complex—made of only the core subunits—formed in etioplasts in dark-grown mustard seedlings, a 700-kDa PEP complex was enriched in chloroplasts in the light (Pfannschmidt and Link 1994). Both complexes were shown to be transcriptionally active (Pfannschmidt and Link 1994, 1997). The assembly of a 1 MDa PEP complex could also be induced by light in Arabidopsis, though no smaller intermediate complex was detected (Yang et al. 2019; Yoo et al. 2019). The discrepancies in the presence of the PEP core complex may be attributable to differences in the stability of the PEP core between the two plant species.

Light signaling to trigger PEP assembly is initiated by photoreceptors. Genetic analysis of PEP assembly in photoreceptor mutants in Arabidopsis seedlings under monochromatic far-red, red, or blue light conditions demonstrated that phys and CRYs can each individually initiate PEP assembly and PhAPG activation (Yang et al. 2019; Yoo et al. 2019; Hwang et al. 2022). PEP assembly can be blocked in *phyA-211* in far-red light, in *phyB-9* in red light, and in *cry1/cry2* in blue light (Yoo et al. 2019; Hwang et al. 2022). We discuss some of the mechanisms of this light-induced signaling below.

1.4.1.4 Degradation of Nuclear PIFs Triggers PEP Assembly

Chloroplast biogenesis is repressed in darkness by a smaller family of basic/helix-loop-helix transcription factors named PHYTOCHROME-INTERACTING FACTORs (PIFs). Both phys and CRYs regulate nuclear gene expression by inhibiting the stability and activity of PIFs (Al-Sady et al. 2006; Leivar and Quail 2011; Pedmale et al. 2016; Yoo et al. 2021). In Arabidopsis, PIFs include eight members (PIF1-8), of which PIF1, PIF3, PIF4, and PIF5 collectively repress PhANGs in the nucleus (Leivar et al. 2008b, 2009; Shin et al. 2009). In parallel, nuclear PIFs remotely repress the assembly of the PEP complex (Yang et al. 2019; Yoo et al. 2019). As such, PIFs act as master repressors of chloroplast biogenesis by synchronously turning off the expression of both PhANGs and PhAPGs. Consistent with this model, a dark-grown *pif1pif3pif4pif5* quadruple mutant (*pifq*) becomes de-etiolated, morphologically mimicking wild-type Col-0 seedlings grown in the light. In *pifq*, plants show early signs of chloroplast biogenesis, such as the disappearance of the prolamellar bodies, the development of rudimentary pro-thylakoid membranes, and the activation of PhANGs and PhAPGs (Leivar et al. 2008b, 2009; Shin et al. 2009; Yang et al. 2019; Yoo et al. 2019; Hwang et al. 2022).

How do phys orchestrate massive transcriptional reprogramming in both the nuclear and plastid genomes? To trigger light responses, phytochromes must first change conformation and subcellular location. In the dark, phys locate to the cytosol, where they are present in an inactive, red-light absorbing Pr form. With light, phys convert to an active Pfr form that moves into the nucleus (Yamaguchi et al. 1999; Kircher et al. 1999; Fankhauser and Chen 2008). In the nucleus, photoactivated phys, primarily phyB, initiate chloroplast biogenesis by releasing the PIF-dependent repression of PhANGs and PhAPGs (Fig. 1.2). The active form of phyB interacts directly with all PIFs to promote their degradation (Legris et al. 2019; Cheng et al. 2021), attenuate their DNA binding (Park et al. 2012, 2018), and repress their transactivation activity (Yoo et al. 2021).

PIF1, PIF3, PIF4, and PIF5 are rapidly degraded during the dark-to-light transition in a phyB-dependent manner (Shen et al. 2005, 2007; Al-Sady et al. 2006; Lorrain et al. 2008; Pham et al. 2018). Binding of phyB promotes PIF3 phosphorylation by PHOTOREGULATORY PROTEIN KINASEs (PPKs) and subsequent ubiquitin-proteasome-dependent degradation (Al-Sady et al. 2006; Ni et al. 2014, 2017; Dong et al. 2017). Two E3 ubiquitin ligases have been identified that

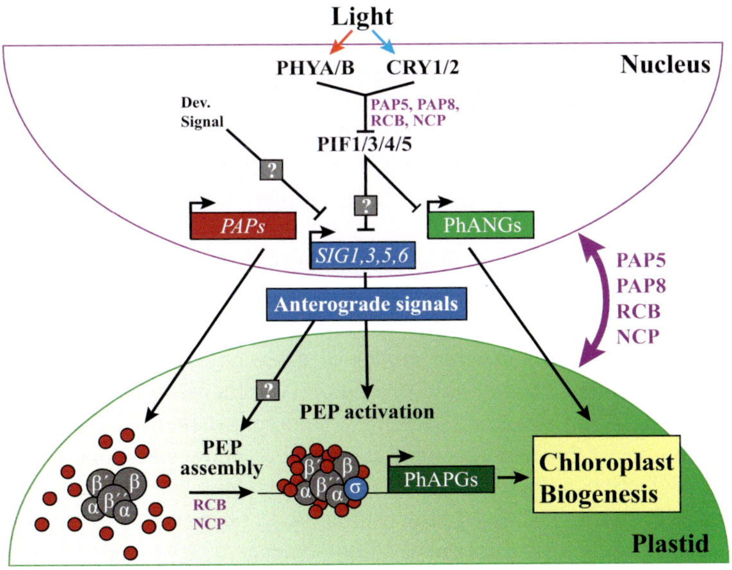

Fig. 1.2 A model of chloroplast biogenesis during light-dependent seedling de-etiolation in dicots. Photoreceptors trigger a cascade of light-induced signaling that leads to the degradation of PIFs and the activation of PhANGs and PhAPGs. To activate PhAPGs, PEP assembly and activation are major regulatory steps that are informed by anterograde signals and the accumulation of plastid-localized, nuclear-encoded proteins. Once PhANGs and PhAPGs are expressed, chloroplast biogenesis is triggered

ubiquitylate PIF3, including the **c**ullin3-**R**ING E3 ubiquitin **l**igase (CRL), with LIGHT-RESPONSE BRIC-A-BRAC/TRAMTRACK/BROADs (LRBs) as the substrate recognition subunits ($CRL3^{LRB}$), and the cullin1 E3, with EIN3-BINDING F BOX PROTEINs (EBFs) as the substrate recognition subunits ($CRL1^{EBF}$) (Ni et al. 2014; Dong et al. 2017). In addition to regulating PIFs via direct interaction, phyB also promotes PIF degradation indirectly by inhibiting factors that stabilize PIFs, such as the CONSTITUTIVE PHOTOMORPHOGENIC 1 (COP1), SUPPRESSOR OF PHYTOCHROME A-105 proteins (SPAs), and DE-ETIOLATED 1(DET1), which constitute the $CRL4^{COP1/SPA}$ and $CRL4^{COP1/DET1}$ E3 ubiquitin ligases (Saijo et al. 2003; Sheerin et al. 2015; Lu et al. 2015). Phys, and also CRYs, interact directly with COP1 and SPAs to block the formation of the COP1-SPA complex in the light (Lian et al. 2011; Liu et al. 2011; Sheerin et al. 2015; Lu et al. 2015). The classic *constitutive photomorphogenic/de-etiolated/fusca* (*cop/det/fus*) mutants, such as *cop1-4*, *det1-1*, and *spa*, exhibit similar phenotypes as *pifq,* including constitutively expressed PhANGs and PhAPGs and assembled PEP complexes in darkness (Reed and Chory 1994; Wei and Deng 1996; Hwang et al. 2022). Therefore, phyB-mediated degradation of PIFs is a major switch to initiate chloroplast biogenesis and serves as a synchronization mechanism that coordinates the expression of nuclear and plastid photosynthesis genes (Fig. 1.2).

1.4.1.5 Nucleus-and-Plastid Dual-Localized Regulators

The anterograde signaling mechanism downstream of PIFs to control the activity of the PEP remains elusive. The anterograde signal is likely encoded by a light-inducible gene repressed in darkness by PIFs, either directly or indirectly. PIFs do not regulate PAPs at the transcript level (Yoo et al. 2019; Hwang et al. 2022); therefore, it is unlikely that PIFs regulate PEP assembly by transcriptionally regulating the expression of PAPs.

The discovery of positive regulators of chloroplast biogenesis has been hindered by the difficulty to distinguish chloroplast-deficient regulator mutants from other albino mutants for genes encoding essential components of the chloroplast (Taylor 1989). Fortunately, recent genetic studies of early phy signaling have serendipitously uncovered a new class of light mutants in Arabidopsis that exhibit a combination of albino and long-hypocotyl seedling phenotypes (Chen and Chory 2011). The founding member of this new mutant class, *hemera* (*hmr*), was identified originally as a mutant defective in the localization of phyB to subnuclear foci called photobodies (Chen et al. 2010). The *hmr* mutant turns out to be defective in phyB signaling and chloroplast biogenesis (Pfalz et al. 2006; Chen et al. 2010). Surprisingly, HMR encodes the PAP5 subunit of the PEP, and it is dual-targeted to both plastids and the nucleus (Chen et al. 2010; Nevarez et al. 2017). In the nucleus, HMR/PAP5 is an acidic-type transcriptional activator that interacts directly with phys as well as PIFs to regulate the stability and activity of PIF1, PIF3, and PIF4 (Galvão et al. 2012; Qiu et al. 2015, 2019). The discovery of HMR/PAP5 revealed an unexpected link between nuclear phyB signaling and plastid transcription, showing that these two spatially separated processes share common functional components (Fig. 1.2). In addition to HMR/PAP5, PAP8 is also dual-targeted to the nucleus and plastids. Similar to HMR/PAP5, PAP8 is required for phyB signaling in the nucleus (Liebers et al. 2020). Intriguingly, PAP8 forms a nuclear protein complex that may contain other PEP components (Liebers et al. 2020; Chambon et al. 2022).

Albino mutants had been ignored historically in the context of light signaling because chlorophyll-deficient mutants were thought to retain normal phy-mediated hypocotyl responses (Borthwick et al. 1951; Jabben and Deitzer 1979). As a result, the entire class of tall-and-albino mutants, like *hmr,* was overlooked (Chen and Chory 2011). A forward genetic screen for *hmr*-like tall-and-albino mutants identified two additional, previously uncharacterized phy signaling components: REGULATOR OF CHLOROPLAST BIOGENESIS (RCB) and NUCLEAR-CONTROL-OF-PEP-ACTIVITY (NCP) (Yang et al. 2019; Yoo et al. 2019). RCB and NCP are two paralogous thioredoxin-like proteins that lack thioredoxin reductase activity (Yang et al. 2019). Intriguingly, like HMR, RCB and NCP also localize to both the nucleus and plastids. Although the biochemical functions of RCB and NCP remain unclear, nuclear RCB and NCP likely work in concert with HMR to regulate photobody formation and the PIFs (Yang et al. 2019; Yoo et al. 2019; Qiu et al. 2021). The key difference between RCB and NCP is that RCB regulates the PEP primarily from the nucleus, whereas NCP is required for both PIF degradation in the nucleus as well as PEP assembly in plastids. This conclusion is supported by the genetic evidence that the *rcb-10/pifq* mutant could rescue the albino phenotype of *rcb-10*, whereas the *ncp-10/pifq* mutant remained albino, like *hmr-5/pifq* (Qiu et al. 2015; Yang et al. 2019; Yoo et al. 2019). Therefore, the *rcb* mutant represents a unique type of albino mutant whose chloroplast defect is due to the accumulation of nuclear repressors of chloroplast biogenesis.

The dual localization of HMR/PAP5, PAP8, RCB, and NCP directly links the nuclear and plastid signaling mechanisms to the control of PhAPG transcription. All four proteins possess putative nuclear localization signals (NLSs) and transit peptides for chloroplast targeting. Intriguingly, the nuclear fractions of HMR/PAP5, PAP8, RCB, and NCP share the same molecular masses as their respective chloroplast counterparts, which are smaller than the mature proteins—i.e., both nuclear and chloroplast fractions lack the transit peptides, which are presumably cleaved upon chloroplast import—suggesting that they localize to the plastids first before being translocated to the nucleus (Nevarez et al. 2017; Yang et al. 2019; Yoo et al. 2019; Liebers et al. 2020). This dual targeting mechanism has been demonstrated for another nuclear and plastidial localized protein, Whirly1 (WHY1) (Grabowski et al. 2008), which is loosely associated with PEP in the nucleoid (Pfalz et al. 2006; Melonek et al. 2010) and required for chloroplast biogenesis and plastid genome repair (Prikryl et al. 2008; Cappadocia et al. 2010, 2012). Plastome-expressed WHY1 in tobacco could translocate to the nucleus, providing strong evidence supporting the existence of a plastid-to-nucleus protein transport mechanism (Isemer et al. 2012). One possible mechanism is via stromules—narrow tubular structures that emanate from plastids (Caplan et al. 2015; Kumar et al. 2018; Lee et al. 2024). Further investigations are needed to elucidate the dual nucleus-plastid protein-targeting mechanism and the functional significance of these dual-localized proteins in PEP assembly and activation by light.

1.4.1.6 Other Nucleus-to-Plastid Signals: Sigma Factors

Beyond proteins that are dual-localized to the nucleus and the plastid, sigma factors—due to their function in promoter selection—are critical nucleus-encoded regulators of PEP activity. Each sigma factor demonstrates unique roles in fine-tuning PEP recruitment, including serving as nucleus-to-plastid anterograde signals during chloroplast biogenesis. This chapter will only briefly review the sigma factors and discuss their role in anterograde signaling. For more extensive details, refer to the following reviews (Lerbs-Mache 2011; Börner et al. 2015; Chi et al. 2015; Ortelt and Link 2021; Puthiyaveetil et al. 2021).

1.4.1.6.1 The Role of SIG1 in Photosystem Stoichiometry

Expression of the sigma factors is controlled by light (Allison 2000) and sigma factors can display a circadian rhythm (Oikawa et al. 2000; Noordally et al. 2013; Schneider et al. 2019). The interplay between the sigma factors and light could serve to protect the plastid during changing light conditions. Though not the only sigma factor involved in delivering light-based information, SIG1 specifically plays a role in adjusting photosystem stoichiometry during fluctuating light (Lerbs-Mache 2011; Chi et al. 2015). SIG1 rapidly accumulates in light (Onda et al. 2008) and regulates the *psaAB* operon (Tozawa et al. 2007) to balance the abundance of the two photosystems, thereby maintaining proper electron transfer. Shorter wavelengths of light favor photosystem II, creating reducing conditions, while longer wavelengths of light favor photosystem I, creating oxidizing conditions. The redox conditions generated by this imbalance influence the phosphorylation status of SIG1 to direct it in the regulation of *psaA* (Shimizu et al. 2010; Macadlo et al. 2020). In addition to its regulation of the *psaAB* operon, SIG1 displays a light-influenced binding pattern with a preference for the promoters of the *psbEFLJ* and *psbBT* operons (Hanaoka et al. 2012). Overall, this demonstrates that, along with being an important general housekeeping sigma factor, SIG1 is a light-controlled and redox-responsive sigma factor that might specifically intervene to influence the transcription of photosystem core subunits and protect the photosynthetic apparatus from light-induced damage.

1.4.1.6.2 The Developmental Roles of SIG2 and SIG6

SIG2 and SIG6 are seen as critical developmental sigma factors that are involved in the transcription of a wide assortment of PEP-dependent genes. The loss of even one of the two sigma factors results in an initial, PEP-deficient-like, pale green phenotype, signifying the importance of the factors for proper chloroplast development and functioning (Hanaoka et al. 2003; Ishizaki et al. 2005). SIG2 is associated with the transcription of tRNA genes such as *trnV-UAC*, *trnM-CAU* and *trnE-UUC*, as well as photosynthesis-associated genes such as *psaJ* and *psbD* from the −256 promoter (Kanamaru et al. 2001; Hanaoka et al. 2003; Nagashima et al. 2004a; Kanamaru and Tanaka 2004). Meanwhile, SIG6 activates many PEP-dependent photosynthesis-associated genes, especially early in development (Ishizaki et al. 2005; Loschelder et al. 2006). Comparison of PEP binding in *sig2* and *sig6* revealed their widespread presence along the plastome, but simultaneously displayed their specificity, with the *sig2* mutation causing a particularly strong reduction for *trnE*, *trnY*, *trnD*, *trnL2*, *trnV1*, and *ndhC* (Palomar et al. 2022). On the other hand, the *sig6* mutation caused an overall strong reduction for most genes, but particularly *trnI1*.

1.4.1.6.3 The Narrow Roles of SIG3 and SIG4

Two of the more recently evolved sigma factors, SIG3 and SIG4, have specific roles in the transcription of *psbN* and *ndhF*, respectively (Favory et al. 2005; Zghidi et al. 2007). For SIG3, its role in the transcription of *psbN* has implicated it in being involved in the regulation of the *psbB* operon (*psbB-psbT-psbH-petB-petD*). Interestingly, *psbN* is located within *psbT* but on the antisense strand, leading to the generation of RNA antisense to *psbT* (Zghidi et al. 2007). The frequent occurrence of antisense transcription in the plastid is unusual, but could have regulatory roles that remain to be deeply explored (Sharwood et al. 2011; Mitschke et al. 2011; Zghidi-Abouzid et al. 2011; Hotto et al. 2011). As for SIG4, the specificity of this sigma factor for *ndhF* can be highlighted by tracing occurrences of SIG4 gene loss and concurrent absence or truncations of the *ndhF* gene (Pfannschmidt et al. 2015; Fu et al. 2021). The activity of SIG4 on *ndhF* may regulate the amount of NDH complexes present in the plastid (Favory et al. 2005).

1.4.1.6.4 The Stress-Responsive Roles of SIG5

SIG5 is induced by a variety of stresses (Nagashima et al. 2004b; Cano-Ramirez et al. 2023). In particular, SIG5 is involved in monitoring both the intensity and quality of light, which is demonstrated by its regulation of *psbD*. SIG5 controls alternative promoter usage of *psbD* during high light and other stress conditions (Christopher 1996; Nagashima et al. 2004b; Kanamaru and Tanaka 2004; Cano-Ramirez et al. 2023). *psbD* can be transcribed from several promoters, including two promoters −256 and −948 from the translation initiation site (Hoffer and Christopher 1997; Christopher and Hoffer 1998; Nagashima et al. 2004b). While SIG2 is involved in transcription initiation at the −256 position (Hanaoka et al. 2003), SIG5 controls initiation from the −948 site, known as the *psbD* light-responsive promoter (*psbD*-LRP). In support of this, transcription from this site is abolished in *sig5* (Tsunoyama et al. 2004; Nagashima et al. 2004b; Shimmura et al. 2017). The extended 5′ untranslated region that is transcribed in high light could be involved in bolstering photosystem functioning during stress by boosting the replacement of the critical D2 protein of photosystem II. Further, not only does SIG5 direct alternative promoter recognition of *psbD*, it also informs the circadian rhythm of this operon and others by promoting oscillating transcription patterns (Noordally et al. 2013; Belbin et al. 2017; Cano-Ramirez et al. 2023).

1.4.1.6.5 Sigma Factors Are Anterograde Signals During Chloroplast Biogenesis

During de-etiolation, four sigma factors—SIG1, SIG3, SIG5, and SIG6—serve as nucleus-to-plastid signals for triggering PEP activation (Fig. 1.2) (Hwang et al. 2022). In etiolated seedlings, the expression of these sigma factors is repressed by PIFs. PhyB-dependent proteolysis of PIFs induces their expression in the light to promote PEP activity (Fig. 1.2). Current evidence indicates that PIFs are transcription activators (Yoo et al. 2021), so it is likely that PIFs repress sigma factors indirectly by transactivating unknown transcription repressor(s). In addition to their involvement in light-induced anterograde signaling, sigma factors could also influence developmentally-regulated anterograde signaling. During wheat leaf development, the genes of multiple sigma factors, including *SIG1*, *SIG2-1*, *SIG2-2*, *SIG3*, *SIG5*, and *SIG6*, are acutely induced at defined leaf segments, indicating that sigma factors also play pivotal roles in the anterograde control of plastid transcription during the proplastid-to-chloroplast developmental transition (Loudya et al. 2021).

1.4.1.7 Modeling Chloroplast Biogenesis in Light

An overview of the steps of chloroplast biogenesis in light can be seen in a pluripotent inducible cell line of single Arabidopsis cells (Dubreuil et al. 2018). Upon light exposure, these cells demonstrate two key regulatory phases as they shift from carrying proplastids (plastid progenitors) to containing mature chloroplasts. The first phase is associated with large-scale gene expression changes in response to light, as the photoreceptors trigger a cascade of downstream signals, with more than 3000 genes showing differential expression. The second phase revolves around activation and development of the photosynthetic chloroplast, which similarly displays a massive shift of gene expression of more than 12,000 genes. During this time, nuclear signals control PEP assembly and, thereby, plastid transcription to achieve chloroplast biogenesis.

The pivotal communication between the nucleus and the chloroplast throughout this process is highlighted by plastid positioning during the second regulatory phase. Plastids start this phase in clusters around the nucleus to, presumably, maximize efficient communication between the two organelles. As chloroplasts develop, they shift away from the nucleus to the cell cortex. The anterograde signaling that takes place to initiate the process of chloroplast biogenesis is critical to give plants their characteristic green color. However, light is not the only factor that regulates chloroplast biogenesis. Developmental cues can play an even greater role in this process. Our final section will briefly discuss progress in this field.

1.4.2 Control of Plastid Transcription During Monocot Leaf Development

The linear developmental gradient of monocot leaves provides an ideal model to investigate the proplastid-to-chloroplast developmental transition. Across the developmental gradient of each leaf blade, there are basal, proliferating progenitor cells near the shoot apical meristem and distal, differentiated cells towards the tip of the leaf (Leech et al. 1973; Chotewutmontri and Barkan 2016). In a single leaf blade, plastids gradually differentiate from non-photosynthetic proplastids at the leaf base, to developing chloroplasts, then to young chloroplasts, and finally to mature chloroplasts near the leaf tip. In contrast to the environmental control of chloroplast biogenesis during de-etiolation, because monocot leaf development occurs in the presence of light, the proplastid-to-chloroplast transition across the leaf blade represents a developmental control of chloroplast biogenesis.

In all monocot models examined, including maize, barley, and wheat, the proplastid-to-chloroplast transition follows two distinct developmental phases (Baumgartner et al. 1989, 1993; Chotewutmontri and Barkan 2016; Loudya et al. 2021). In the initial phase—where the host cell transitions from proliferation to differentiation—plastids also differentiate from proplastids to developing chloroplasts, replicating their DNA, dividing rapidly and growing in size. At this point, transcription in plastids is carried out mainly by the NEP, RPOTp (Loudya et al. 2021), which activates plastid genes associated with plastid transcription and translation. These include *rpoB-rpoC1-rpoC2* and genes encoding ribosomal proteins (Baumgartner et al. 1989, 1993; Chotewutmontri and Barkan 2016; Loudya et al. 2021). Concurrently, nuclear-encoded sigma factors and PAPs, such as PAP2, PAP3, PAP5, and PAP6, are activated to promote plastid gene expression by the PEP (Loudya et al. 2021). After the initial phase—when cells undergo a transition from cell differentiation to expansion—developing plastids fully differentiate to young chloroplasts, which is accompanied by build-up of the photosynthesis machinery. The PEP gradually replaces the NEP as the major RNAP in plastids and transcripts of PhAPGs, the primary targets of the PEP, peak in this phase. However, as chloroplasts continue to mature, plastome transcription declines.

Intriguingly, RCB and NCP are activated at distinct stages during wheat leaf development (Loudya et al. 2021). RCB is activated in the initial phase of chloroplast biogenesis, corroborating its central role in the nucleus to initiate chloroplast biogenesis (Yoo et al. 2019; Loudya et al. 2021). In contrast, NCP is activated only at the leaf tip, suggesting a role during chloroplast maturation (Yang et al. 2019;

Loudya et al. 2021). Together, the studies using monocot models highlight the importance of controlling plastid transcription during proplastid-to-chloroplast differentiation and reveal that developmental cues control plastid transcription by orchestrating two distinct anterograde mechanisms to activate the NEP and PEP sequentially during the two phases of chloroplast biogenesis.

1.5 Concluding Thoughts

Plastid transcription is a key part of energy production, biosynthesis, and development in photosynthetic organisms. As such, the nucleus strictly regulates the plastid genome. Using anterograde signaling, the nucleus directs chloroplast biogenesis during de-etiolation and leaf development through the comprehensive regulation of plastid transcription itself. To do this, PEP assembly and activity are key points of control to establish plastome expression.

We have only begun to listen in on the light-based conversations that take place during the etioplast-to-chloroplast conversion and the developmental proplastid-to-chloroplast conversion. Many questions remain surrounding both PEP structure and overall nucleus-plastid communication. The cryo-EM structure of the PEP super-complex revealed a complicated web of interactions between the PEP core and PAP subunits, but further research is needed to determine the specific roles of each PAP and how these roles support PEP activation and nuclear control of chloroplast function. The order of assembly of the massive PEP complex also remains a mystery, as does the existence of alternate assemblies during different stages of transcription. Overall, the construction of such a large complex for the transcription of a relatively small genome underscores the conclusion that the assembly of PEP itself is a major point of control for plastid transcription. Additionally, the identities and mechanisms of anterograde signals that trigger light-based and developmentally-controlled PEP assembly remain unresolved. An array of nucleus-encoded proteins, such as sigma factors, PAP5, PAP8, RCB and NCP, seem to be involved in the communication pathways that lead to chloroplast biogenesis, but their specific roles are unknown.

On its surface, transcription of the relatively small plastome appears deceptively simple. In actuality, an exciting array of regulatory pathways are required for the activation of these PhAPGs to spur chloroplast biogenesis. Future work in this field will continue to unravel inter-organellar signaling mechanisms to broaden our understanding of a critical, bacterial transcription system within eukaryotic cells.

Acknowledgements This work was supported by National Institute of General Medical Sciences grant R01GM132765 to M.C. and National Science Foundation grant IOS-2034015 to X.C. and M.C. N.F. was supported by the US Department of Education GAANN grant P200A210029, National Science Foundation Research Traineeship grant 1922642, and a Howard Hughes Medical Institute bridge fund to X.C.

References

Abdallah F, Salamini F, Leister D (2000) A prediction of the size and evolutionary origin of the proteome of chloroplasts of *Arabidopsis*. Trends Plant Sci 5:141–142

Allison LA (2000) The role of sigma factors in plastid transcription. Biochimie 82:537–548

Allison LA, Simon LD, Maliga P (1996) Deletion of *rpoB* reveals a second distinct transcription system in plastids of higher plants. EMBO J 15:2802–2809

Al-Sady B, Ni W, Kircher S et al (2006) Photoactivated phytochrome induces rapid PIF3 phosphorylation prior to proteasome-mediated degradation. Mol Cell 23:439–446

Arabidopsis Genome Initiative (2000) Analysis of the genome sequence of the flowering plant *Arabidopsis thaliana*. Nature 408:796–815

Armstrong GA (1998) Greening in the dark: light-independent chlorophyll biosynthesis from anoxygenic photosynthetic bacteria to gymnosperms. J Photochem Photobiol B 43:87–100

Arsova B, Hoja U, Wimmelbacher M et al (2010) Plastidial thioredoxin z interacts with two fructokinase-like proteins in a thiol-dependent manner: evidence for an essential role in chloroplast development in *Arabidopsis* and *Nicotiana benthamiana*. Plant Cell 22:1498–1515

Artsimovitch I, Svetlov V, Murakami KS, Landick R (2003) Co-overexpression of *Escherichia coli* RNA polymerase subunits allows isolation and analysis of mutant enzymes lacking lineage-specific sequence insertions. J Biol Chem 278:12344–12355

Baba K, Schmidt J, Espinosa-Ruiz A et al (2004) Organellar gene transcription and early seedling development are affected in the *rpoT;2* mutant of *Arabidopsis*. Plant J 38:38–48

Barkan A (2011) Expression of plastid genes: organelle-specific elaborations on a prokaryotic scaffold. Plant Physiol 155:1520–1532

Barkan A, Small I (2014) Pentatricopeptide repeat proteins in plants. Annu Rev Plant Biol 65:415–442

Baumgartner BJ, Rapp JC, Mullet JE (1989) Plastid transcription activity and DNA copy number increase early in barley chloroplast development. Plant Physiol 89:1011–1018

Baumgartner BJ, Rapp JC, Mullet JE (1993) Plastid genes encoding the transcription/translation apparatus are differentially transcribed early in barley (*Hordeum vulgare*) chloroplast development. Plant Physiol 101:781–791

Belbin FE, Noordally ZB, Wetherill SJ et al (2017) Integration of light and circadian signals that regulate chloroplast transcription by a nuclear-encoded sigma factor. New Phytol 213:727–738

Bennett J, Jenkins GI, Hartley MR (1984) Differential regulation of the accumulation of the light-harvesting chlorophyll a/b complex and ribulose bisphosphate carboxylase/oxygenase in greening pea leaves. J Cell Biochem 25:1–13

Bergsland KJ, Haselkorn R (1991) Evolutionary relationships among eubacteria, cyanobacteria, and chloroplasts: evidence from the *rpoC1* gene of *Anabaena sp.* strain PCC 7120. J Bacteriol 173:3446–3455

Bock S, Ortelt J, Link G (2014) *AtSIG6* and other members of the sigma gene family jointly but differentially determine plastid target gene expression in *Arabidopsis thaliana*. Front Plant Sci 5:667

Börner T, Aleynikova AY, Zubo YO, Kusnetsov VV (2015) Chloroplast RNA polymerases: role in chloroplast biogenesis. Biochim Biophys Acta 1847:761–769

Borthwick HA, Hendricks SB, Parker MW (1951) Action spectrum for inhibition of stem growth in dark-grown seedlings of albino and nonalbino barley (*Hordeum vulgare*). Bot Gaz 113:95–105

Bottomley W, Smith HJ, Bogorad L (1971) RNA polymerases of maize: partial purification and properties of the chloroplast enzyme. Proc Natl Acad Sci USA 68:2412–2416

Briat JF, Laulhere JP, Mache R (1979) Transcription activity of a DNA-protein complex isolated from spinach plastids. Eur J Biochem 98:285–292

Burgess RR, Travers AA, Dunn JJ, Bautz EK (1969) Factor stimulating transcription by RNA polymerase. Nature 221:43–46

Burgie ES, Vierstra RD (2014) Phytochromes: an atomic perspective on photoactivation and signaling. Plant Cell 26:4568–4583

Cahoon AB, Harris FM, Stern DB (2004) Analysis of developing maize plastids reveals two mRNA stability classes correlating with RNA polymerase type. EMBO Rep 5:801–806
Cahoon AB, Takacs EM, Sharpe RM, Stern DB (2008) Nuclear, chloroplast, and mitochondrial transcript abundance along a maize leaf developmental gradient. Plant Mol Biol 66:33–46
Cano-Ramirez DL, Panter PE, Takemura T et al (2023) Low-temperature and circadian signals are integrated by the sigma factor SIG5. Nat Plants 9:661–672
Caplan JL, Kumar AS, Park E et al (2015) Chloroplast stromules function during innate immunity. Dev Cell 34:45–57
Cappadocia L, Maréchal A, Parent J-S et al (2010) Crystal structures of DNA-Whirly complexes and their role in *Arabidopsis* organelle genome repair. Plant Cell 22:1849–1867
Cappadocia L, Parent J-S, Zampini E et al (2012) A conserved lysine residue of plant Whirly proteins is necessary for higher order protein assembly and protection against DNA damage. Nucleic Acids Res 40:258–269
Castandet B, Germain A, Hotto AM, Stern DB (2019) Systematic sequencing of chloroplast transcript termini from Arabidopsis thaliana reveals >200 transcription initiation sites and the extensive imprints of RNA-binding proteins and secondary structures. Nucleic Acids Res 47:11889–11905
Chambon L, Gillet F-X, Chieb M et al (2022) PAP8/pTAC6 is part of a nuclear protein complex and displays RNA recognition motifs of viral origin. Int J Mol Sci 23. https://doi.org/10.3390/ijms23063059
Chang SH, Lee S, Um TY et al (2017) pTAC10, a key subunit of plastid-encoded RNA polymerase, promotes chloroplast development. Plant Physiol 174:435–449
Chang C-Y, Chen L-J, Li H-M (2023) Chloroplast import motor subunits FtsHi1 and FtsHi2 are located on opposite sides of the inner envelope membrane. Proc Natl Acad Sci USA 120:e2307747120
Chateigner-Boutin A-L, Ramos-Vega M, Guevara-García A et al (2008) CLB19, a pentatricopeptide repeat protein required for editing of rpoA and clpP chloroplast transcripts. Plant J 56:590–602
Chen M, Chory J (2011) Phytochrome signaling mechanisms and the control of plant development. Trends Cell Biol 21:664–671
Chen M, Schwab R, Chory J (2003) Characterization of the requirements for localization of phytochrome B to nuclear bodies. Proc Natl Acad Sci USA 100:14493–14498
Chen M, Chory J, Fankhauser C (2004) Light signal transduction in higher plants. Annu Rev Genet 38:87–117
Chen M, Tao Y, Lim J et al (2005) Regulation of phytochrome B nuclear localization through light-dependent unmasking of nuclear-localization signals. Curr Biol 15:637–642
Chen M, Galvão RM, Li M et al (2010) Arabidopsis HEMERA/pTAC12 initiates photomorphogenesis by phytochromes. Cell 141:1230–1240
Cheng M-C, Kathare PK, Paik I, Huq E (2021) Phytochrome signaling networks. Annu Rev Plant Biol 72:217–244
Chi W, He B, Mao J et al (2015) Plastid sigma factors: their individual functions and regulation in transcription. Biochim Biophys Acta 1847:770–778
Chibani K, Tarrago L, Schürmann P et al (2011) Biochemical properties of poplar thioredoxin z. FEBS Lett 585:1077–1081
Christopher DA (1996) Leaf development and phytochrome modulate the activation of *psbD-psbC* transcription by high-fluence blue light in barley chloroplasts. Photosynth Res 47:239–251
Chotewutmontri P, Barkan A (2016) Dynamics of chloroplast translation during chloroplast differentiation in maize. PLoS Genet 12:e1006106
Christopher DA, Hoffer PH (1998) DET1 represses a chloroplast blue light-responsive promoter in a developmental and tissue-specific manner in *Arabidopsis thaliana*. Plant J 14:1–11
Courtois F, Merendino L, Demarsy E et al (2007) Phage-type RNA polymerase RPOTmp transcribes the *rrn* operon from the PC promoter at early developmental stages in *Arabidopsis*. Plant Physiol 145:712–721

De Santis-Maclossek G, Kofer W, Bock A et al (1999) Targeted disruption of the plastid RNA polymerase genes *rpoA*, *B* and *C1*: molecular biology, biochemistry and ultrastructure. Plant J 18:477–489

Deng XW, Gruissem W (1987) Control of plastid gene expression during development: the limited role of transcriptional regulation. Cell 49:379–387

Deng XW, Stern DB, Tonkyn JC, Gruissem W (1987) Plastid run-on transcription. Application to determine the transcriptional regulation of spinach plastid genes. J Biol Chem 262:9641–9648

dePamphilis CW, Palmer JD (1990) Loss of photosynthetic and chlororespiratory genes from the plastid genome of a parasitic flowering plant. Nature 348:337–339

Díaz MG, Hernández-Verdeja T, Kremnev D et al (2018) Redox regulation of PEP activity during seedling establishment in *Arabidopsis thaliana*. Nat Commun 9:50

Dong J, Ni W, Yu R et al (2017) Light-dependent degradation of PIF3 by SCFEBF1/2 promotes a photomorphogenic response in *Arabidopsis*. Curr Biol 27:2420–2430.e6

Dubreuil C, Jin X, de Dios Barajas-López J et al (2018) Establishment of photosynthesis through chloroplast development is controlled by two distinct regulatory phases. Plant Physiol 176:1199–1214

Emanuel C, Weihe A, Graner A et al (2004) Chloroplast development affects expression of phage-type RNA polymerases in barley leaves. Plant J 38:460–472

Fankhauser C, Chen M (2008) Transposing phytochrome into the nucleus. Trends Plant Sci 13:596–601

Favory J-J, Kobayshi M, Tanaka K et al (2005) Specific function of a plastid sigma factor for *ndhF* gene transcription. Nucleic Acids Res 33:5991–5999

Feklístov A, Sharon BD, Darst SA, Gross CA (2014) Bacterial sigma factors: a historical, structural, and genomic perspective. Ann Rev Microbiol 68:357–376

Fu X-X, Zhang J, Zhang G-Q et al (2021) Insights into the origin and evolution of plant sigma factors. J Syst Evol 59:326–340

Fujiwara M, Nagashima A, Kanamaru K et al (2000) Three new nuclear genes, *sigD*, *sigE* and *sigF*, encoding putative plastid RNA polymerase sigma factors in *Arabidopsis thaliana*. FEBS Lett 481:47–52

Galvão RM, Li M, Kothadia SM et al (2012) Photoactivated phytochromes interact with HEMERA and promote its accumulation to establish photomorphogenesis in *Arabidopsis*. Genes Dev 26:1851–1863

Gao Z-P, Yu Q-B, Zhao T-T et al (2011) A functional component of the transcriptionally active chromosome complex, Arabidopsis pTAC14, interacts with pTAC12/HEMERA and regulates plastid gene expression. Plant Physiol 157:1733–1745

Garcia M, Myouga F, Takechi K et al (2008) An *Arabidopsis* homolog of the bacterial peptidoglycan synthesis enzyme MurE has an essential role in chloroplast development. Plant J 53:924–934

Gatenby AA, Castleton JA, Saul MW (1981) Expression in *E. coli* of maize and wheat chloroplast genes for large subunit of ribulose bisphosphate carboxylase. Nature 291:117–121

Gilkerson J, Perez-Ruiz JM, Chory J, Callis J (2012) The plastid-localized pfkB-type carbohydrate kinases FRUCTOKINASE-LIKE 1 and 2 are essential for growth and development of *Arabidopsis thaliana*. BMC Plant Biol 12:102

Gould SB, Waller RF, McFadden GI (2008) Plastid evolution. Annu Rev Plant Biol 59:491–517

Grabowski E, Miao Y, Mulisch M, Krupinska K (2008) Single-stranded DNA-binding protein Whirly1 in barley leaves is located in plastids and the nucleus of the same cell. Plant Physiol 147:1800–1804

Grübler B, Merendino L, Twardziok SO et al (2017) Light and plastid signals regulate different sets of genes in the albino mutant Pap7-1. Plant Physiol 175:1203–1219

Hajdukiewicz PT, Allison LA, Maliga P (1997) The two RNA polymerases encoded by the nuclear and the plastid compartments transcribe distinct groups of genes in tobacco plastids. EMBO J 16:4041–4048

Hallick RB, Lipper C, Richards OC, Rutter WJ (1976) Isolation of a transcriptionally active chromosome from chloroplasts of *Euglena gracilis*. Biochemistry 15:3039–3045

Hanaoka M, Kanamaru K, Takahashi H, Tanaka K (2003) Molecular genetic analysis of chloroplast gene promoters dependent on SIG2, a nucleus-encoded sigma factor for the plastid-encoded RNA polymerase, in *Arabidopsis thaliana*. Nucleic Acids Res 31:7090–7098

Hanaoka M, Kato M, Anma M, Tanaka K (2012) SIG1, a sigma factor for the chloroplast RNA polymerase, differently associates with multiple DNA regions in the chloroplast chromosomes in vivo. Int J Mol Sci 13:12182–12194

He L, Zhang S, Qiu Z et al (2018) Fructokinase-like protein 1 interacts with TRXz to regulate chloroplast development in rice. J Integr Plant Biol 60:94–111

Hedtke B, Börner T, Weihe A (1997) Mitochondrial and chloroplast phage-type RNA polymerases in *Arabidopsis*. Science 277:809–811

Hedtke B, Börner T, Weihe A (2000) One RNA polymerase serving two genomes. EMBO Rep 1:435–440

Hess WR, Prombona A, Fieder B et al (1993) Chloroplast *rps15* and the *rpoB/C1/C2* gene cluster are strongly transcribed in ribosome-deficient plastids: evidence for a functioning non-chloroplast-encoded RNA polymerase. EMBO J 12:563–571

Hoffer PH, Christopher DA (1997) Structure and blue-light-responsive transcription of a chloroplast *psbD* promoter from *Arabidopsis thaliana*. Plant Physiol 115:213–222

Hotto AM, Schmitz RJ, Fei Z et al (2011) Unexpected diversity of chloroplast noncoding RNAs as revealed by deep sequencing of the *Arabidopsis* transcriptome G3(1):559–570

Hricová A, Quesada V, Micol JL (2006) The *SCABRA3* nuclear gene encodes the plastid RpoTp RNA polymerase, which is required for chloroplast biogenesis and mesophyll cell proliferation in *Arabidopsis*. Plant Physiol 141:942–956

Hu J, Bogorad L (1990) Maize chloroplast RNA polymerase: the 180-, 120-, and 38-kilodalton polypeptides are encoded in chloroplast genes. Proc Natl Acad Sci USA 87:1531–1535

Hu J, Troxler RF, Bogorad L (1991) Maize chloroplast RNA polymerase: the 78-kilodaiton polypeptide is encoded by the plastid *rpoC1* gene. Nucleic Acids Res 19:3431–3434

Hu W, Franklin KA, Sharrock RA et al (2013) Unanticipated regulatory roles for Arabidopsis phytochromes revealed by null mutant analysis. Proc Natl Acad Sci USA 110:1542–1547

Huang C, Yu Q-B, Lv R-H et al (2013) The reduced plastid-encoded polymerase-dependent plastid gene expression leads to the delayed greening of the *Arabidopsis fln2* mutant. PLoS One 8:e73092

Hwang Y, Han S, Yoo CY et al (2022) Anterograde signaling controls plastid transcription via sigma factors separately from nuclear photosynthesis genes. Nat Commun 13:7440

Imamura S, Asayama M (2009) Sigma factors for cyanobacterial transcription. Gene Regul Syst Biol 3:65–87

Isemer R, Mulisch M, Schäfer A et al (2012) Recombinant Whirly1 translocates from transplastomic chloroplasts to the nucleus. FEBS Lett 586:85–88

Ishizaki Y, Tsunoyama Y, Hatano K et al (2005) A nuclear-encoded sigma factor, *Arabidopsis* SIG6, recognizes sigma-70 type chloroplast promoters and regulates early chloroplast development in cotyledons. Plant J 42:133–144

Jabben M, Deitzer GF (1979) Effects of the herbicide San 9789 on photomorphogenic responses. Plant Physiol 63:481–485

Jeon Y, Jung HJ, Kang H et al (2012) S1 domain-containing STF modulates plastid transcription and chloroplast biogenesis in *Nicotiana benthamiana*. New Phytol 193:349–363

Kanamaru K, Tanaka K (2004) Roles of chloroplast RNA polymerase sigma factors in chloroplast development and stress response in higher plants. Biosci Biotechnol Biochem 68:2215–2223

Kanamaru K, Nagashima A, Fujiwara M et al (2001) An Arabidopsis sigma factor (SIG2)-dependent expression of plastid-encoded tRNAs in chloroplasts. Plant Cell Physiol 42:1034–1043

Kanno A, Hirai A (1993) A transcription map of the chloroplast genome from rice (*Oryza sativa*). Curr Genet 23:166–174

Kendrick R, Chotewutmontri P, Belcher S, Barkan A (2022) Correlated retrograde and developmental regulons implicate multiple retrograde signals as coordinators of chloroplast development in maize. Plant Cell 34:4897–4919
Kessler F, Schnell D (2009) Chloroplast biogenesis: diversity and regulation of the protein import apparatus. Curr Opin Cell Biol 21:494–500
Kindgren P, Strand Å (2015) Chloroplast transcription, untangling the Gordian Knot. New Phytol 206:889–891
Kindgren P, Kremnev D, Blanco NE et al (2012) The plastid redox insensitive 2 mutant of Arabidopsis is impaired in PEP activity and high light-dependent plastid redox signalling to the nucleus. Plant J 70:279–291
Kircher S, Kozma-Bognar L, Kim L et al (1999) Light quality-dependent nuclear import of the plant photoreceptors phytochrome A and B. Plant Cell 11:1445–1456
Kirk JT (1964) Studies on RNA synthesis in chloroplast preparations. Biochem Biophys Res Commun 16:233–238
Kirk JTO, Tilney-Bassett RAE (1978) The plastids: their chemistry, structure. Growth Inheritance 6
Klein RR, Mullet JE (1990) Light-induced transcription of chloroplast genes. *psbA* transcription is differentially enhanced in illuminated barley. J Biol Chem 265:1895–1902
Klein S, Schiff JA (1972) The correlated appearance of prolamellar bodies, protochlorophyll(ide) species, and the shibata shift during development of bean etioplasts in the dark. Plant Physiol 49:619–626
Kleine T, Maier UG, Leister D (2009) DNA transfer from organelles to the nucleus: the idiosyncratic genetics of endosymbiosis. Annu Rev Plant Biol 60:115–138
Koussevitzky S, Nott A, Mockler TC et al (2007) Signals from chloroplasts converge to regulate nuclear gene expression. Science 316:715–719
Krause K, Maier RM, Kofer W et al (2000) Disruption of plastid-encoded RNA polymerase genes in tobacco: expression of only a distinct set of genes is not based on selective transcription of the plastid chromosome. Mol Gen Genet 263:1022–1030
Kremnev D, Strand A (2014) Plastid encoded RNA polymerase activity and expression of photosynthesis genes required for embryo and seed development in Arabidopsis. Front Plant Sci 5:385
Kühn K, Bohne A-V, Liere K et al (2007) Arabidopsis phage-type RNA polymerases: accurate in vitro transcription of organellar genes. Plant Cell 19:959–971
Kumar AS, Park E, Nedo A et al (2018) Stromule extension along microtubules coordinated with actin-mediated anchoring guides perinuclear chloroplast movement during innate immunity. elife 7:e23625. https://doi.org/10.7554/eLife.23625
Lane WJ, Darst SA (2010) Molecular evolution of multisubunit RNA polymerases: sequence analysis. J Mol Biol 395:671–685
Lee J-H, Doan TM, Senthilkumar S, Yoo CY (2024) Mechanism of nucleus-chloroplast communication by alternative promoter usage and stromules to establish photomorphogenesis in *Arabidopsis*. bioRxiv 2024.05.13.593997
Leech RM, Rumsby MG, Thomson WW (1973) Plastid differentiation, acyl lipid, and fatty acid changes in developing green maize leaves. Plant Physiol 52:240–245
Lefebvre-Legendre L, Merendino L, Rivier C, Goldschmidt-Clermont M (2014) On the complexity of chloroplast RNA metabolism: *psaA* trans-splicing can be bypassed in *Chlamydomonas*. Mol Biol Evol 31:2697–2707
Legen J, Kemp S, Krause K et al (2002) Comparative analysis of plastid transcription profiles of entire plastid chromosomes from tobacco attributed to wild-type and PEP-deficient transcription machineries. Plant J 31:171–188
Legris M, Ince YÇ, Fankhauser C (2019) Molecular mechanisms underlying phytochrome-controlled morphogenesis in plants. Nat Commun 10:5219. https://doi.org/10.1038/s41467-019-13045-0
Leivar P, Quail PH (2011) PIFs: pivotal components in a cellular signaling hub. Trends Plant Sci 16:19–28

Leivar P, Monte E, Al-Sady B et al (2008a) The *Arabidopsis* phytochrome-interacting factor PIF7, together with PIF3 and PIF4, regulates responses to prolonged red light by modulating phyB levels. Plant Cell 20:337–352

Leivar P, Monte E, Oka Y et al (2008b) Multiple phytochrome-interacting bHLH transcription factors repress premature seedling photomorphogenesis in darkness. Curr Biol 18:1815–1823

Leivar P, Tepperman JM, Monte E et al (2009) Definition of early transcriptional circuitry involved in light-induced reversal of PIF-imposed repression of photomorphogenesis in young *Arabidopsis* seedlings. Plant Cell 21:3535–3553

Lerbs-Mache S (2011) Function of plastid sigma factors in higher plants: regulation of gene expression or just preservation of constitutive transcription? Plant Mol Biol 76:235–249

Li L, Ljung K, Breton G et al (2012) Linking photoreceptor excitation to changes in plant architecture. Genes Dev 26:785–790

Lian H-L, He S-B, Zhang Y-C et al (2011) Blue-light-dependent interaction of cryptochrome 1 with SPA1 defines a dynamic signaling mechanism. Genes Dev 25:1023–1028

Liebers M, Gillet F-X, Israel A et al (2020) Nucleo-plastidic PAP8/pTAC6 couples chloroplast formation with photomorphogenesis. EMBO J 39:e104941

Liere K, Börner T (2007) Transcription and transcriptional regulation in plastids. In: Bock R (ed) Cell and molecular biology of plastids. Springer, Berlin, Heidelberg, pp 121–174

Liere K, Weihe A, Börner T (2011) The transcription machineries of plant mitochondria and chloroplasts: composition, function, and regulation. J Plant Physiol 168:1345–1360

Link G (1982) Phytochrome control of plastid mRNA in mustard (*Sinapis alba* L.). Planta 154:81–86

Little MC, Hallick RB (1988) Chloroplast *rpoA*, *rpoB*, and *rpoC* genes specify at least three components of a chloroplast DNA-dependent RNA polymerase active in tRNA and mRNA transcription. J Biol Chem 263:14302–14307

Liu B, Zuo Z, Liu H et al (2011) *Arabidopsis* cryptochrome 1 interacts with SPA1 to suppress COP1 activity in response to blue light. Genes Dev 25:1029–1034

Lorrain S, Allen T, Duek PD et al (2008) Phytochrome-mediated inhibition of shade avoidance involves degradation of growth-promoting bHLH transcription factors. Plant J 53:312–323

Loschelder H, Schweer J, Link B, Link G (2006) Dual temporal role of plastid sigma factor 6 in *Arabidopsis* development. Plant Physiol 142:642–650

Loudya N, Mishra P, Takahagi K et al (2021) Cellular and transcriptomic analyses reveal two-staged chloroplast biogenesis underpinning photosynthesis build-up in the wheat leaf. Genome Biol 22:151

Lu X-D, Zhou C-M, Xu P-B et al (2015) Red-light-dependent interaction of phyB with SPA1 promotes COP1-SPA1 dissociation and photomorphogenic development in *Arabidopsis*. Mol Plant 8:467–478

Luo Q, Lian H-L, He S-B et al (2014) COP1 and phyB physically interact with PIL1 to regulate its stability and photomorphogenic development in *Arabidopsis*. Plant Cell 26:2441–2456

Macadlo LA, Ibrahim IM, Puthiyaveetil S (2020) Sigma factor 1 in chloroplast gene transcription and photosynthetic light acclimation. J Exp Bot 71:1029–1038

Mapleston RE, Griffiths WT (1980) Light modulation of the activity of protochlorophyllide reductase. Biochem J 189:125–133

Martin W, Herrmann RG (1998) Gene transfer from organelles to the nucleus: how much, what happens, and why? Plant Physiol 118:9–17

Martin W, Kowallik KV (1999) Annotated English translation of Mereschkowsky's 1905 paper "Über Natur und Ursprung der Chromatophoren im Pflanzenreiche". Eur J Phycol 34:287–295

Martín M, Sabater B (2010) Plastid *ndh* genes in plant evolution. Plant Physiol Biochem 48:636–645

Martin W, Rujan T, Richly E et al (2002) Evolutionary analysis of *Arabidopsis*, cyanobacterial, and chloroplast genomes reveals plastid phylogeny and thousands of cyanobacterial genes in the nucleus. Proc Natl Acad Sci USA 99:12246–12251

Mazumder A, Lin M, Kapanidis AN, Ebright RH (2020) Closing and opening of the RNA polymerase trigger loop. Proc Natl Acad Sci USA 117:15642–15649
Melonek J, Mulisch M, Schmitz-Linneweber C et al (2010) Whirly1 in chloroplasts associates with intron containing RNAs and rarely co-localizes with nucleoids. Planta 232:471–481
Mereschkowsky C (1905) Über natur und ursprung der chromatophoren im pflanzenreiche. Biol Centralbl 25:593
Mitschke J, Georg J, Scholz I et al (2011) An experimentally anchored map of transcriptional start sites in the model cyanobacterium *Synechocystis* sp. PCC6803. Proc Natl Acad Sci USA 108:2124–2129
Mohanta TK, Mohanta YK, Sharma N (2023) Anticodon table of the chloroplast genome and identification of putative quadruplet anticodons in chloroplast tRNAs. Sci Rep 13:760
Morden CW, Wolfe KH, dePamphilis CW, Palmer JD (1991) Plastid translation and transcription genes in a non-photosynthetic plant: intact, missing and pseudo genes. EMBO J 10:3281–3288
Mullet JE, Klein RR (1987) Transcription and RNA stability are important determinants of higher plant chloroplast RNA levels. EMBO J 6:1571–1579
Murakami KS, Masuda S, Campbell EA et al (2002) Structural basis of transcription initiation: an RNA polymerase holoenzyme-DNA complex. Science 296:1285–1290
Myouga F, Hosoda C, Umezawa T et al (2008) A heterocomplex of iron superoxide dismutases defends chloroplast nucleoids against oxidative stress and is essential for chloroplast development in *Arabidopsis*. Plant Cell 20:3148–3162
Nagashima A, Hanaoka M, Motohashi R et al (2004a) DNA microarray analysis of plastid gene expression in an *Arabidopsis* mutant deficient in a plastid transcription factor sigma, SIG2. Biosci Biotechnol Biochem 68:694–704
Nagashima A, Hanaoka M, Shikanai T et al (2004b) The multiple-stress responsive plastid sigma factor, SIG5, directs activation of the *psbD* blue light-responsive promoter (BLRP) in *Arabidopsis thaliana*. Plant Cell Physiol 45:357–368
Nawaz G, Kang H (2017) Chloroplast- or mitochondria-targeted DEAD-Box RNA helicases play essential roles in organellar RNA metabolism and abiotic stress responses. Front Plant Sci 8:871
Nevarez PA, Qiu Y, Inoue H et al (2017) Mechanism of dual targeting of the phytochrome signaling component HEMERA/pTAC12 to plastids and the nucleus. Plant Physiol 173:1953–1966
Ni W, Xu S-L, Tepperman JM et al (2014) A mutually assured destruction mechanism attenuates light signaling in *Arabidopsis*. Science 344:1160–1164
Ni W, Xu S-L, González-Grandío E et al (2017) PPKs mediate direct signal transfer from phytochrome photoreceptors to transcription factor PIF3. Nat Commun 8:15236
Noordally ZB, Ishii K, Atkins KA et al (2013) Circadian control of chloroplast transcription by a nuclear-encoded timing signal. Science 339:1316–1319
Nott A, Jung H-S, Koussevitzky S, Chory J (2006) Plastid-to-nucleus retrograde signaling. Annu Rev Plant Biol 57:739–759
Oelmüller R (1989) Photooxidative destruction of chloroplasts and its effect on nuclear gene expression and extraplastidic enzyme levels. Photochem Photobiol 49:229–239
Ohyama K, Fukuzawa H, Kohchi T et al (1986) Chloroplast gene organization deduced from complete sequence of liverwort *Marchantia polymorpha* chloroplast DNA. Nature 322:572–574
Oikawa K, Fujiwara M, Nakazato E et al (2000) Characterization of two plastid sigma factors, SigA1 and SigA2, that mainly function in matured chloroplasts in *Nicotiana tabacum*. Gene 261:221–228
Onda Y, Yagi Y, Saito Y et al (2008) Light induction of Arabidopsis SIG1 and SIG5 transcripts in mature leaves: differential roles of cryptochrome 1 and cryptochrome 2 and dual function of SIG5 in the recognition of plastid promoters. Plant J 55:968–978
Ortelt J, Link G (2014) Plastid gene transcription: promoters and RNA polymerases. In: Maliga P (ed) Chloroplast biotechnology: methods and protocols. Humana Press, Totowa, NJ, pp 47–72

Ortelt J, Link G (2021) Plastid gene transcription: an update on promoters and RNA polymerases. In: Maliga P (ed) Chloroplast biotechnology: methods and protocols. Springer, New York, NY, pp 49–76

Palomar VM, Jaksich S, Fujii S et al (2022) High-resolution map of plastid-encoded RNA polymerase binding patterns demonstrates a major role of transcription in chloroplast gene expression. Plant J 111:1139–1151

Park E, Park J, Kim J et al (2012) Phytochrome B inhibits binding of phytochrome-interacting factors to their target promoters. Plant J 72:537–546

Park E, Kim Y, Choi G (2018) Phytochrome B requires PIF degradation and sequestration to induce light responses across a wide range of light conditions. Plant Cell 30:1277–1292

Parks BM, Quail PH (1993) *hy8*, a new class of arabidopsis long hypocotyl mutants deficient in functional phytochrome A. Plant Cell 5:39–48

Patel UR, Gautam S, Chatterji D (2020) Validation of omega subunit of RNA polymerase as a functional entity. Biomol Ther 10:1588. https://doi.org/10.3390/biom10111588

Pearce SF, Rebelo-Guiomar P, D'Souza AR et al (2017) Regulation of mammalian mitochondrial gene expression: recent advances. Trends Biochem Sci 42:625–639

Pedmale UV, Huang S-SC, Zander M et al (2016) Cryptochromes Interact directly with PIFs to control plant growth in limiting blue light. Cell 164:233–245

Peltier G, Aro E-M, Shikanai T (2016) NDH-1 and NDH-2 plastoquinone reductases in oxygenic photosynthesis. Annu Rev Plant Biol 67:55–80

Pfalz J, Pfannschmidt T (2013) Essential nucleoid proteins in early chloroplast development. Trends Plant Sci 18:186–194

Pfalz J, Liere K, Kandlbinder A et al (2006) pTAC2, -6, and -12 are components of the transcriptionally active plastid chromosome that are required for plastid gene expression. Plant Cell 18:176–197

Pfalz J, Holtzegel U, Barkan A et al (2015) ZmpTAC12 binds single-stranded nucleic acids and is essential for accumulation of the plastid-encoded polymerase complex in maize. New Phytol 206:1024–1037

Pfannschmidt T, Link G (1994) Separation of two classes of plastid DNA-dependent RNA polymerases that are differentially expressed in mustard (*Sinapis alba* L.) seedlings. Plant Mol Biol 25:69–81

Pfannschmidt T, Link G (1997) The A and B forms of plastid DNA-dependent RNA polymerase from mustard (*Sinapis alba* L.) transcribe the same genes in a different developmental context. Mol Gen Genet 257:35–44

Pfannschmidt T, Blanvillain R, Merendino L et al (2015) Plastid RNA polymerases: orchestration of enzymes with different evolutionary origins controls chloroplast biogenesis during the plant life cycle. J Exp Bot 66:6957–6973

Pham VN, Kathare PK, Huq E (2018) Dynamic regulation of PIF5 by COP1-SPA complex to optimize photomorphogenesis in *Arabidopsis*. Plant J 96:260–273

Polya GM, Jagendorf AT (1971a) Wheat leaf RNA polymerases. I. Partial purification and characterization of nuclear, chloroplast and soluble DNA-dependent enzymes. Arch Biochem Biophys 146:635–648

Polya GM, Jagendorf AT (1971b) Wheat leaf RNA polymerases. II. Kinetic characterization and template specificities of nuclear, chloroplast, and soluble enzymes. Arch Biochem Biophys 146:649–657

do Prado PFV, Ahrens FM, Liebers M et al (2024) Structure of the multi-subunit chloroplast RNA polymerase. Mol Cell 84:910–925.e5

Prikryl J, Watkins KP, Friso G et al (2008) A member of the Whirly family is a multifunctional RNA- and DNA-binding protein that is essential for chloroplast biogenesis. Nucleic Acids Res 36:5152–5165

Puthiyaveetil S, McKenzie SD, Kayanja GE, Ibrahim IM (2021) Transcription initiation as a control point in plastid gene expression. Biochim Biophys Acta Gene Regul Mech 1864:194689

Qayyum MZ, Imashimizu M, Leanca M et al (2024) Structure and function of the Si3 insertion integrated into the trigger loop/helix of cyanobacterial RNA polymerase. Proc Natl Acad Sci USA 121:e2311480121

Qiu Y, Li M, Pasoreck EK et al (2015) HEMERA couples the proteolysis and transcriptional activity of PHYTOCHROME INTERACTING FACTORs in *Arabidopsis* photomorphogenesis. Plant Cell 27:1409–1427

Qiu Y, Li M, Kim RJ-A et al (2019) Daytime temperature is sensed by phytochrome B in *Arabidopsis* through a transcriptional activator HEMERA. Nat Commun 10:140

Qiu Y, Pasoreck EK, Yoo CY et al (2021) RCB initiates *Arabidopsis* thermomorphogenesis by stabilizing the thermoregulator PIF4 in the daytime. Nat Commun 12:2042

Reed JW, Chory J (1994) Mutational analyses of light-controlled seedling development in *Arabidopsis*. Semin Cell Biol 5:327–334

Reed JW, Nagpal P, Poole DS et al (1993) Mutations in the gene for the red/far-red light receptor phytochrome B alter cell elongation and physiological responses throughout *Arabidopsis* development. Plant Cell 5:147–157

Ris H, Plaut W (1962) Ultrastructure of DNA-containing areas in the chloroplast of *Chlamydomonas*. J Cell Biol 13:383–391

Rockwell NC, Su Y-S, Lagarias JC (2006) Phytochrome structure and signaling mechanisms. Annu Rev Plant Biol 57:837–858

Ruckle ME, Burgoon LD, Lawrence LA et al (2012) Plastids are major regulators of light signaling in *Arabidopsis*. Plant Physiol 159:366–390

Ruedas R, Muthukumar SS, Kieffer-Jaquinod S et al (2022) Three-dimensional envelope and subunit interactions of the plastid-encoded RNA polymerase from *Sinapis alba*. Int J Mol Sci 23:9922. https://doi.org/10.3390/ijms23179922

Sagan L (1967) On the origin of mitosing cells. J Theor Biol 14:255–274

Sager R, Ishida MR (1963) Chloroplast DNA in *Chlamydomonas*. Proc Natl Acad Sci USA 50:725–730

Saijo Y, Sullivan JA, Wang H et al (2003) The COP1-SPA1 interaction defines a critical step in phytochrome A-mediated regulation of HY5 activity. Genes Dev 17:2642–2647

Sakai A, Kawano S, Kuroiwa T (1992) Conversion of proplastids to amyloplasts in tobacco cultured cells is accompanied by changes in the transcriptional activities of plastid genes. Plant Physiol 100:1062–1066

Sakamoto K, Nagatani A (1996) Nuclear localization activity of phytochrome B. Plant J 10:859–868

Sanitá Lima M, Smith DR (2017) Pervasive, genome-wide transcription in the organelle genomes of diverse plastid-bearing protists. G3 7:3789–3796

Santel HJ, Apel K (1981) The protochlorophyllide holochrome of barley (*Hordeum vulgare* L.). The effect of light on the NADPH:protochlorophyllide oxidoreductase. Eur J Biochem 120:95–103

Sato S, Nakamura Y, Kaneko T et al (1999) Complete structure of the chloroplast genome of Arabidopsis thaliana. DNA Res 6:283–290

Schneider T, Bolger A, Zeier J et al (2019) Fluctuating light interacts with time of day and leaf development stage to reprogram gene expression. Plant Physiol 179:1632–1657

Schröter Y, Steiner S, Matthäi K, Pfannschmidt T (2010) Analysis of oligomeric protein complexes in the chloroplast sub-proteome of nucleic acid-binding proteins from mustard reveals potential redox regulators of plastid gene expression. Proteomics 10:2191–2204

Schweer J, Türkeri H, Kolpack A, Link G (2010) Role and regulation of plastid sigma factors and their functional interactors during chloroplast transcription – recent lessons from Arabidopsis thaliana. Eur J Cell Biol 89:940–946

Shahar N, Weiner I, Stotsky L et al (2019) Prediction and large-scale analysis of primary operons in plastids reveals unique genetic features in the evolution of chloroplasts. Nucleic Acids Res 47:3344–3352

Sharrock RA, Quail PH (1989) Novel phytochrome sequences in *Arabidopsis thaliana*: structure, evolution, and differential expression of a plant regulatory photoreceptor family. Genes Dev 3:1745–1757

Sharwood RE, Hotto AM, Bollenbach TJ, Stern DB (2011) Overaccumulation of the chloroplast antisense RNA AS5 is correlated with decreased abundance of 5S rRNA in vivo and inefficient 5S rRNA maturation in vitro. RNA 17:230–243

Sheerin DJ, Menon C, zur Oven-Krockhaus S et al (2015) Light-activated phytochrome A and B interact with members of the SPA family to promote photomorphogenesis in Arabidopsis by reorganizing the COP1/SPA complex. Plant Cell 27:189–201

Shen H, Moon J, Huq E (2005) PIF1 is regulated by light-mediated degradation through the ubiquitin-26S proteasome pathway to optimize photomorphogenesis of seedlings in *Arabidopsis*. Plant J 44:1023–1035

Shen Y, Khanna R, Carle CM, Quail PH (2007) Phytochrome induces rapid PIF5 phosphorylation and degradation in response to red-light activation. Plant Physiol 145:1043–1051

Shi C, Wang S, Xia E-H et al (2016) Full transcription of the chloroplast genome in photosynthetic eukaryotes. Sci Rep 6:30135

Shiina T, Ishizaki Y, Yagi Y, Nakahira Y (2009) Function and evolution of plastid sigma factors. Plant Biotechnol 26:57–66

Shikanai T (2016) Chloroplast NDH: a different enzyme with a structure similar to that of respiratory NADH dehydrogenase. Biochim Biophys Acta 1857:1015–1022

Shimizu M, Kato H, Ogawa T et al (2010) Sigma factor phosphorylation in the photosynthetic control of photosystem stoichiometry. Proc Natl Acad Sci USA 107:10760–10764

Shimmura S, Nozoe M, Kitora S et al (2017) Comparative analysis of chloroplast *psbD* promoters in terrestrial plants. Front Plant Sci 8:1186

Shin J, Kim K, Kang H et al (2009) Phytochromes promote seedling light responses by inhibiting four negatively-acting phytochrome-interacting factors. Proc Natl Acad Sci USA 106:7660–7665

Shinozaki K, Ohme M, Tanaka M et al (1986) The complete nucleotide sequence of the tobacco chloroplast genome: its gene organization and expression. EMBO J 5:2043–2049

Silhavy D, Maliga P (1998) Mapping of promoters for the nucleus-encoded plastid RNA polymerase (NEP) in the *iojap* maize mutant. Curr Genet 33:340–344

Smith HJ, Bogorad L (1974) The polypeptide subunit structure of the DNA-dependent RNA polymerase of *Zea mays* chloroplasts. Proc Natl Acad Sci USA 71:4839–4842

Smith AC, Purton S (2002) The transcriptional apparatus of algal plastids. Eur J Phycol 37:301–311

Song S, Wang Y, Ding X et al (2023) A plastid RNA polymerase-associated protein is involved in early chloroplast development in rice. Agronomy 13:1424

de Souza A, Wang J-Z, Dehesh K (2017) Retrograde signals: integrators of interorganellar communication and orchestrators of plant development. Annu Rev Plant Biol 68:85–108

Steiner S, Schröter Y, Pfalz J, Pfannschmidt T (2011) Identification of essential subunits in the plastid-encoded RNA polymerase complex reveals building blocks for proper plastid development. Plant Physiol 157:1043–1055

Stern DB, Goldschmidt-Clermont M, Hanson MR (2010) Chloroplast RNA metabolism. Annu Rev Plant Biol 61:125–155

Strasser B, Sánchez-Lamas M, Yanovsky MJ et al (2010) *Arabidopsis thaliana* life without phytochromes. Proc Natl Acad Sci USA 107:4776–4781

Strittmatter G, Gozdzicka-Jozefiak A, Kössel H (1985) Identification of an rRNA operon promoter from *Zea mays* chloroplasts which excludes the proximal $tRNA^{Val}_{GAC}$ from the primary transcript. EMBO J 4:599–604

Sugita M, Sugiura M (1996) Regulation of gene expression in chloroplasts of higher plants. Plant Mol Biol 32:315–326

Sugiura M, Wakasugi T, Kung S (1989) Compilation and comparison of transfer RNA genes from tobacco chloroplasts. CRC Crit Rev Plant Sci 8:89–101

Susek RE, Ausubel FM, Chory J (1993) Signal transduction mutants of *Arabidopsis* uncouple nuclear *CAB* and *RBCS* gene expression from chloroplast development. Cell 74:787–799
Sutherland C, Murakami KS (2018) An introduction to the structure and function of the catalytic core enzyme of *Escherichia coli* RNA polymerase. EcoSal Plus 8. https://doi.org/10.1128/ecosalplus.ESP-0004-2018
Suzuki JY, Ytterberg AJ, Beardslee TA et al (2004) Affinity purification of the tobacco plastid RNA polymerase and in vitro reconstitution of the holoenzyme. Plant J 40:164–172
Swiatecka-Hagenbruch M, Liere K, Börner T (2007) High diversity of plastidial promoters in *Arabidopsis thaliana*. Mol Gen Genomics 277:725–734
Swiatecka-Hagenbruch M, Emanuel C, Hedtke B et al (2008) Impaired function of the phage-type RNA polymerase RpoTp in transcription of chloroplast genes is compensated by a second phage-type RNA polymerase. Nucleic Acids Res 36:785–792
Takano M, Inagaki N, Xie X et al (2009) Phytochromes are the sole photoreceptors for perceiving red/far-red light in rice. Proc Natl Acad Sci USA 106:14705–14710
Taylor WC (1989) Regulatory interactions between nuclear and plastid genomes. Annu Rev Plant Physiol Plant Mol Biol 40:211–233
Thompson WF, Everett M, Polans NO et al (1983) Phytochrome control of RNA levels in developing pea and mung-bean leaves. Planta 158:487–500
Tozawa Y, Teraishi M, Sasaki T et al (2007) The plastid sigma factor SIG1 maintains photosystem I activity via regulated expression of the *psaA* operon in rice chloroplasts. Plant J 52:124–132
Tsunoyama Y, Ishizaki Y, Morikawa K et al (2004) Blue light-induced transcription of plastid-encoded *psbD* gene is mediated by a nuclear-encoded transcription initiation factor, AtSig5. Proc Natl Acad Sci USA 101:3304–3309
Turmel M, Otis C, Lemieux C (2002) The chloroplast and mitochondrial genome sequences of the charophyte *Chaetosphaeridium globosum*: insights into the timing of the events that restructured organelle DNAs within the green algal lineage that led to land plants. Proc Natl Acad Sci USA 99:11275–11280
Van Buskirk EK, Decker PV, Chen M (2012) Photobodies in light signaling. Plant Physiol 158:52–60
Vassylyev DG, Sekine S-I, Laptenko O et al (2002) Crystal structure of a bacterial RNA polymerase holoenzyme at 2.6 A resolution. Nature 417:712–719
Vergara-Cruces Á, Pramanick I, Pearce D et al (2024) Structure of the plant plastid-encoded RNA polymerase. Cell 187:1145–1159.e21
Wang Q, Zuo Z, Wang X et al (2018) Beyond the photocycle-how cryptochromes regulate photoresponses in plants? Curr Opin Plant Biol 45:120–126
Wang J, Miguel Palomar V, Min J-H, Wierzbicki AT (2024) pTAC3 and pTAC14 are required for binding of plastid-encoded RNA polymerase to DNA. bioRxiv 2024.03.01.582956
Wei N, Deng XW (1996) The role of the COP/DET/FUS genes in light control of arabidopsis seedling development. Plant Physiol 112:871–878
Weihe A, Börner T (1999) Transcription and the architecture of promoters in chloroplasts. Trends Plant Sci 4:169–170
Whitelam GC, Johnson E, Peng J et al (1993) Phytochrome A null mutants of *Arabidopsis* display a wild-type phenotype in white light. Plant Cell 5:757–768
Wicke S, Schneeweiss GM, dePamphilis CW et al (2011) The evolution of the plastid chromosome in land plants: gene content, gene order, gene function. Plant Mol Biol 76:273–297
Wimmelbacher M, Börnke F (2014) Redox activity of thioredoxin z and fructokinase-like protein 1 is dispensable for autotrophic growth of *Arabidopsis thaliana*. J Exp Bot 65:2405–2413
Windgassen TA, Mooney RA, Nayak D et al (2014) Trigger-helix folding pathway and SI3 mediate catalysis and hairpin-stabilized pausing by *Escherichia coli* RNA polymerase. Nucleic Acids Res 42:12707–12721
Wollman FA (2001) State transitions reveal the dynamics and flexibility of the photosynthetic apparatus. EMBO J 20:3623–3630

Woodson JD, Perez-Ruiz JM, Schmitz RJ et al (2013) Sigma factor-mediated plastid retrograde signals control nuclear gene expression. Plant J 73:1–13

Wu X-X, Mu W-H, Li F et al (2024) Cryo-EM structures of the plant plastid-encoded RNA polymerase. Cell 187:1127–1144.e21

Yagi Y, Shiina T (2014) Recent advances in the study of chloroplast gene expression and its evolution. Front Plant Sci 5:61

Yagi Y, Shiina T (2016) The chloroplast gene-expression system. In: Khan MS, Khan IA, Barh D (eds) Applied molecular biotechnology. CRC Press, pp 91–125

Yagi Y, Ishizaki Y, Nakahira Y et al (2012) Eukaryotic-type plastid nucleoid protein pTAC3 is essential for transcription by the bacterial-type plastid RNA polymerase. Proc Natl Acad Sci USA 109:7541–7546

Yamaguchi R, Nakamura M, Mochizuki N et al (1999) Light-dependent translocation of a phytochrome B-GFP fusion protein to the nucleus in transgenic *Arabidopsis*. J Cell Biol 145:437–445

Yang EJ, Yoo CY, Liu J et al (2019) NCP activates chloroplast transcription by controlling phytochrome-dependent dual nuclear and plastidial switches. Nat Commun 10:2630

Yoo CY, Pasoreck EK, Wang H et al (2019) Phytochrome activates the plastid-encoded RNA polymerase for chloroplast biogenesis via nucleus-to-plastid signaling. Nat Commun 10:2629

Yoo CY, Han S, Chen M (2020) Nucleus-to-plastid phytochrome signalling in controlling chloroplast biogenesis. Ann Plant Rev online 251–280

Yoo CY, He J, Sang Q et al (2021) Direct photoresponsive inhibition of a p53-like transcription activation domain in PIF3 by *Arabidopsis* phytochrome B. Nat Commun 12:5614

Yu Q-B, Lu Y, Ma Q et al (2013) TAC7, an essential component of the plastid transcriptionally active chromosome complex, interacts with FLN1, TAC10, TAC12 and TAC14 to regulate chloroplast gene expression in *Arabidopsis thaliana*. Physiol Plant 148:408–421

Zeng C, Jiao Q, Jia T, Hu X (2022) Updated progress on group II intron splicing factors in plant chloroplasts. Curr Issues Mol Biol 44:4229–4239

Zghidi W, Merendino L, Cottet A et al (2007) Nucleus-encoded plastid sigma factor SIG3 transcribes specifically the *psbN* gene in plastids. Nucleic Acids Res 35:455–464

Zghidi-Abouzid O, Merendino L, Buhr F et al (2011) Characterization of plastid *psbT* sense and antisense RNAs. Nucleic Acids Res 39:5379–5387

Zhang G, Campbell EA, Minakhin L et al (1999) Crystal structure of *Thermus aquaticus* core RNA polymerase at 3.3 A resolution. Cell 98:811–824

Zhang Y, Tian L, Lu C (2023) Chloroplast gene expression: recent advances and perspectives. Plant Commun 4:100611

Zhao X, Huang J, Chory J (2020) Unraveling the linkage between retrograde signaling and RNA metabolism in plants. Trends Plant Sci 25:141–147

Zhelyazkova P, Sharma CM, Förstner KU et al (2012) The primary transcriptome of barley chloroplasts: numerous noncoding RNAs and the dominating role of the plastid-encoded RNA polymerase. Plant Cell 24:123–136

Zoschke R, Bock R (2018) Chloroplast translation: structural and functional organization, operational control, and regulation. Plant Cell 30:745–770

Chapter 2
RNA Binding Proteins Regulating Chloroplast RNA Metabolism

Andrea A. Zanini, Mohammad F. Azim, Tyra N. McCray, and Tessa M. Burch-Smith

Abstract Nucleus-encoded RNA-binding proteins and ribonucleases have crucial roles in calibrating the transcriptional output of chloroplast genes. Primary chloroplast transcripts are processed through a series of regulated steps including polycistronic transcript processing, splicing to remove introns, cytidine-to-uridine (C-to-U) editing, and controlled mRNA turnover. Key players include pentatricopeptide repeat (PPR) proteins, which stabilize transcripts and guide their processing, RNA helicases, which unwind double-stranded RNA structures by utilizing the ATP hydrolysis energy, and a variety of RNases such as PNPase (a 3′ → 5′ exonuclease), RNase E (endoribonuclease), and RNase J (endoribonuclease and 5′ → 3′ exonuclease). These enzymes coordinate the maturation and degradation of RNA, ensuring the removal of defective or unnecessary transcripts and maintaining RNA stability. The intricate interplay between RNA-binding proteins and ribonucleases shapes the complex transcript patterns within chloroplasts and provides the precise regulation of gene expression. Given that chloroplasts play a pivotal role in cellular processes such as photosynthesis, plant defense, and primary and specialized metabolism, chloroplast RNA metabolism is therefore an essential component of plant development. This chapter discusses the roles of various nucleus-encoded, chloroplast-localized RBPs in chloroplast RNA metabolism and gene expression.

Keywords Chloroplast · RNA metabolism · RNA maturation · RNA degradation · Helicases · Ribonucleases · Exonucleases

---

A. A. Zanini · T. M. Burch-Smith (✉)
Donald Danforth Plant Science Center, St. Louis, MO, USA
e-mail: tburch-smith@danforthcenter.org

M. F. Azim
Donald Danforth Plant Science Center, St. Louis, MO, USA

Department of Biochemistry & Cellular and Molecular Biology, University of Tennessee, Knoxville, TN, USA

T. N. McCray
10x Genomics, Pleasanton, CA, USA

T. M. Burch-Smith (ed.), *Chloroplast Gene Expression: Regulation, Stress Signaling and Biotechnology*, Nucleic Acids and Molecular Biology 37,
https://doi.org/10.1007/978-3-031-70098-9_2

Abbreviations

ABA	Abscisic acid
MORF	Multiple Organellar RNA editing Factor
NTP	nucleoside triphosphate
ORRM	Organelle RNA Recognition Motif protein
OZ	Organelle Zinc finger protein
PEP	plastid-encoded RNA polymerase
PNPase	polynucleotide phosphorylase
PPR	Pentatricopeptide repeat
RBP	RNA-binding proteins
RH	RNA helicase
RIP	RNA-editing factor interacting protein
RNase	ribonuclease

2.1 Introduction

Chloroplasts play a pivotal role in cellular processes such as photosynthesis, plant defense, and primary and specialized metabolism. Chloroplast genomes, remnants of their cyanobacterial progenitors, exhibit prokaryotic features such as genes organized into operons, short intergenic regions, and limited transcriptional regulation. This prokaryote-derived genome requires extensive post-transcriptional RNA processing to generate mature transcripts for translation within this organelle to produce chloroplast-encoded proteins (Fig. 2.1). Therefore, the processes and machinery of chloroplast gene expression essential for expression of the chloroplast genome are distinct from those of nuclear gene expression. Post-transcriptional RNA metabolism in chloroplasts is complex and includes polycistronic transcript processing, splicing to remove introns, cytidine-to-uridine (C-to-U) editing, and controlled mRNA turnover. Thus, as it is necessary for the production of essential chloroplast proteins, including components of the photosynthetic electron transport chain and carbon fixation, chloroplast RNA processing is crucial for cellular function (reviewed by Dreyfuss et al. 1993; Simpson and Filipowicz 1996; Jung et al. 2013; MacIntosh and Castandet 2020; Small et al. 2023).

The chloroplast genome (often called the plastome) is essential for the semi-autonomy of the organelle and it encodes a highly conserved set of about 120–200 genes in land plants (Palmer 1985). The genes encode polypeptides mainly related to photosynthetic proteins, and the plastid gene expression machinery, including ribosomal proteins, rRNAs, and tRNAs (Sugiura 1992). Each chloroplast contains multiple copies of the plastome which are structured into DNA-protein complexes known as nucleoids. These nucleoids serve as a core for DNA replication, gene expression, rRNA processing, and ribosome assembly and transcription (Kuroiwa 1991; Majeran et al. 2012). Plastid nucleoids, localized in the stroma and anchored to the thylakoid membranes, can vary in shape, number, and distribution across plant and algal groups, and they have a dynamic role in chloroplast development

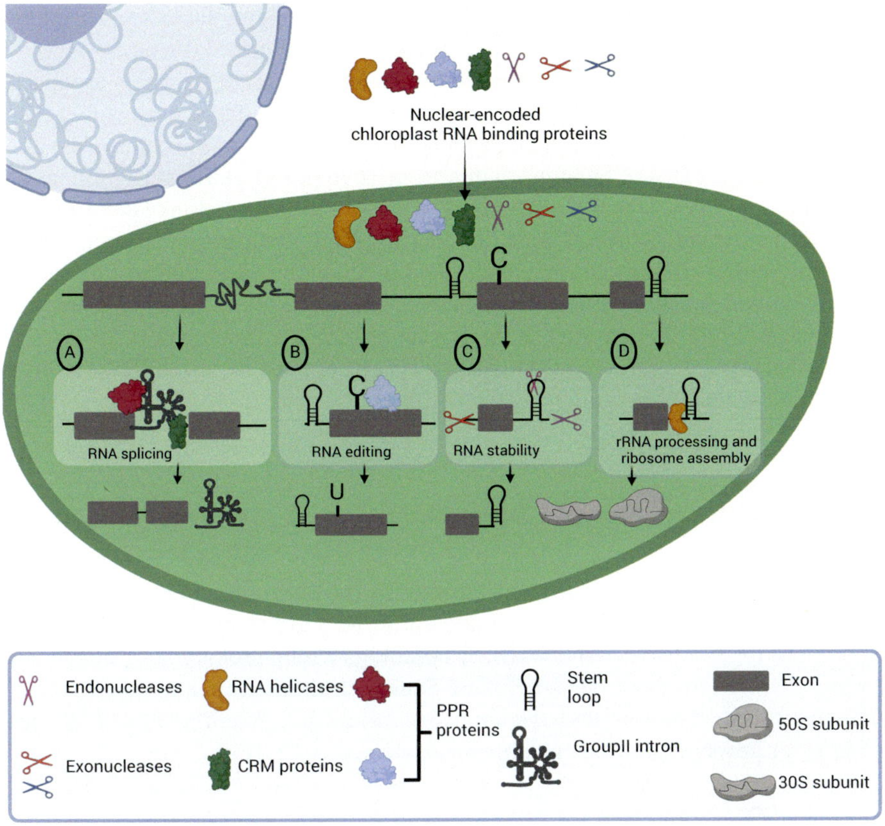

Fig. 2.1 Overview of chloroplast RNA maturation. The post-transcriptional processing of chloroplast RNA transcripts includes intron splicing (**a**), C-to-U RNA editing (**b**), RNA stability (**c**), rRNA processing, and ribosomal assembly (**d**). Nuclear-encoded chloroplast RNA-binding proteins, including endonucleases, exonucleases, and pentatricopeptide proteins (PPR), are involved in chloroplast RNA metabolism. (**a**) Splicing factors are involved in removing introns and fusing the exons. (**b**) PPR proteins facilitate C-to-U RNA editing by targeting editing sites. (**c**) Endonucleases cleave the RNA transcript intercistronically, and exonucleases process the RNA from the 5′- or 3′-end to make mature RNA. (**d**) RNA helicases facilitate chloroplast rRNA processing and assembly of 30S and 50S ribosomal subunits. Image generated with BioRender

(Kuroiwa 1991; Pfalz et al. 2012; Greiner et al. 2020). Recent proteome analysis of proplastids and mature chloroplasts isolated from maize have facilitated the identification of proteins enriched in nucleoids (Majeran et al. 2012). Around 131 proteins, known or predicted to be involved in plastid RNA metabolism, supported the importance of nucleoids in RNA processing, translation, and ribosome assembly (Majeran et al. 2012). The subunits of the plastid-encoded RNA polymerase (PEP) complex, DEAD-box RNA helicases (RH3 homologs and RH39), nucleases

(RNaseJ and PNPase), a few PPR proteins and many splicing factors were, among others, the most abundant RNA-related proteins in the nucleoid fractions.

The chloroplast proteome comprises almost 3000 proteins, of which around 80 are encoded by the plastome (Chotewutmontri et al. 2017). Thus, most chloroplast-resident proteins are encoded in the nucleus and imported into the chloroplast after translation in the cytosol. Included in this group of imported chloroplast proteins are RNA-binding proteins (RBPs) with essential roles in the expression of the plastome. These RBPs belong to several large families of proteins, such as the pentatricopeptide (PPR) and helicase families, and many of them are essential for chloroplast maturation and function, evidenced by embryonic or seedling lethality when they are mutated or lost (Tzafrir et al. 2004; Bryant et al. 2011; Belcher et al. 2015). This chapter discusses the roles of nucleus-encoded, chloroplast-localized RBPS in chloroplast RNA metabolism and gene expression.

2.2 RNA Processing in Chloroplast Gene Expression

The plastid genome reflects its cyanobacterial (prokaryotic) origins. First, most plastid genes are organized in clusters and transcribed as polycistrons. Second, most chloroplast ribonucleases (RNases) have homologs within prokaryotes (Stoppel and Meurer 2012; Germain et al. 2013; Dos Santos et al. 2018). Chloroplast primary RNA transcripts therefore undergo extensive processing, requiring both RBPs and RNases for proper maturation (Stoppel and Meurer 2012) (Fig. 2.1).

2.2.1 Polycistronic RNA Processing

Chloroplast 70S ribosomes consist of two subunits: the small 30S subunit and the large 50S subunit. The 30S subunit comprises 16S rRNA, and the 50S subunit comprises 23S, 4.5S, and 5S rRNAs. Multiple RNA helicases play a unique role in the assembly of functional chloroplasts by facilitating the expression of the chloroplast genome (Goldschmidt-Clermont 1998; Tanner and Linder 2001; Stern et al. 2010) through ribosomal RNA (rRNA) processing (Asakura et al. 2012; Gu et al. 2014; Fig. 2.2). Chloroplast rRNAs are transcribed as a polycistronic transcript which is then post-transcriptionally processed by the coordinated action of RNA helicases and multiple other enzymes, including endonucleases and exonucleases (Stern et al. 2010).

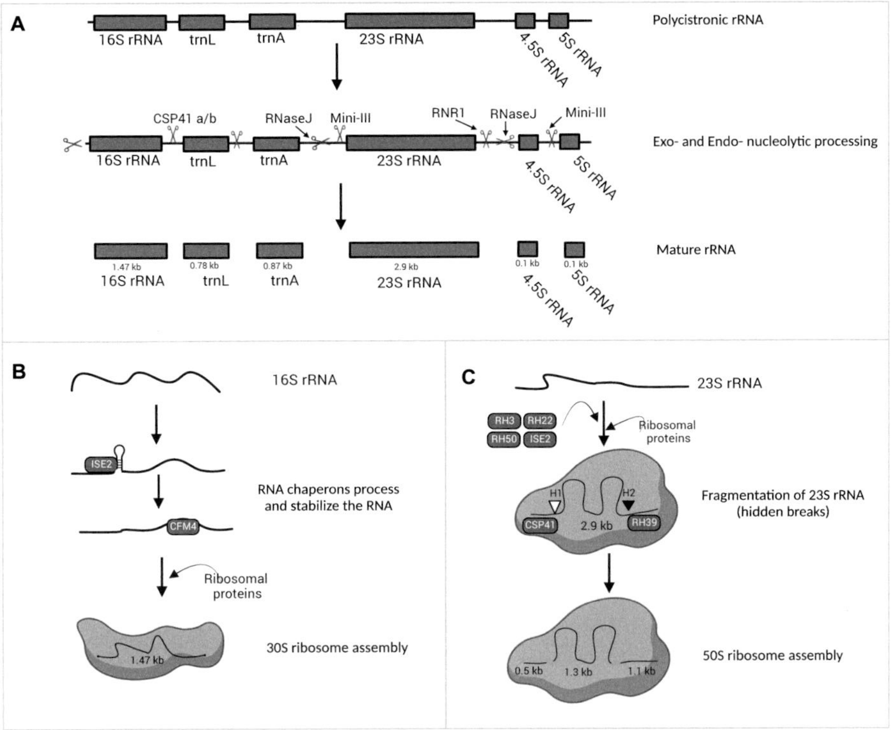

Fig. 2.2 Chloroplast rRNA processing and ribosomal assembly. (**a**) Full-length (7.4 kb) polycistronic rRNA transcript is processed through endonucleolytic (CSP41 a/b, Mini-III) and exonucleolytic (RNase J, PNPase) action to produce the final processed form. (**b**) 16S rRNA is processed and stabilized by RNA chaperones (RNA helicase ISE2 and CRM domain-containing protein CFM4). (**c**) Post-maturation processing of 23S rRNA. 23S rRNA (2.9 kb) is cleaved into 0.5, 1.3, and 1.1 kb fragments during 50S ribosomal assembly. RH39 is required for processing at the 3′-end (black triangles). Cleavage reactions at the 3′- and 5′-ends (white triangles) require CSP41. The length of the fragments is indicated by the number below the fragments. Image generated with BioRender

2.2.2 *RNA Splicing and Intron Removal*

Interestingly, plant organelles contain an abundance of introns per kilobase of genetic information and require RNA splicing and editing, processes that are lacking in prokaryotes (Oldenkott et al. 2014; Small et al. 2020; Miura et al. 2022). RNA splicing, in which intron sequences are removed from precursor RNAs and exons joined to produce mature RNAs, is a critical step for chloroplast gene expression. The chloroplast genome in land plants has 17–21 introns. Most of them are group II introns, and only one of them is the group I intron, that of *trnL* (de Longevialle et al. 2010; Germain et al. 2013). In contrast, *C. reinhardtii* has five group I introns- one in *rrnL*, and four in *psbA*. Group I introns contain a secondary structure of 10 domains, while group II is composed of six domains. Group I intron splicing requires guanosine as a cofactor, whereas group II uses adenosine during the two-step phosphoryl transfer reaction of splicing. During group I intron splicing, a nucleophilic attack by a guanine nucleotide on the 5′-end of the intron is followed by an attack of now free 3′-OH of the upstream exon on the 5′-end of the downstream exon (Kruger et al. 1982). Group II intron splicing occurs in two *trans*-esterifications: first, nucleophilic attack by the 2′-OH of a bulged adenosine in the helix of domain 6 on the 5′-end of the intron gives rise to a branched structure, which results in releasing the intron as a lariat structure after the second *trans*-esterification (Podar et al. 1998). Group II introns are further classified into the IIA and IIB subgroups based on their structural features. The major structural differences between the two subgroups are their exon-binding sites (EBSs), the internal loop of domain III, and the connection between domains I–IV (Michel et al. 1989; Toor et al. 2001). Considering their conserved structural elements and splicing mechanism, most chloroplast introns can be classified as group II introns. Although group II introns of chloroplast genes can undergo autocatalytic splicing *in vitro*, they require splicing factors for splicing *in vivo* (de Longevialle et al. 2010). Several RBPs are required for chloroplast intron splicing.

Over evolution, chloroplast group II introns lost the ability to self-splice due to the degeneration of RNA structure and loss of conserved intron-encoded maturase. MATURASE K (matK) is encoded by the intron of the *trnK* gene, and its location within the *trnK* gene implies involvement in splicing the *trnK* precursor (Zoschke et al. 2010). It has been reported to coimmunoprecipitate with group IIA introns, including the introns of tRNAs (*trnA, trnI, trnV,* and *trnK),* ribosomal proteins (*rpl2* and second intron of *rps12*), and one subunit of the ATP synthase (*atpF*) (Zoschke et al. 2010).

Noticeably, RNA splicing of some chloroplast genes is required for environmental stress responses in plants, especially responses related to abscisic Acid (ABA) signaling; however, it is still unknown how this regulation occurs (Asakura et al. 2012; Gu et al. 2014). Additionally, defects in chloroplast RNA splicing lead to impaired photosynthetic function, disrupting chloroplast homeostasis and making the organism more vulnerable to environmental stressors (Zhu et al. 2020; An et al. 2023).

2.2.3 Cytidine to Uridine RNA Editing

RNA editing in plants mainly involves deamination reactions that convert cytidines (C) in organellar RNAs into uridines (U), called C-to-U RNA editing (Chateigner-Boutin and Small 2010; Takenaka et al. 2013; Yan et al. 2018; Fig. 2.3). In chloroplasts of flowering plants, RNA editing occurs at 20–60 sites, predominantly in the coding regions, and with only a few sites in introns, non-coding regions, UTRs, and tRNAs (Ichinose and Sugita 2016; Edera et al. 2018). RNA editing in organelles can alter the amino acid sequences of proteins and plays a critical role in post-transcriptional regulation. This process can involve changing specific nucleotides in the mRNA, which can result in the introduction of translation start sites (like AUG codons) or the elimination of premature stop codons. These changes directly affect the protein sequence synthesized from the edited mRNA. Additionally, RNA editing can influence the secondary structures of transcripts. By altering base pairs, RNA editing can affect the stability of mRNA molecules and their ability to be properly spliced (Shikanai 2015; Small et al. 2020). RNA editing occurs in chloroplast and mitochondria across all major plant lineages, including bryophytes, gymnosperms, and angiosperms (Wakasugi et al. 1996; Steinhauser et al. 1999; Sugita et al. 2006; Takenaka et al. 2012), except certain species of liverworts in the Marchantiales branch that lack mRNA editing in plastids (Rudinger et al. 2008). The frequency of RNA editing events varies among different lineages. In flowering plants, 30–40 editing events occur in the plastid (Takenaka et al. 2013), and 400–600 editing events occur in mitochondria (Mower 2009). On the other hand, the moss *Physcomitrella patens* (now *Physcomitrium patens*) exhibits only two editing events in the plastid and eleven editing events in mitochondria (Ichinose et al. 2014). Since no RNA editing has been observed in any algae, RNA editing in plant organelles evolved independently from RNA processing when the first plants moved from the aquatic environment onto the land (Takenaka et al. 2013). Notably, no cytoplasmic RNA editing has been observed in plants, suggesting this process is exclusive to the plastid and mitochondria (Takenaka et al. 2013).

2.2.4 RNA Stability

RNA transcripts undergo extensive processing, requiring both RNA-binding proteins (RBPs) and RNases for proper maturation (Stoppel and Meurer 2012). This interplay results in complex transcript patterns within plant organelles, with RBPs contributing to the stability of organellar transcripts over extended periods (Germain et al. 2013). After transcription, intercistronic cleavage, RNA 5′ and 3′ processing, decay is required for the maintenance of RNA stability, ensuring proper gene expression and functional plastid biogenesis. This stability is regulated by nucleus-encoded and chloroplast-localized ribonucleases with exo- and endonucleolytic activity such as polynucleotide phosphorylase (PNPase), RNR1, RNase J, RNase E, CSP41a, and CSP41b (Fig. 2.4).

2.3 RNA Binding Proteins in Chloroplast Gene Expression

RBPs contain one or more RNA-binding domains, such as RNA recognition motifs (RRM), chloroplast RNA-splicing and ribosome maturation (CRM) domains, and K homology (KH) domains (Jacobs and Kuck 2011). These RNA-binding domains determine the binding specificity of RNA interactions in eukaryotes.

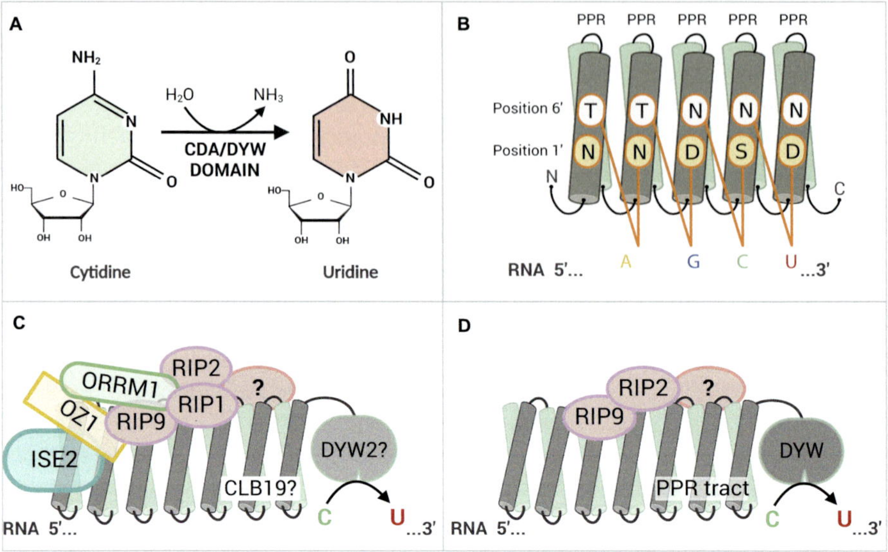

Fig. 2.3 C-to-U RNA editing and editosomes in chloroplasts. (**a**) The DYW domain acts as a hydrolytic deaminase to change cytosine to uracil; (CDA, CYTIDINE DEAMINASE). (**b**) Basis of the PPR recognition code in RNA binding. Each PPR motif's amino acid position 6′ and 1′ is denoted in white and yellow circles, respectively. Tyrosine, asparagine, aspartic acid, and serine are each represented by the letters T, N, D, and S, respectively. The specific binding site of mRNA bases is recognized by amino acid combinations at positions 6′ and 1′ of PPRs (T, N) (T at 6′, N at 1′) binding to guanine (G), adenine (A), cytidine (C), cytidine (D), uridine (U), and (N, N) to either C or U. (**c**) and (**d**) are two models of editosomes proposed by Sandoval et al. (2019) based on size separation of chloroplast extracts and data from Arabidopsis mutants (Bentolila et al. 2012; Takenaka et al. 2012; Sun et al. 2013; Bobik et al. 2017). Members of non-PPR families: RNA editing factor interacting proteins (RIPs), organelle RNA recognition motif (ORRM) proteins, and organelle zinc-finger (OZ) proteins, Ski-like RNA helicase ISE2 and PPR proteins interact to form the editosome. Image generated with BioRender

The RRM motif, also referred to as the RNA Binding Domain (RBD) or ribonucleoprotein (RNP) domain, was first identified in nuclear heterogeneous RNP particles (hnRNPs) of eukaryotes. It is also present in prokaryotes, primarily cyanobacteria, and with lower frequency in viruses (Adam et al. 1986; Maris et al. 2005). An RRM is composed of 90 amino acids and adopts a conserved structure known as an α/β sandwich ($\beta_1\alpha_1\beta_2\beta_3\alpha_2\beta_4$), comprising a four-stranded β-sheet and two α-helices (Burd and Dreyfuss 1994; Clery et al. 2008). The RRM defining characteristics include two canonical consensus sequences: [RK]-G-[FY]-[GA]-[FY]-[ILV]-X-[FY] (RNP1) located in β3 and [ILV]-[FY]-[ILV]-X-N-L (RNP2) situated in β1 (Maris et al. 2005). RRM-containing RBPs interact with other proteins to modulate their RNA-binding properties, protein-protein interactions, or enzymatic activity (Lunde et al. 2007).

The CRM domain is restricted to archaea, bacteria, and plants (Ostheimer et al. 2003). In the two former groups, it is represented by the YhbY protein while in

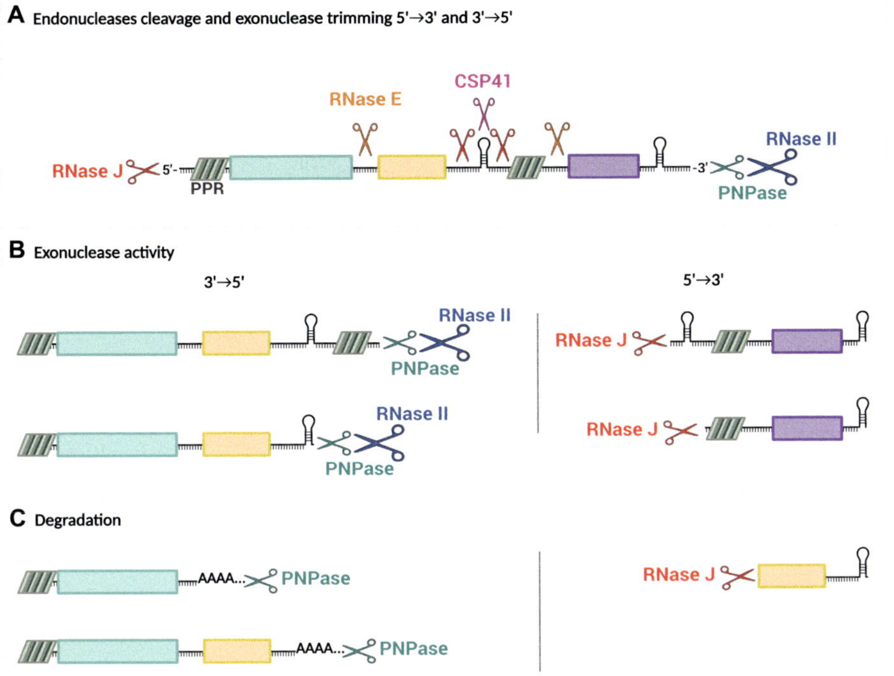

Fig. 2.4 RNA stability and ribonucleases. (**a**) The primary mRNA is subject to cleavage by the endoribonucleases RNase J, RNase E, and CSP41 which cleave intercistronically or between protein binding sites, like PPR proteins. RNase J may also function as a 5′ → 3′ exonuclease, primarily responsible for defining the 5′-ends of mRNA. PNPase, a 3′ → 5′ exonuclease, trims 3′-ends but is stopped by RNA secondary structures or bound PPR proteins. (**b**) Exoribonucleases produce the final transcript termini. Mature ends are protected from degradation by sequence-specific RNA-binding proteins and secondary structures. While PNPase and RNase II may occasionally overcome secondary structures, they typically stop, requiring 5′ → 3′ decay or another endonucleolytic cleavage. (**c**) Polynucleotide tail addition by PNPase stimulates decay of illegitimate or misprocessed transcripts. RNase J is also involved in RNA degradation and surveillance, mainly by preventing double-stranded RNA formation. Figure inspired by MacIntosh and Castandet (2020). Image generated with BioRender

plants CRMs are RNA binding domains present in a family of plant-specific proteins whose characterized members play important roles in the splicing of group I and group II introns in chloroplasts (Till et al. 2001; Ostheimer et al. 2003; Asakura and Barkan 2006, 2007; Barkan et al. 2007; Asakura et al. 2008; Zhang et al. 2019, 2020; Shen et al. 2020; Feiz et al. 2021) or mitochondria. All bacterial YhbY orthologues harbor the conserved sequence G-X-X-G that is present at least in one of the CRM domains of multi-CRM-domain proteins, responsible for the RNA binding ability of CRM domain-containing proteins (Sager et al. 2020).

The K homology (KH) domain was first identified in the human hnRNP K protein and is the second most widespread motif in RBPs after the RRM (Lorkovic and

Barta 2002). The domain is 60 amino acids long with hydrophobic residues and a conserved consensus sequence (VIGXXGXXI) inserted into the middle of the domain (Burd and Dreyfuss 1994).

The PENTRATRICOPEPTIDE REPEAT (PPR) protein family is a highly conserved group of RBPs in eukaryotes that play crucial roles in organellar function, plant physiology, and development (Lurin et al. 2004; Rovira and Smith 2019). They are characterized by tandem 35-amino acid repeats that have the same core structure although differing sequence. These PPR motifs bind RNA molecules and participate in post-transcriptional processes (Chen et al. 2018). Curiously, while animal genomes encode 3–7 PPR proteins, the angiosperm genome, on average, encodes more than 400 PPR proteins (Lurin et al. 2004). PPR proteins can be categorized into P and PLS subfamilies according to the variants of the P motif, including L (long) motif and S (small) motif (Lurin et al. 2004; more details are provided later in this chapter).

The major classes of chloroplast localized RBPs and their roles in chloroplast gene expression will be discussed in this chapter.

2.3.1 RNA Helicases

A general feature of RNAs is the ability to form complex secondary structures important for biological activities (Zhu et al. 2021). Moreover, as biological activity is dynamic, the interactions between RNA-RNA, RNA-DNA, and RNA-protein complexes must be transient and dissociate in a specific order and in an efficient manner to ensure function. In this context, RNA Helicases (RHs) are key enzymes in RNA metabolism and, among their functions, they modulate RNA secondary structure, allowing RNA molecules to adopt the conformations required for interaction with their target proteins. RHs, characterized by seven to eight conserved motifs, are crucial in nucleic acid metabolism (de la Cruz et al. 1999). These motifs play a pivotal role in binding to an NTP, typically ATP, and utilizing the hydrolysis energy to carry out their activities which can include unwinding double-stranded RNA structures. RHs are present in almost all organisms, from bacteria and viruses to humans. RHs have been classified into six superfamilies, namely SF1 through SF6, based on their sequence, occurrence, and motif characteristics (Singleton et al. 2007). Among these superfamilies, SF1 and SF2 are the largest, each further subdivided into various groups. Superfamily I (SF1) contains the best structurally-characterized helicases and includes bona fide helicases and α enzymes, often harboring essential inserts in each helicase domain (Singleton et al. 2007; Jankowsky and Fairman 2007). Superfamily II (SF2), the largest helicase family, encompasses members sharing a conserved core that includes 12 identified sequence motifs that function in ATP hydrolysis, RNA binding, and intramolecular interactions (Gorbalenya and Koonin 1993; Fairman-Williams et al. 2010; Linder and Fuller-Pace 2013). The DEAD-box and related families such as DEAH, DExH, and DExD, collectively known as the DExD/H helicase family, belong to SF2 (Tanner and Linder 2001; Caruthers and McKay 2002).

SF2 helicase members are categorized into three sub-groups based on their structures: Ski2, DEAD-box, and DEAH-box (Caruthers and McKay 2002; Cordin et al. 2006; Linder and Owttrim 2009). The two DExD-box and DExH-box helicases families, the largest within SF2, are named after their Asp-Glu-Ala-Asp (DEAD) or Asp-Glu-Ala-His (DEAH) amino acid sequences and share evolutionarily conserved motifs within their core (Linder et al. 1989). DEAD-box proteins form the largest RH family, with 37 members in humans and 26 in *Saccharomyces cerevisiae* (Fairman-Williams et al. 2010). In plants, the Arabidopsis genome contains 58 members while the rice genome holds more than 50 (Mingam et al. 2004; Umate et al. 2010). They harbor nine motifs called Q, I, Ia, Ib, II, III, IV, V, and VI (Caruthers and McKay 2002; Silverman et al. 2003; Tuteja and Tuteja 2004; Kwong et al. 2005; Singleton et al. 2007) organized into two helicase domains, domain I and domain II. Each motif has a specific contribution to helicase activity such as ATP binding and hydrolysis, RNA binding, and ATP-dependent intramolecular RNA remodeling (reviewed by Nawaz and Kang 2017). Furthermore, based on divergent sequence motifs, other related families were defined after characterizing the DEAD-box family. Thus, alterations in motif II have defined the Ski2, DEAD-box, and DEAH-box families (Caruthers and McKay 2002; Cordin et al. 2006; Linder and Owttrim 2009). In plants, most DEAD-box members possess the helicase core but lack additional domains in their C-terminal regions compared to DEAH-box and Ski2-like RHs. In human and yeast cells the function of DExD/H-box RHs is to participate in ribosome biogenesis and regulate cellular proliferation and stress responses, while in plants they are involved in most cellular events associated with RNA, playing an important role in plant development and abiotic stress tolerance by modulating pre-rRNA processing (Jarmoskaite and Russell 2014; Liu and Imai 2018).

2.3.1.1 DEAD-Box RNA Helicases

RNA molecules require precise folding for interactions with proteins, with DEAD-box helicases playing a crucial role in restructuring RNA secondary structures by locally separating duplexes. These helicases initially bind to the duplex, engage one end, and use ATP energy to translocate directionally along one strand while displacing the other strand (Jarmoskaite and Russell 2011). DEAD-box proteins function as "motor proteins" and are likely activated and regulated by their interacting partners in specific biological contexts. Although not universally applicable, it is plausible that a given DEAD-box protein may be activated differently by various partners during different cellular processes (Linder and Jankowsky 2011; Jarmoskaite and Russell 2014).

Out of 58 DEAD Arabidopsis RHs, 11 have predicted N-terminal chloroplast-targeting sequences. Proteomic studies reported six Arabidopsis DEAD-box RHs (RH3, RH22, RH26, RH39, RH47, RH50) localized to the chloroplast (Olinares et al. 2010). Maize homologs of these RHs are found in plastids and were enriched in plastid nucleoids (Majeran et al. 2012). The tobacco RH58, named VDL helicase, is also found in the chloroplast (Wang et al. 2000). Maize RH58 was found at low abundance in nucleoids (Majeran et al. 2012), but the Arabidopsis homolog of RH58 was not detected. In phylogenetic analysis four RNA helicases (RH22, RH39, RH47, RH50) form one clade, implying their common ancestry. In contrast, RH3 belongs to a different clade and it was not a member of the GreenCut2 group of conserved chloroplast proteins (Bobik et al. 2017), which suggests it has distinct functions from the other four RNA helicases. RH3 and RH26 have all the conserved RH motifs, compared to the other four RNA helicases that lack several conserved motifs of DEAD-box RNA helicases (Asakura et al. 2012).

RH3 is required for the splicing of group II introns in Arabidopsis and maize chloroplasts (Asakura et al. 2012). RH3 binds to *rps12, rpl2, trnA,* and *trnl* type II introns and is involved in splicing *trnA* and *rpl12* introns (Asakura et al. 2012). Phylogenetic analyses found that RH3 is closely related to mitochondrial RNA helicase AtRH53/PMH2, which also exhibits function in intron II splicing (Köhler et al. 2010). The *rh3* null mutant is embryo-lethal, whereas a weak allele resulted in pale-green seedlings (Asakura et al. 2012). *RH3* is predominantly expressed in leaves, although detected in flowers and siliques, but not in stems and roots in Arabidopsis (Asakura et al. 2012; Gu et al. 2014). Indeed, RH3 shows peak accumulation during chloroplast formation in the sink-source transition zone of developing maize leaves. RH3 is also involved in assembling 50S ribosomal subunits. RH3 in maize co-immunoprecipitated with pre-23S and mature 23S rRNA but not with 16S rRNA. Arabidopsis *rh3* mutants accumulate reduced amounts of 23S, 4.5S, and 5S rRNA (Asakura et al. 2012), and 16S rRNA accumulation was also affected in a development-dependent manner (Lee et al. 2013).

RH22 is involved in processing the rRNA of the 50S ribosome subunit (Chi et al. 2012). Reduced expression of *RH22* led to the accumulation of the precursors of 23S and 4.5S rRNA (Chi et al. 2012). RH22 can bind to a fragment of domain I of 23S rRNA without processing. So, the rRNA processing defect in the *rh22* mutant did not result from the processing but was caused by the defective assembly of the 50S ribosomal subunit. RH22, L24, and the 5′ end of the 23S rRNA form a protein/RNA complex involved in the early assembly of the 50S subunit. RH22 may unwind and/or rearrange the structure of the 23S precursor rRNA to allow it to contact L24. Interestingly, RH22 is not associated with 30S particles (Chi et al. 2012), but *rh22* mutants exhibit defects in the maturation of 16S rRNA without precursor rRNA accumulation. These defects may result either from the indirect consequence of the defect in 50S ribosome subunit biogenesis or RH22 may have a separate role in other processes like rRNA gene expression such as rRNA transcription or the stabilization of rRNA. Mutants with knockdown of *RH22* expression displayed delayed cotyledon greening, while complete loss of function was lethal (Chi et al. 2012).

RH39 is involved in the post-maturation processing of 23S rRNA in monocots and dicots (Nishimura et al. 2010). Arabidopsis RH39 is involved in processing a stem-loop formed around the hidden break site between nucleotides 1744 and 1762 of the 23S rRNA. The nucleotide sequence and the stem-loop structures at the 3′ end processing regions are highly conserved between Arabidopsis and maize 23S rRNAs. RH39 may remodel the rRNA/RNase complex and/or release proteins bound in the processing region to allow proper positioning of RNase on the stem-loom and to facilitate the introduction of the nick. The *rh39* mutant, also known as *nara12-1 (necessary for the achievement of RuBisCO accumulation)*, exhibits defects in the elongation steps of chloroplast transition without affecting initiation and termination stages (Nishimura et al. 2010).

RH50 associates with transcriptionally active plastid chromosomes together with plastid ribosomal proteins, PEP core enzyme, and proteins involved in transcription, translation, and RNA metabolism, including RNases and other DEAD-box RNA helicases (Olinares et al. 2010; Majeran et al. 2012). Arabidopsis AtRH50 is required for the maturation of the 23S and 4.5S rRNA. The *rh50* mutants accumulated transcripts from the 23S-4.5S rRNA intergenic region. This suggests that RH50 may unwind double-stranded rRNA of the 23S-4.5S intergenic region to promote rRNA processing. Rice RH50 exhibits RNA helicase activity *in vitro*, but no direct target has yet been identified (Li et al. 2008). The *rh50* mutants, unlike *rh3*, *rh22*, and *rh39* mutants, were of normal size and photosynthetic performance. However, *rh50* mutant plants were sensitive to erythromycin and low temperature, with polysome loading disrupted in low temperatures (Paieri et al. 2018).

2.3.1.2 SKI2-Like RNA Helicases

ISE2 is a Ski-2-like chloroplast RNA helicase, the only protein from this family that is found in the chloroplast. Ski2-like helicases unwind RNA substrates with high processivity, compared to DEAD-box RNA helicases, which unwind RNA with low processivity. Despite their sequence similarity, Arabidopsis ISE2 and its orthologues differ from other Ski-like proteins (Bobik et al. 2017). Unlike ISE2, other Ski2-like proteins localize to the cytoplasm and degrade RNA (Dorcey et al. 2012). ISE2 was identified as a member of GreenCut2 (Karpowicz et al. 2011; Bobik et al. 2017)—a set of conserved proteins specific to the plant lineage—whereas other Ski2-like proteins do not belong to the GreenCut2. ISE2 is also a member of PlastidCut2, a subset of GreenCut2 proteins. PlastidCut2 includes 124 nuclear-encoded proteins of photosynthetic eukaryotes regardless of their plant lineage (Karpowicz et al. 2011).

ISE2 is required for the splicing of *rpl2, atpF, clpP1,* and *rps12* type II introns (Carlotto et al. 2016; Bobik et al. 2017). Indeed, ISE2 associates with 20 of 21 introns encoded by the chloroplast genome (Bobik et al. 2017). For group II intron splicing, ISE2 and RH3 may be involved in folding or unfolding the RNA secondary structure of the targeted chloroplast transcripts. ISE2 is also required for site-specific C-to-U RNA editing. The role of ISE2 in the editing of several chloroplast transcripts has been determined in Arabidopsis (Bobik et al. 2017) as in maize (Sandoval et al. 2019) and it has been found to interact with multiple editing factors, including RIP1/MORF8, RIP9, and ORRM to form an editosome complex (Sandoval et al. 2019) (Fig. 2.3). Editosome complexes comprise PPR and RIP/

MORFs proteins, which are required for RNA editing (Takenaka et al. 2012). ISE2 may facilitate the annealing of target RNA to editosome complex or may unwind the local RNA secondary structure that allows the editosome to access the substrate or may remove the editosome complex after RNA editing. In addition to intron splicing and C-to-U RNA editing, ISE2 is also involved in rRNA processing, ribosome assembly, and protein accumulation in chloroplasts (Bobik et al. 2017). Thus, ISE2 may also play a role as an RNA chaperone to bind in a non-specific manner to promote the function of other processing factors (Bobik et al. 2017). Indeed, ISE2 interacts with a PPR protein, EMB175/PPR103/IPI1 that is involved in rRNA processing and stabilization, stabilizing the *rpl16* mRNA by impeding 5′→3' RNA degradation in both maize (Hammani et al. 2016) and *N. benthamiana* (McCray et al. 2024). ISE2's multiple roles in chloroplast are supported by identifying 111 chloroplast RNA features as possible substrates for ISE2 (Bobik et al. 2017).

2.3.2 *Endoribonucleases*

Endoribonucleases are enzymes that cleave RNA molecules internally (Fig. 2.4). These RNases are essential for regulating the abundance and quality of RNA in plant cells, ensuring proper gene expression, and maintaining cellular and genomic integrity by cleaving mRNA internally at specific sites. Several endoribonucleases are key enzymes involved in the processing and degradation of chloroplast transcripts, helping to generate mature and functional RNA by precisely processing precursor transcripts and removing unwanted RNA sequences (Germain et al. 2013), also contributing to gene expression regulation and maintenance of chloroplast function.

2.3.2.1 RNase E

The involvement of RNase E in RNA processing and degradation, primarily studied in *E. coli*, extends to higher plants where homologs are encoded in the nuclear genome. Chloroplast RNA degradation occurs via the polyadenylation-assisted degradation pathway (Schuster and Stern 2009). Arabidopsis RNase E, localized to the chloroplast stroma, exhibits endonucleolytic activity similar to its *E. coli* counterpart, preferring A/U-rich sequences and forming an oligomeric complex of ~680 kDa (Schein et al. 2008). This suggests that plant RNase E likely participates in the initial endonucleolytic cleavage of chloroplast RNA degradation, possibly alongside other chloroplast endonucleases such as RNase J and CSP41. Analysis of *rne* mutants revealed that, unlike in bacteria, RNase E is not essential for survival. However, although other enzymes can partially compensate for its absence, deficiency still leads multiple defects in chloroplast RNA metabolism (Walter et al.

2010; Stoppel and Meurer 2012). RNase E appears to work in concert with the RBPs and transcriptional regulator RHON1 (Stoppel et al. 2012; Yang et al. 2020). The knockout plants exhibit a notable accumulation of polycistronic precursor transcripts and a significant reduction in several mature monocistronic mRNAs. This indicates a critical role for RNase E in the intercistronic processing of primary transcripts from chloroplast operons (Walter et al. 2010).

2.3.2.2 CSP41

The two homologous CHLOROPLAST STEM-LOOP BINDING PROTEINS (CSP)41a and CSP41b are widely distributed and highly conserved endoribonucleases found exclusively in photosynthetic organisms. Analysis of mutants of the two Arabidopsis homologs showed that while *csp41a* and *csp41b* single mutant plants resembled wild-type plants, the *csp41a csp41b* double mutant plant exhibited elevated levels of 23S rRNA precursors, with impaired chloroplast polysome assembly (Beligni and Mayfield 2008), suggesting potential overlapping functions of CSP41a and CSP41b in the plastid rRNA maturation pathway (Fig. 2.2). Early evidence suggested that CSP41 initiates the degradation of chloroplast transcripts by cleaving them internally, serving as the initial stage in the poly(A)-stimulated degradation pathway (Yang and Stern 1997). Further, reduced expression of *CSP41a* in *Nicotiana tabacum* suggested a role in initiating mRNA turnover through endonucleolytic cleavages (Bollenbach et al. 2003). However, CSP41 proteins are more likely involved in RNA stabilization rather than cleavage (Qi et al. 2012). While they exhibit strong endoribonuclease activity on stem-loops *in vitro* (Yang et al. 1996; Yang and Stern 1997; Bollenbach et al. 2003), recent RNA immunoprecipitation-microarray (RIP-Chip) analyses suggest their primary function is RNA stabilization (Qi et al. 2012). It has been proposed that the endoribonuclease activity of CSP41 could be regulated *in vivo*, with catalytically active forms behaving like ribonucleases, and inactive forms acting as stabilization factors (Leister 2014).

2.3.3 Exoribonucleases

The general function of exonucleases is to degrade RNA from its ends (5′→3′ or 3′→5′), determining the length of time an mRNA exists in the cell before it is degraded (mRNA turnover) and quality control. During RNA processing, exoribonucleases trim RNA precursors to produce mature transcripts and remove fragments generated (Fig. 2.4). These enzymes are therefore essential for maintaining the chloroplast's integrity and functionality and here we describe some of the major players.

2.3.3.1 Polynucleotide Phosphorylase (PNPase)

In contrast to the stabilizing role observed in the poly(A) tail of nucleus-encoded mRNA, the poly(A) tail present on bacterial or chloroplast RNA usually acts as a signal for exonucleolytic degradation of the RNA molecule. The main role of the PNPase exoribonuclease is to degrade RNA through phosphorolytic 3′→5′ activity. However, the enzyme is also able to add poly(A)-rich tails to the 3′-ends of some RNAs (Walter et al. 2002; Yehudai-Resheff et al. 2001). The balance between these two activities is regulated by nucleoside diphosphates and inorganic phosphate (Pi). While ADP inhibited RNA degradation and enhanced polymerization of the poly(A) tail, the addition of Pi enhanced degradation (Yehudai-Resheff et al. 2001). In *E. coli,* PNPase is part of the degradosome complex (Symmons et al. 2002; Liou et al. 2002; Yehudai-Resheff 2003), whereas chloroplast PNPase consists of a homotrimeric complex and lacks any known interactions with other proteins (Baginsky et al. 2001; Yehudai-Resheff 2003). The structure of the three monomers forms a central channel that can accommodate a single-stranded RNA molecule, and it has been hypothesized that the structural regulation of both degradation and polymerization activities are sensitive to a 3′-end stem-loop structure (Yehudai-Resheff 2003).

Loss or repression of PNPase activity leads to defects in mRNA and rRNA processing within chloroplasts (Walter et al. 2002; Marchive et al. 2009). However, Arabidopsis mutants can grow to full maturity, although displaying chlorotic young leaves and smaller size than wild type (Marchive et al. 2009; Germain et al. 2011). Analysis using RNA-seq coupled with conventional molecular biology techniques revealed that PNPase mutants accumulate numerous RNAs with 3′ extensions, alongside additional abnormalities in RNA splicing and editing (Hotto et al. 2011; Castandet et al. 2013, 2019).

2.3.3.2 RNase II, RNR1

The RNR exoribonuclease family members, typified by *E. coli* RNase II and RNase R, are hydrolytically processive 3′→5′ exoribonucleases that release 5′ monophosphate nucleotides. In Arabidopsis, the RNase II/R homolog RNR1 is dually localized to chloroplasts and mitochondria, has 3′→5′ exoribonuclease activity, and cleaves RNA during mRNA turnover and rRNA maturation (Germain et al. 2012). Arabidopsis RNR1 activity is blocked both *in vitro* and *in vivo* due to 3′-end secondary structures and RNA-binding proteins, contributing to the shortness of 3′-end extensions generated by chloroplast PNPase (Bollenbach et al. 2005; Germain et al. 2012). *rnr1* null mutants only germinated on sucrose-containing media and had a pale leaf phenotype associated with chloroplast development defects, implying impaired photosynthesis (Bollenbach et al. 2005). In addition, analysis of *rnr1* mutants revealed its role in rRNA accumulation and for final rRNA 3′-end maturation. A more in-depth examination of *rnr1* RNA profiles highlighted the enzyme's essential role in mRNA 3′-end maturation, potentially overlapping with the function of PNPase. In the *rnr1* mutant, many mRNAs displayed short 3′ extensions, suggesting the enzyme's proximity to RNA secondary structures and RNA-binding protein sites at mature mRNA 3′ ends (Germain et al. 2012). This observation suggests a collaborative role between RNR1 and PNPase in RNA processing pathways.

2.3.3.3 RNase J

The evolutionarily conserved RNase J belongs to the metallo-β-lactamase family of ribonucleases (Condon and Gilet 2011). With both 5′→3′ exoribonuclease and endoribonuclease function, RNase J has been shown to trim the 5′-end of several transcripts and degrade long antisense RNA. The role of RNase J in the chloroplast genome has been recently reviewed (Hotto et al. 2020). Arabidopsis null mutants for RNase J are embryo-lethal and, although is still not clear how the lack of RNase J activity results in lethality, it is suggested the cause could be the broadly impaired chloroplast function, or due to its specific function in the processing or degradation of a particular transcript (Hotto et al. 2020). Analysis of plants with reduced *RNase J* expression in *N. benthamiana* revealed that the mature transcript 5′ ends are generated by endonucleolytic cleavage followed by 5′→3′ exonuclease degradation until is blocked by the corresponding PPR protein (Luro et al. 2013). Interestingly, in plants with reduced expression of RNase J, the accumulation of antisense RNA species, likely forming duplexes, blocked translation (Sharwood et al. 2011). These results imply that RNase J likely acts *in vivo* to degrade antisense RNAs before duplex formation, indicating its crucial role in chloroplast RNA quality control (Sharwood et al. 2011; Halpert et al. 2019).

2.3.4 Other Ribonucleases

Chloroplasts harbor a diverse array of RNases beyond the well-characterized major players described so far, each with distinct roles in RNA processing and metabolism. Mini-RNase III, a homolog of *Bacillus subtilis* mini-III endoribonuclease, processes specific regions of rRNA, the 5′ end of the 23S rRNA and the 3′ end of the 4.5S rRNA and plays a role in intron degradation following splicing in Arabidopsis (Redko et al. 2008; Hotto et al. 2015). YbeY, analogous to the endoribonuclease in bacteria, is essential for chloroplast development and specifically involved in rRNA operon maturation (Liu et al. 2015). ERIL1, a conserved 3′→5′ exoribonuclease from the DEDDh family (Graw et al. 2014), is implicated in chloroplast RNA processing, although its precise molecular defects are yet to be fully elucidated (Mermigka et al. 2016).

2.3.5 *Chloroplast RNA Splicing and Ribosome Maturation (CRM) Domain-Containing Proteins*

Splicing of chloroplast group II introns requires nucleus-encoded splicing factors, including the CHLOROPLAST RNA SPLICING AND RIBOSOME MATURATION (CRM) domain-containing proteins. The CRM domain is an RNA recognition and binding domain that can bind to chloroplast introns. While prokaryotes have only one CRM domain protein, plant genomes encode multiple CRM domain proteins (Barkan et al. 2007). Most of the CRM domain proteins are predicted to be localized in the plastid, and a few in the mitochondria or nuclei. They contain 1–4 repeating CRM domains that can bind to chloroplast intron RNA for splicing.

2.3.5.1 Chloroplast RNA Splicing 1 (CRS1), CRS2

The first CRM gene, *CRS1,* was identified in maize, and the protein is involved in the splicing of the *atpF* intron (Jenkins et al. 1997; Till et al. 2001; Ostersetzer et al. 2005). ZmCRS1 has been reported to co-immunoprecipitate with ZmWHY in an RNA-dependent manner, which suggests ZmWHY binds to CRS1's RNA ligand, *atpF* intron (Prikryl et al. 2008). ZmWHY belongs to the 'Whirly' protein family, which are plant-specific DNA-binding proteins, necessary for the splicing of *atpF* (Prikryl et al. 2008). Rice has three CRS1 orthologs that are involved in the splicing of introns of the *ndhA, ndhB, petD ycf3,* and *trnL* transcripts.

2.3.5.2 CRS2-Associated Factor 1 (CAF1) and CAF2

Maize CAF1 and CAF2 interact with CRS2 and are involved in the splicing of group II introns. CRS2 is a peptidyl-tRNA hydrolase associated with CAF1 and CAF2 to form a splicing complex (Ostheimer et al. 2006). In plants, CRS2 lacks amino acids that, in *E. coli*, are required for the hydrolase enzyme to cleave the ester bond between the tRNA and nascent peptide. However, CRS2 in plants has evolved the ability to perform group II intron splicing. The CRS2-CAF1 complex is required for the splicing of group IIB introns in maize *petD, rpl16, rps16, trnG, ycf3* transcripts. The *rpoC1* intron *and ClpP* intron 1 are spliced by the CRS2-CAF1 complex in Arabidopsis, but these introns are absent in maize (Asakura and Barkan 2006). The CRS2-CAF2 complex is required for the splicing of five group IIB introns in the *ndhB*, the *petB*, *the rps12* (intron 1), the *ndhA*, and the *ycf3* (intron 1) transcripts in maize (Ostheimer et al. 2003).

2.3.5.3 CRM Family Members 1 (CFM1), CFM2 and CFM3

CFM1 is necessary for the splicing of the introns of *trnI, trnV,* and *trnA*, and its function is conserved between maize and *Setaria virdis* (Feiz et al. 2021). CFM1 interacts with other splicing factors (THA1, WTF1, RNC1, and mTERF4) that bind to overlapping subsets of introns, as shown by co-immunoprecipitation experiments (Feiz et al. 2021). These findings suggest that these proteins do not act sequentially in a linear pathway but rather interact simultaneously with their respective intron ligands, allowing efficient splicing by coordinated activity of multiple splicing factors at the same time (Feiz et al. 2021). CFM2 is part of large ribonucleoprotein complexes with CAF1 and/or CAF2 and is required for the splicing of group I introns, *trnL,* and group II introns *ndhA, ycf3-1,* and *Clp-P2* in Arabidopsis. In maize CAF1 was found to co-immunoprecipitate with APO1—a member of the ACCUMULATION OF PHOTOSYSTEM (APO) protein family. Arabidopsis AtAPO1 is involved in the splicing of *petD, ndhB, ndhA, ycf3* (intron 2), and *clpP* (intron 2) (Watkins et al. 2011). CFM3 has three CRM domains and localizes to both chloroplast and mitochondria. It is required for the group II intron splicing of *petB, petD, ndhB, rpl16, rps16,* and *trnG* in the chloroplast.

CAF1 and CAF2 were found to co-immunoprecipitate with the ribonuclease III domain protein RNC1 (Barkan et al. 2007). RNC1 facilitates the splicing of group IIA introns (*trnV, trnk, trnL, trnA, atpF, rps12)* and IIB introns (*ndhB, petB, petD, trnG).* The PLANT ORGANELLAR RNA RECOGNITION (PORR) protein family can indirectly participate in RNA splicing (Kroeger et al. 2009). The maize PORR protein WHAT'S THIS FACTOR (ZmWTF)1 interacts with RNC1 to splice *petD, petB, ndhB, rps12-int2, trnl, trnA, trnG, trnV,* and *trnK* introns. LEFKOTHEA is a PORR domain-containing protein involved in group II intron splicing of *rpl2* and *petB* (Daras et al. 2019). mTERF4 is a member of the mitochondrial transcription termination factor protein family and is involved in the splicing of *trnl, trnA, rpl2, ndhB, atpF,* and *ycf3-2* introns (Hammani and Barkan 2014) (See Chapter in this volume). CMF2 interacts with the PPR protein EMBRYO DEFECTIVE 1270 (EMB1270). This interaction is required for the splicing of the *clpP1* intron, the *ycf3* intron, *ndhA*, and *ndhB*. Since PPR proteins bind to RNA in a sequence-specific manner, the interaction between PPR proteins and general splicing factors like CFM2 may help target their constituent complexes to the corresponding intron to form the full splicing complexes (discussed below). In addition, several RHs are also involved in intron splicing, as described previously.

2.3.6 *Pentatricopeptide Repeat Proteins*

The appearance of PPR proteins and the appearance of chloroplast RNA editing occurred at about the same time during plants' transition from water to land (Tillich and Krause 2010). It has been hypothesized that RNA editing may have evolved during this transition to protect plants from UV light (Takenaka et al. 2013).

Plant PPR proteins are targeted to either the mitochondria or chloroplast and bind to one or several single-stranded RNA molecules via 35-amino acid sequences present in helical tandem arrays of 2–30 repeat motifs each separated by a sharp turn (Shikanai 2006; Barkan et al. 2012; Okuda et al. 2007; Tavares-Carreon et al. 2008). PPR proteins are involved in post-transcriptional processing and the maturation of RNA (Schallenberg-Rudinger et al. 2013a) by influencing RNA splicing (Khrouchtchova et al. 2012), RNA cleavage (Hattori et al. 2007), RNA stability/turnover (Liu et al. 2010), and the site-specific sequence alteration of RNA transcripts through RNA editing (Barkan and Small 2014). Thus, they are site-specific trans-factors involved in group II intron splicing, intergenic processing, stabilization, and RNA editing of organelle transcripts (Nakamura and Sugita 2008; Mach 2009). The absence of PPR proteins affects nuclear gene expression (Ding et al. 2006; Koussevitzky et al. 2007; Liu et al. 2010) via chloroplast-to-nucleus retrograde signaling. In keeping with their important roles in chloroplast gene expression *ppr* mutants often exhibit chlorotic or albino phenotypes and show defective chloroplast structures coupled to defects in RNA metabolism (Tseng et al. 2010; Hammani et al. 2016). Indeed, loss of PPR proteins can lead to embryonic arrest or severe developmental defects (Barkan and Small 2014).

PPR proteins are classified into the P subfamily or the plant-specific PLS subfamily. The PLS subfamily is further classified according to the following C-terminal domains: (i) E group members are classified as those containing an E domain, (ii) E+ group proteins are classified as those containing an E and an E+ domain, and (iii) DYW group members are classified as those containing an E, E+ and a DYW domain, which is named after the C-terminal amino acid triplet of Asp, Tyr and Trp (Shikanai 2006). The tandem array of PPR motifs facilitates the recognition and interaction with RNA sequences (Barkan et al. 2012). The PPR motif likely evolved from a tetracopeptide repeat (TPR) motif early in the evolution of eukaryotes (Lurin et al. 2004). The TPR motif is a 34 amino acid repeating unit that generally serves as a protein-interaction scaffold (D'Andrea and Regan 2003). Nucleotide-binding probabilities for the motifs (P, L, and S) are based on the amino acids found at the 6 and 1′ position (first amino acid of the subsequent C terminal PPR motif) of each PPR motif (Barkan et al. 2012).

2.3.6.1 PPR Proteins in Splicing

Several PPR proteins are involved in group II intron splicing. Thylakoid assembly 8 (THA8) (Khrouchtchova et al. 2012) is a PPR protein that participates in the splicing of *ycf3-2* and *trnA* in maize and Arabidopsis. It has been reported to co-immunoprecipitate with WTF1 and RNC1 splicing factor. The crystal structure of THA8 from *Brachypodium distachyon* reveals that RNA binding induces the dimerization of THA8 (Ke et al. 2013). Another PPR protein, PPR5, binds to the 50-nucleotide region chloroplast *trnG* intron in maize (Beick et al. 2008; Williams-Carrier et al. 2008), which may prevent hairpin formation and facilitate the function of a key *cis*-element for splicing. It also protects the *trnG* intron from endonucleolytic cleavage. HCF152 is a well-studied Arabidopsis PPR protein that binds a site between *psbH* and *psbB* and is involved in the splicing of the *petB* intron (Nakamura et al. 2003). EMB1270 is a P-type PPR protein, and its mutant exhibits defective intron splicing of *clpP1-2*, *ycf3-1*, *ndhA*, and *ndhB* (Zhang et al. 2021). EMB1270 also interacts with another chloroplast splicing factor, CFM2. PPR proteins including PPR4 (Lee et al. 2019), EMBRYO DEFECTIVE 2654 (EMB2654) (Aryamanesh et al. 2017), PBF2 and PDM4 (Iswanto et al. 2020), and ECD2 (Wang et al. 2021) were reported to function as splicing factors. PDM1/SEL1 is a PLS-type protein that participates in editing *accD-1* and splicing of group II intron in *trnK* and *ndhA* in Arabidopsis (Zhang et al. 2015). The editing of approximately 27 sites was affected in *pdm1/sel1* mutants (Pyo et al. 2013). Additionally, PDM1/SEL1 was found to interact with the MORF proteins (MORF9, MORF2, and MORF8) (Zhang et al. 2015), indicating that PDM1 is a crucial component of the editosomes in plastids. PPR4 and the PPR protein EMB2654 bind to the 5′-end of *rps12* intron 1b and the 3′-end of *rps12* intron 1a, respectively, for proper folding and formation of the splicing complex with CRS2-CAF2 complexes (Schmitz-Linneweber et al. 2006; Aryamanesh et al. 2017; Lee et al. 2019). The *rps12* intron 1 is the only intron that is trans-spliced in land plants. Before splicing together into one functional RNA, two parts of *rps12* are transcribed from different parts of the chloroplast genome. One part of *rps12* intron 1—*rps12 1a*—is approximately 30 kb away from another part, *rps12* intron *1b* (Glanz and Kuck 2009). These precursors are cleaved and brought together before regular splicing.

2.3.6.2 PPR Proteins in C-TO-U RNA Editing

Chloroplast transcript editing defects were first discovered in Arabidopsis due to mutations in genes encoding PPR proteins (Kotera et al. 2005). Subsequent research has found that the proteins most often associated with RNA editing are PLS Class PPR proteins, which contain a characteristic repeating triplet of P, L (long variants of P), and S (short variants of P) motifs (Lurin et al. 2004; Cheng et al. 2016) (Fig. 2.3). More specifically, PPR proteins that belong to the DYW subclass are involved in the RNA editing of organelle transcripts. Approximately half of the PLS-class proteins harbor a C-terminal DYW domain (Lurin et al. 2004). DYW domain contains a conserved HxE(x)nCxxX catalytic-site signature similar to known cytidine deaminases from other kingdoms. This suggests the DYW domain may serve as a cytidine deaminase during RNA editing (Salone et al. 2007; Schallenberg-Rudinger et al. 2013b). Recent findings support and confirm this suggestion.

The moss DYW protein PPR56 can effectively edit C-to-U of *ccmFC* transcript without additional cofactors (Oldenkott et al. 2019; Hayes and Santibanez 2020). Further, the crystal structure of the DYW domain of the chloroplast editing factor ORGANELLE TRANSCRIPT PROCESSING 86 (OTP86 DYW) of Arabidopsis was resolved at 2.5Å resolution (Takenaka et al. 2021). This structure provided invaluable insight into the mechanisms used by PPR proteins in the deamination of target sites to cause editing. An inserted gating domain interrupts the deaminase of OTP86DYW and restrains the entry of the substrate cytidine into the catalytic site. The OTP86DYW gating domain alters its conformation upon activation and recruits a water molecule as a fourth ligand between a zinc ion and the catalytic residue E894 to attach the cytidine for deamination (Takenaka et al. 2021). The DYW domain remains in the autoinhibited state without the water molecule, which prevents off-target editing events.

Interestingly, some proteins carrying a variant of the DYW motif were also implicated in C-to-U RNA editing although it is not known whether they have roles in the deamination. EMB175/PPR103/IPI1 (ISE2 PROTEIN INTERACTOR 1) carries a DLW motif instead of a DYW motif and lacks a glutamate residue critical for coordination with one of the catalytic Zn^{2+} ions but editing at 18 of 23 tested sites was reduced in *N. benthamiana* plants where *IPI1* was silenced (McCray et al. 2024). Further, there are PPR proteins that do not possess the DYW motif that nonetheless have been implicated in RNA editing. These include CRR4, CRR21, and CLB19 (Robbins et al. 2009). These proteins, however, may interact with DYW-domain-containing proteins to carry out editing, as was revealed by CRR4 and its association with DYW1 to edit the *ndhD-1* site in Arabidopsis chloroplast transcripts (Boussardon et al. 2012). These results support the existence of protein complexes, called editosomes, to carry out editing (Fig. 2.3).

CRR2 has 9 PPR motifs, and it was initially thought to be an endoribonuclease and responsible for the *ndhB* 5′-end processing, but later research suggested it protects the *ndhB* 5′-end from degradation (Nakamura and Sugita 2008; Okuda et al. 2009; Ruwe et al. 2019). The *crr2* mutant in Arabidopsis revealed its role in RNA processing (Hashimoto et al. 2003). SOT1 and PPR53 are also PPR proteins identified in Arabidopsis and maize, respectively, and are vital for chloroplast development and rRNA maturation (Zoschke et al. 2016; Zhou et al. 2017). They protect RNA transcripts and facilitate rRNA maturation, hinting at their potential role as sequence-specific endoribonucleases in chloroplasts (Zhou et al. 2017).

Although mRNA editing in flowering plants is generally performed as C-to-U alternation, reverse editing (U-to-C) was also detected in the organelles of three land plant clades, including hornworts, lycophytes, and fern (Small et al. 2020). This U-to-C editing is phylogenetically correlated with a subgroup of DYW domains, referred to as DYW:KP, that contains conserved Lys-Pro (KP) dipeptide at the start of the domain (Gerke et al. 2020; Gutmann et al. 2020). These differences in amino acid sequences may affect the ability of the catalytic pocket to perform the C-to-U or U-to-C reactions.

2.3.7 *RIP/MORF Proteins*

PPR proteins and non-PPR editing factors interact to form large multiprotein complexes called editosomes that perform C-to-U RNA editing (Sun et al. 2016; Sandoval et al. 2019) (Fig. 2.3). Flowering plants have many essential non-PPR editing factors such as MULTIPLE ORGANELLAR RNA EDITING FACTOR (MORF) proteins, ORGANELLE RNA RECOGNITION MOTIF PROTEIN (ORRM) proteins, the RNA helicase ISE2, and ORGANELLE ZINC FINGER (OZ) proteins (Takenaka et al. 2012; Sun et al. 2013, 2015; Bobik et al. 2017; Gipson et al. 2022). The RNA EDITING FACTOR INTERACTING PROTEINS (RIP)/ MORF proteins are a small family of proteins required for C-to-U editing in both chloroplasts and mitochondria (Takenaka et al. 2012; Bentolila et al. 2013). RIP/MORF proteins lack the DYW domain but are essential components of the RNA editing machinery. The RIP/MORF proteins possess a conserved domain of approximately 100 amino acids called the MORF box. This domain is required for multimerization and interaction with PPR proteins (Takenaka et al. 2012; Bayer-Csaszar et al. 2017; Haag et al. 2017; Yan et al. 2017; Yang et al. 2018). For example, MORF9 binds to the motif close to the concave surface of the PLS triplet that changes the angle between two helicases of motif I of the PPR protein. This altered structure led to the compression of the PPR superhelix (Haag et al. 2017; Yan et al. 2017, 2018) that ultimately increases RNA-binding activity of the PPR protein (Yan et al. 2017; Royan et al. 2021).

While RIP/MORF2 and 9 function in the chloroplasts, the other RIP/MORF proteins are involved in mitochondrial editing (Takenaka et al. 2012), and RIP1/MORF8 is found in both chloroplasts and mitochondria (Bentolila et al. 2012). RIP1/MORF8 is involved in the editing of 75% of mitochondrial sites and around 20% of plastid sites (Bentolila et al. 2013). RIP3 and RIP8, together with RIP1, are also involved in the editing of over 85% of mitochondrial sites (Bentolila et al. 2013). The importance of the RIP/MORF proteins to C-to-U editing is exemplified by the finding that loss of some RIP/MORF proteins results in defective editing of all chloroplast sites (RIP/MORF2 and 9) (Takenaka et al. 2012) and hundreds of mitochondrial sites (Bentolila et al. 2012). This contrasts with PPR proteins, which are usually involved in editing one or a few sites. Other chloroplast RBPs have also been implicated in C-to-U RNA editing. Mutations in the exoribonuclease PNPase, which functions in 3′-end maturation and tRNA precursor degradation (Ruwe et al. 2013), were found to cause editing defects. Additionally, cp31, a common chloroplast RNA-binding protein, was found to be required for editing several sites (Hirose and Sugiura 2001). It should also be mentioned that stress or defects in chloroplast function and mutations in members of the tetrapyrrole biosynthetic pathway were found to also cause RNA editing defects (Castandet et al. 2013; Kakizaki et al. 2012; Tseng et al. 2010; Zhang et al. 2014), although the particular pattern of editing defects are unique.

2.3.8 *mTERF Proteins*

MITOCHONDRIAL TRANSCRIPTION TERMINATION FACTORs (mTERFs) are modular proteins belonging to the family of helical-repeat proteins, the same as PPR proteins (Roberti et al. 2009). These proteins are capable of binding nucleic acids and help in the regulation of organellar gene expression machinery. In plants, mTERFs are localized in chloroplasts and/or mitochondria and have been characterized primarily through analysis of Arabidopsis (Kindgren et al. 2012) and maize (Zhao et al. 2014) mutants. See Chap. 3 in this volume.

2.4 Engineering RBPs for Biotechnological Application

Characterizing the structure and binding mechanism of RBPs to their targets has opened new possibilities for their application in synthetic biology. The bacterial RBP Cas13 fused with functional domains like adenosine deaminase has been used to modify or degrade RNAs (Kavuri et al. 2022). While Cas13 is effective at targeting nuclear and cytoplasmic transcripts, there are limitations in targeting organelle transcripts due to challenges in importing guide RNAs across organelle membranes (Sager et al. 2020).

PPR proteins emerge as strong candidates for engineered tools to manipulate chloroplast gene expression due to their RNA binding code, modular structure, and natural ability to deliver a catalytic C-terminal domain. The advantage of synthetic PPR proteins is that they determine their RNA binding specificity without requiring guide RNA (McDowell et al. 2022). The modular, base-specific, and predictable binding of PPR protein to single-stranded RNAs make them attractive as designable tools for transcript manipulation. The PPR code is developed based on the PPR motifs, which recognize RNA bases according to the identities of two key amino acids in the PPR motif (Barkan et al. 2012; Yagi et al. 2013; Kobayashi et al. 2019; Yan et al. 2019). This development has allowed for the design of synthetic PPR proteins with predictable and modifiable RNA targets (Coquille et al. 2014; Shen et al. 2015; Yan et al. 2019; Royan et al. 2021). Synthetic PPR protein based on consensus motif sequences can recognize a specific RNA target with high affinity and specificity *in vitro* (Coquille et al. 2014; Gully et al. 2015; Shen et al. 2015; Yan et al. 2019) and *in vivo* (McDermott et al. 2019). Synthetic P-class PPR proteins have been engineered to stabilize and protect transcripts from 5′ or 3′ exonucleases. They may also be designed to modify the RNA secondary structure, allowing the ribosome to access the transcript and facilitate translation of the target. Additionally, synthetic PLS-class PPR proteins have been designed to induce C-to-U editing via a DYW domain (Royan et al. 2021) or U-to-C editing via a DYW:KP domain (Ichinose et al. 2022).

The ability to design PPR protein to manipulate specified RNA targets is exciting and potentially of great importance for application in biotechnology, medicine, and basic research. However, there are currently no published demonstrations that synthetic RNA editing factors can be designed to target a completely novel editing site. This ability would generate exciting possibilities for controlling gene expression.

2.5 Conclusions and Perspectives

The plant cell expends a lot of energy to generate the large number of RBPs essential for chloroplast biogenesis and function, which is justified as RBPs ensure the precise regulation and expression of photosynthetic proteins through their roles in RNA processing for chloroplast gene expression. Thus, RBPs coordinate gene expression between the nuclear and chloroplast genomes, essential for coordinated production of photosynthetic proteins encoded by both genomes. The essential role of RBPs in chloroplast function and plant survival is highlighted by the overrepresentation of chloroplast RBP mutants in collection of Arabidopsis embryo-defective (emb) mutants (Hsu et al. 2010; Bryant et al. 2011) and maize seedling-lethal mutants (Belcher et al. 2015). PPR proteins and chloroplast ribosomal proteins are particularly well represented in these groups of genes. Therefore, investing energy in encoding the high number of RBPs that facilitate chloroplast biogenesis and maintenance benefits the plant.

While much has been learnt about chloroplast RNA processing and the RBPs involved, many issues remain. These include the order and interdependence of RNA processing steps. Interestingly, RNA editing is required for high splicing efficiency (Kurihara-Yonemoto and Handa 2001). Suppressing an editing site results in an accumulation of unspliced precursors in wheat. However, editing sites close to the intron-exon junction are edited only after the mRNA splicing. Additionally, in maize, Arabidopsis, and tobacco, some sites are edited, and others are not within the same transcript; therefore, specific trans-factors for each editing site have been postulated (Peeters and Hanson 2002; Tseng et al. 2010). While some trans-factors may be specific to one site, other trans-factors may edit several sites (Robbins et al. 2009). The tissue specificity of RNA processing is also of interest, as differences in plastid function can be expected to dictate distinct RNA processing to fine tune chloroplast gene expression for maximal function. For example, transcripts of the NADH dehydrogenase subunits showed drastically reduced editing in roots compared to tobacco leaves, illustrating that the presence and/or the activity of trans-factors that are required for RNA editing depend on the tissue type (Chateigner-Boutin and Hanson 2002).

As a better understanding of chloroplast RNA processing and the RBP players is gained, the potential to manipulate chloroplast gene expression to achieve desired outcomes emerges. Together with advances in chloroplast transformation (see Chap. 6), this holds the promise of using the chloroplast as a factory for production of high-value products, for manipulating photosynthesis to better cope with environmental and biotic stresses and for other possible uses that will become apparent only as the technologies are developed. Excitingly, the vast number and functionalities of chloroplast RBPs suggests endless possibilities for engineering.

Acknowledgements This work and other work on chloroplast gene expression in the Burch-Smith lab is supported by the National Science Foundation, grant MCB-221027.

References

Adam SA, Nakagawa T, Swanson MS, Woodruff TK, Dreyfuss G (1986) mRNA polyadenylate-binding protein: gene isolation and sequencing and identification of a ribonucleoprotein consensus sequence. Mol Cell Biol 6:2932–2943

An H, Ke X, Li L, Liu Y, Yuan S, Wang Q, Hou X, Zhao J (2023) ALBINO EMBRYO AND SEEDLING is required for RNA splicing and chloroplast homeostasis in Arabidopsis. Plant Physiol 193:483–501

Aryamanesh N, Ruwe H, Sanglard LV, Eshraghi L, Bussell JD, Howell KA, Small I, des Francs-Small CC (2017) The pentatricopeptide repeat protein EMB2654 is essential for trans-splicing of a chloroplast small ribosomal subunit transcript. Plant Physiol 173:1164–1176

Asakura Y, Barkan A (2006) Arabidopsis Orthologs of maize chloroplast splicing factors promote splicing of orthologous and species-specific group II introns. Plant Physiol 142:(4)1656–1663

Asakura Y, Barkan A (2007) A CRM domain protein functions dually in group I and group II intron splicing in land plant chloroplasts. Plant Cell 19:3864–3875

Asakura Y, Bayraktar OA, Barkan A (2008) Two CRM protein subfamilies cooperate in the splicing of group IIB introns in chloroplasts. RNA 14:2319–2332

Asakura Y, Galarneau E, Watkins KP, Barkan A, van Wijk KJ (2012) Chloroplast RH3 DEAD box RNA helicases in maize and Arabidopsis function in splicing of specific group II introns and affect chloroplast ribosome biogenesis. Plant Physiol 159:961–974

Baginsky S, Shteiman-Kotler A, Liveanu V, Yehudai-Resheff S, Bellaoui M, Settlage RE, Shabanowitz J, Hunt DF, Schuster G, Gruissem W (2001) Chloroplast PNPase exists as a homo-multimer enzyme complex that is distinct from the Escherichia coli degradosome. RNA 7:1464–1475

Barkan A, Small I (2014) Pentatricopeptide repeat proteins in plants. Annu Rev Plant Biol 65:415–442

Barkan A, Klipcan L, Ostersetzer O, Kawamura T, Asakura Y, Watkins KP (2007) The CRM domain: an RNA binding module derived from an ancient ribosome-associated protein. RNA 13:55–64

Barkan A, Rojas M, Fujii S, Yap A, Chong YS, Bond CS, Small I (2012) A combinatorial amino acid code for RNA recognition by pentatricopeptide repeat proteins. PLoS Genet 8:e1002910

Bayer-Csaszar E, Haag S, Jorg A, Glass F, Hartel B, Obata T, Meyer EH, Brennicke A, Takenaka M (2017) The conserved domain in MORF proteins has distinct affinities to the PPR and E elements in PPR RNA editing factors. Biochim Biophys Acta Gene Regul Mech 1860:813–828

Beick S, Schmitz-Linneweber C, Williams-Carrier R, Jensen B, Barkan A (2008) The pentatricopeptide repeat protein PPR5 stabilizes a specific tRNA precursor in maize chloroplasts. Mol Cell Biol 28:5337–5347

Belcher S, Williams-Carrier R, Stiffler N, Barkan A (2015) Large-scale genetic analysis of chloroplast biogenesis in maize. Biochim Biophys Acta 1847:1004–1016

Beligni MV, Mayfield SP (2008) Arabidopsis thaliana mutants reveal a role for CSP41a and CSP41b, two ribosome-associated endonucleases, in chloroplast ribosomal RNA metabolism. Plant Mol Biol 67:389–401

Bentolila S, Heller WP, Sun T, Babina AM, Friso G, van Wijk KJ, Hanson MR (2012) RIP1, a member of an Arabidopsis protein family, interacts with the protein RARE1 and broadly affects RNA editing. Proc Natl Acad Sci USA 109:E1453–E1461

Bentolila S, Oh J, Hanson MR, Bukowski R (2013) Comprehensive high-resolution analysis of the role of an Arabidopsis gene family in RNA editing. PLoS Genet 9:e1003584

Bobik K, McCray TN, Ernest B, Fernandez JC, Howell KA, Lane T, Staton M, Burch-Smith TM (2017) The chloroplast RNA helicase ISE2 is required for multiple chloroplast RNA processing steps in Arabidopsis thaliana. Plant J 91:114–131

Bollenbach TJ, Tatman DA, Stern DB (2003) CSP41a, a multifunctional RNA-binding protein, initiates mRNA turnover in tobacco chloroplasts. Plant J 36:842–852

Bollenbach TJ, Lange H, Gutierrez R, Erhardt M, Stern DB, Gagliardi D (2005) RNR1, a 3′-5′ exoribonuclease belonging to the RNR superfamily, catalyzes 3′ maturation of chloroplast ribosomal RNAs in Arabidopsis thaliana. Nucleic Acids Res 33:2751–2763

Boussardon C, Salone V, Avon A, Berthome R, Hammani K, Okuda K, Shikanai T, Small I, Lurin C (2012) Two interacting proteins are necessary for the editing of the NdhD-1 site in Arabidopsis plastids. Plant Cell 24:3684–3694

Bryant N, Lloyd J, Sweeney C, Myouga F, Meinke D (2011) Identification of nuclear genes encoding chloroplast-localized proteins required for embryo development in Arabidopsis. Plant Physiol 155:1678–1689

Burd CG, Dreyfuss G (1994) RNA binding specificity of hnRNP A1: significance of hnRNP A1 high-affinity binding sites in pre-mRNA splicing. EMBO J 13:1197–1204

Carlotto N, Wirth S, Furman N, Ferreyra Solari N, Ariel F, Crespi M, Kobayashi K (2016) The chloroplastic DEVH-box RNA helicase INCREASED SIZE EXCLUSION LIMIT 2 involved in plasmodesmata regulation is required for group II intron splicing. Plant Cell Environ 39:165–173

Caruthers JM, McKay DB (2002) Helicase structure and mechanism. Curr Opin Struct Biol 12:123–133

Castandet B, Hotto AM, Fei Z, Stern DB (2013) Strand-specific RNA sequencing uncovers chloroplast ribonuclease functions. FEBS Lett 587:3096–3101

Castandet B, Germain A, Hotto AM, Stern DB (2019) Systematic sequencing of chloroplast transcript termini from Arabidopsis thaliana reveals >200 transcription initiation sites and the extensive imprints of RNA-binding proteins and secondary structures. Nucleic Acids Res 47:11889–11905

Chateigner-Boutin AL, Hanson MR (2002) Cross-competition in transgenic chloroplasts expressing single editing sites reveals shared cis elements. Mol Cell Biol 22:8448–8456

Chateigner-Boutin AL, Small I (2010) Plant RNA editing. RNA Biol 7:213–219

Chen G, Zou Y, Hu J, Ding Y (2018) Genome-wide analysis of the rice PPR gene family and their expression profiles under different stress treatments. BMC Genomics 19:720

Cheng S, Gutmann B, Zhong X, Ye Y, Fisher MF, Bai F, Castleden I, Song Y, Song B, Huang J, Liu X, Xu X, Lim BL, Bond CS, Yiu SM, Small I (2016) Redefining the structural motifs that determine RNA binding and RNA editing by pentatricopeptide repeat proteins in land plants. Plant J 85:532–547

Chi W, He B, Mao J, Li Q, Ma J, Ji D, Zou M, Zhang L (2012) The function of RH22, a DEAD RNA helicase, in the biogenesis of the 50S ribosomal subunits of Arabidopsis chloroplasts. Plant Physiol 158:693–707

Chotewutmontri P, Holbrook K, Bruce BD (2017) Plastid protein targeting: preprotein recognition and translocation. Int Rev Cell Mol Biol 330:227–294

Clery A, Blatter M, Allain FH (2008) RNA recognition motifs: boring? Not quite. Curr Opin Struct Biol 18:290–298

Condon C, Gilet L (2011) The metallo-β-lactamase family of ribonucleases. In: Nicholson AW (ed) Ribonucleases. Springer, Berlin, Heidelberg, pp 245–267

Coquille S, Filipovska A, Chia T, Rajappa L, Lingford JP, Razif MF, Thore S, Rackham O (2014) An artificial PPR scaffold for programmable RNA recognition. Nat Commun 5:5729

Cordin O, Banroques J, Tanner NK, Linder P (2006) The DEAD-box protein family of RNA helicases. Gene 367:17–37

de la Cruz J, Kressler D, Linder P (1999) Unwinding RNA in Saccharomyces cerevisiae: DEAD-box proteins and related families. Trends Biochem Sci 24:192–198

de Longevialle AF, Small ID, Lurin C (2010) Nuclearly encoded splicing factors implicated in RNA splicing in higher plant organelles. Mol Plant 3:691–705

D'Andrea LD, Regan L (2003) TPR proteins: the versatile helix. Trends Biochem Sci 28:655–662

Daras G, Rigas S, Alatzas A, Samiotaki M, Chatzopoulos D, Tsitsekian D, Papadaki V, Templalexis D, Banilas G, Athanasiadou AM, Kostourou V, Panayotou G, Hatzopoulos P

(2019) LEFKOTHEA regulates nuclear and chloroplast mRNA splicing in plants. Dev Cell 50:767–779.e767

Ding YH, Liu NY, Tang ZS, Liu J, Yang WC (2006) Arabidopsis GLUTAMINE-RICH PROTEIN23 is essential for early embryogenesis and encodes a novel nuclear PPR motif protein that interacts with RNA polymerase II subunit III. Plant Cell 18:815–830

Dorcey E, Rodriguez-Villalon A, Salinas P, Santuari L, Pradervand S, Harshman K, Hardtke CS (2012) Context-dependent dual role of SKI8 homologs in mRNA synthesis and turnover. PLoS Genet 8:e1002652

Dos Santos RF, Quendera AP, Boavida S, Seixas AF, Arraiano CM, Andrade JM (2018) Major 3′–5′ exoribonucleases in the metabolism of coding and non-coding RNA. Prog Mol Biol Transl Sci 159:101–155

Dreyfuss G, Matunis MJ, Pinol-Roma S, Burd CG (1993) hnRNP proteins and the biogenesis of mRNA. Annu Rev Biochem 62:289–321

Edera AA, Gandini CL, Sanchez-Puerta MV (2018) Towards a comprehensive picture of C-to-U RNA editing sites in angiosperm mitochondria. Plant Mol Biol 97:215–231

Fairman-Williams ME, Guenther UP, Jankowsky E (2010) SF1 and SF2 helicases: family matters. Curr Opin Struct Biol 20:313–324

Feiz L, Asakura Y, Mao L, Strickler SR, Fei Z, Rojas M, Barkan A, Stern DB (2021) CFM1, a member of the CRM-domain protein family, functions in chloroplast group II intron splicing in Setaria viridis. Plant J 105:639–648

Gerke P, Szovenyi P, Neubauer A, Lenz H, Gutmann B, McDowell R, Small I, Schallenberg-Rudinger M, Knoop V (2020) Towards a plant model for enigmatic U-to-C RNA editing: the organelle genomes, transcriptomes, editomes and candidate RNA editing factors in the hornwort Anthoceros agrestis. New Phytol 225:1974–1992

Germain A, Herlich S, Larom S, Kim SH, Schuster G, Stern DB (2011) Mutational analysis of Arabidopsis chloroplast polynucleotide phosphorylase reveals roles for both RNase PH core domains in polyadenylation, RNA 3′-end maturation and intron degradation. Plant J 67:381–394

Germain A, Kim SH, Gutierrez R, Stern DB (2012) Ribonuclease II preserves chloroplast RNA homeostasis by increasing mRNA decay rates, and cooperates with polynucleotide phosphorylase in 3′ end maturation. Plant J 72:960–971

Germain A, Hotto AM, Barkan A, Stern DB (2013) RNA processing and decay in plastids. Wiley Interdiscip Rev RNA 4:295–316

Gipson AB, Hanson MR, Bentolila S (2022) The RanBP2 zinc finger domains of chloroplast RNA editing factor OZ1 are required for protein-protein interactions and conversion of C to U. Plant J 109:215–226

Glanz S, Kuck U (2009) Trans-splicing of organelle introns – a detour to continuous RNAs. BioEssays 31:921–934

Goldschmidt-Clermont M (1998) Coordination of nuclear and chloroplast gene expression in plant cells. Int Rev Cytol 177:115–180

Gorbalenya AE, Koonin EV (1993) Helicases: amino acid sequence comparisons and structure-function relationships. Curr Opin Struc Biol 3:419–429

Graw F, Balagopal A, Kandathil AJ, Ray SC, Thomas DL, Ribeiro RM, Perelson AS (2014) Inferring viral dynamics in chronically HCV infected patients from the spatial distribution of infected hepatocytes. PLoS Comput Biol 10:e1003934

Greiner S, Golczyk H, Malinova I, Pellizzer T, Bock R, Borner T, Herrmann RG (2020) Chloroplast nucleoids are highly dynamic in ploidy, number, and structure during angiosperm leaf development. Plant J 102:730–746

Gu L, Xu T, Lee K, Lee KH, Kang H (2014) A chloroplast-localized DEAD-box RNA helicaseAtRH3 is essential for intron splicing and plays an important role in the growth and stress response in Arabidopsis thaliana. Plant Physiol Biochem 82:309–318

Gully BS, Shah KR, Lee M, Shearston K, Smith NM, Sadowska A, Blythe AJ, Bernath-Levin K, Stanley WA, Small ID, Bond CS (2015) The design and structural characterization of a synthetic pentatricopeptide repeat protein. Acta Crystallogr D Biol Crystallogr 71:196–208

Gutmann B, Royan S, Schallenberg-Rudinger M, Lenz H, Castleden IR, McDowell R, Vacher MA, Tonti-Filippini J, Bond CS, Knoop V, Small ID (2020) The expansion and diversification of pentatricopeptide repeat RNA-editing factors in plants. Mol Plant 13:215–230

Haag S, Schindler M, Berndt L, Brennicke A, Takenaka M, Weber G (2017) Crystal structures of the Arabidopsis thaliana organellar RNA editing factors MORF1 and MORF9. Nucleic Acids Res 45:4915–4928

Halpert M, Liveanu V, Glaser F, Schuster G (2019) The Arabidopsis chloroplast RNase J displays both exo- and robust endonucleolytic activities. Plant Mol Biol 99:17–29

Hammani K, Barkan A (2014) An mTERF domain protein functions in group II intron splicing in maize chloroplasts. Nucleic Acids Res 42:5033–5042

Hammani K, Takenaka M, Miranda R, Barkan A (2016) A PPR protein in the PLS subfamily stabilizes the 5′-end of processed rpl16 mRNAs in maize chloroplasts. Nucleic Acids Res 44:4278–4288

Hashimoto M, Endo T, Peltier G, Tasaka M, Shikanai T (2003) A nucleus-encoded factor, CRR2, is essential for the expression of chloroplast ndhB in Arabidopsis. Plant J 36:541–549

Hattori M, Miyake H, Sugita M (2007) A pentatricopeptide repeat protein is required for RNA processing of clpP Pre-mRNA in moss chloroplasts. J Biol Chem 282:10773–10782

Hayes ML, Santibanez PI (2020) A plant pentatricopeptide repeat protein with a DYW-deaminase domain is sufficient for catalyzing C-to-U RNA editing in vitro. J Biol Chem 295:3497–3505

Hirose T, Sugiura M (2001) Involvement of a site-specific trans-acting factor and a common RNA-binding protein in the editing of chloroplast mRNAs: development of a chloroplast in vitro RNA editing system. EMBO J 20:1144–1152

Hotto AM, Schmitz RJ, Fei Z, Ecker JR, Stern DB (2011) Unexpected diversity of chloroplast non-coding RNAs as revealed by deep sequencing of the Arabidopsis transcriptome. G3 (Bethesda) 1:559–570

Hotto AM, Castandet B, Gilet L, Higdon A, Condon C, Stern DB (2015) Arabidopsis chloroplast mini-ribonuclease III participates in rRNA maturation and intron recycling. Plant Cell 27:724–740

Hotto AM, Stern DB, Schuster G (2020) Plant Ribonuclease J: an essential player in maintaining chloroplast RNA quality control for gene expression. Plants 9:334

Hsu SC, Belmonte MF, Harada JJ, Inoue K (2010) Indispensable roles of plastids in Arabidopsis thaliana embryogenesis. Curr Genomics 11:338–349

Ichinose M, Sugita M (2016) RNA editing and its molecular mechanism in plant organelles. Genes (Basel) 8:5

Ichinose M, Uchida M, Sugita M (2014) Identification of a pentatricopeptide repeat RNA editing factor in Physcomitrella patens chloroplasts. FEBS Lett 588:4060–4064

Ichinose M, Kawabata M, Akaiwa Y, Shimajiri Y, Nakamura I, Tamai T, Nakamura T, Yagi Y, Gutmann B (2022) U-to-C RNA editing by synthetic PPR-DYW proteins in bacteria and human culture cells. Commun Biol 5:968

Iswanto ABB, Shon JC, Liu KH, Vu MH, Kumar R, Kim JY (2020) Sphingolipids modulate secretion of glycosylphosphatidylinositol-anchored plasmodesmata proteins and callose deposition. Plant Physiol 184:407–420

Jacobs J, Kuck U (2011) Function of chloroplast RNA-binding proteins. Cell Mol Life Sci 68:735–748

Jankowsky E, Fairman ME (2007) RNA helicases – one fold for many functions. Curr Opin Struct Biol 17:316–324

Jarmoskaite I, Russell R (2011) DEAD-box proteins as RNA helicases and chaperones. Wiley Interdiscip Rev RNA 2:135–152

Jarmoskaite I, Russell R (2014) RNA helicase proteins as chaperones and remodelers. Annu Rev Biochem 83:697–725

Jenkins BD, Kulhanek DJ, Barkan A (1997) Nuclear mutations that block group II RNA splicing in maize chloroplasts reveal several intron classes with distinct requirements for splicing factors. Plant Cell 9:283–296

Jung HJ, Park SJ, Kang H (2013) Regulation of RNA metabolism in plant development and stress responses. J Plant Biol 56:123–129

Kakizaki T, Yazu F, Nakayama K, Ito-Inaba Y, Inaba T (2012) Plastid signalling under multiple conditions is accompanied by a common defect in RNA editing in plastids. J Exp Bot 63:251–260

Karpowicz SJ, Prochnik SE, Grossman AR, Merchant SS (2011) The GreenCut2 resource, a phylogenomically derived inventory of proteins specific to the plant lineage. J Biol Chem 286:21427–21439

Kavuri NR, Ramasamy M, Qi Y, Mandadi K (2022) Applications of CRISPR/Cas13-based RNA editing in plants. Cells 11:2665

Ke J, Chen RZ, Ban T, Zhou XE, Gu X, Tan MH, Chen C, Kang Y, Brunzelle JS, Zhu JK, Melcher K, Xu HE (2013) Structural basis for RNA recognition by a dimeric PPR-protein complex. Nat Struct Mol Biol 20:1377–1382

Khrouchtchova A, Monde RA, Barkan A (2012) A short PPR protein required for the splicing of specific group II introns in angiosperm chloroplasts. RNA 18:1197–1209

Kindgren P, Kremnev D, Blanco NE, de Dios Barajas Lopez J, Fernandez AP, Tellgren-Roth C, Kleine T, Small I, Strand A (2012) The plastid redox insensitive 2 mutant of Arabidopsis is impaired in PEP activity and high light-dependent plastid redox signalling to the nucleus. Plant J 70:279–291

Kobayashi T, Yagi Y, Nakamura T (2019) Comprehensive prediction of target RNA editing sites for PLS-class PPR proteins in Arabidopsis thaliana. Plant Cell Physiol 60:862–874

Köhler D, Schmidt-Gattung S, Binder S (2010) The DEAD-box protein PMH2 is required for efficient group II intron splicing in mitochondria of Arabidopsis thaliana. Plant Mol Biol 72:459–467

Kotera E, Tasaka M, Shikanai T (2005) A pentatricopeptide repeat protein is essential for RNA editing in chloroplasts. Nature 433:326–330

Koussevitzky S, Nott A, Mockler TC, Hong F, Sachetto-Martins G, Surpin M, Lim J, Mittler R, Chory J (2007) Signals from chloroplasts converge to regulate nuclear gene expression. Science 316:715–719

Kroeger TS, Watkins KP, Friso G, van Wijk KJ, Barkan A (2009) A plant-specific RNA-binding domain revealed through analysis of chloroplast group II intron splicing. Proc Natl Acad Sci USA 106:4537–4542

Kruger K, Grabowski PJ, Zaug AJ, Sands J, Gottschling DE, Cech TR (1982) Self-splicing RNA: autoexcision and autocyclization of the ribosomal RNA intervening sequence of Tetrahymena. Cell 31:147–157

Kurihara-Yonemoto S, Handa H (2001) Low temperature affects the processing pattern and RNA editing status of the mitochondrial cox2 transcripts in wheat. Curr Genet 40:203–208

Kuroiwa T (1991) The replication, differentiation, and inheritance of plastids with emphasis on the concept of organelle nuclei. Int Rev Cytol 128:1–62

Kwong AD, Rao BG, Jeang KT (2005) Viral and cellular RNA helicases as antiviral targets. Nat Rev Drug Discov 4:845–853

Lee KH, Park J, Williams DS, Xiong Y, Hwang I, Kang BH (2013) Defective chloroplast development inhibits maintenance of normal levels of abscisic acid in a mutant of the Arabidopsis RH3 DEAD-box protein during early post-germination growth. Plant J 73:720–732

Lee K, Park SJ, Colas des Francs-Small C, Whitby M, Small I, Kang H (2019) The coordinated action of PPR4 and EMB2654 on each intron half mediates trans-splicing of rps12 transcripts in plant chloroplasts. Plant J 100:1193–1207

Leister D (2014) Complex(iti)es of the ubiquitous RNA-binding CSP41 proteins. Front Plant Sci 5:255

Li D, Liu H, Zhang H, Wang X, Song F (2008) OsBIRH1, a DEAD-box RNA helicase with functions in modulating defence responses against pathogen infection and oxidative stress. J Exp Bot 59:2133–2146

Linder P, Fuller-Pace FV (2013) Looking back on the birth of DEAD-box RNA helicases. Biochim Biophys Acta 1829:750–755
Linder P, Jankowsky E (2011) From unwinding to clamping – the DEAD box RNA helicase family. Nat Rev Mol Cell Biol 1:505–516
Linder P, Owttrim GW (2009) Plant RNA helicases: linking aberrant and silencing RNA. Trends Plant Sci 14:344–352
Linder P, Lasko PF, Ashburner M, Leroy P, Nielsen PJ, Nishi K, Schnier J, Slonimski PP (1989) Birth of the D-E-A-D box. Nature 337:121–122
Liou GG, Chang HY, Lin CS, Lin-Chao S (2002) DEAD box RhlB RNA helicase physically associates with exoribonuclease PNPase to degrade double-stranded RNA independent of the degradosome-assembling region of RNase E. J Biol Chem 277:41157–41162
Liu Y, Imai R (2018) Function of plant DExD/H-box RNA helicases associated with ribosomal RNA biogenesis. Front Plant Sci 9:125
Liu Y, He J, Chen Z, Ren X, Hong X, Gong Z (2010) ABA overly-sensitive 5 (ABO5), encoding a pentatricopeptide repeat protein required for cis-splicing of mitochondrial nad2 intron 3, is involved in the abscisic acid response in Arabidopsis. Plant J 63:749–765
Liu J, Zhou W, Liu G, Yang C, Sun Y, Wu W, Cao S, Wang C, Hai G, Wang Z, Bock R, Huang J, Cheng Y (2015) The conserved endoribonuclease YbeY is required for chloroplast ribosomal RNA processing in Arabidopsis. Plant Physiol 168:205–221
Lorkovic ZJ, Barta A (2002) Genome analysis: RNA recognition motif (RRM) and K homology (KH) domain RNA-binding proteins from the flowering plant Arabidopsis thaliana. Nucleic Acids Res 30:623–635
Lunde BM, Moore C, Varani G (2007) RNA-binding proteins: modular design for efficient function. Nat Rev Mol Cell Biol 8:479–490
Lurin C, Andres C, Aubourg S, Bellaoui M, Bitton F, Bruyere C, Caboche M, Debast C, Gualberto J, Hoffmann B, Lecharny A, Le Ret M, Martin-Magniette ML, Mireau H, Peeters N, Renou JP, Szurek B, Taconnat L, Small I (2004) Genome-wide analysis of Arabidopsis pentatricopeptide repeat proteins reveals their essential role in organelle biogenesis. Plant Cell 16:2089–2103
Luro S, Germain A, Sharwood RE, Stern DB (2013) RNase J participates in a pentatricopeptide repeat protein-mediated 5′ end maturation of chloroplast mRNAs. Nucleic Acids Res 41:9141–9151
Mach J (2009) Chloroplast RNA editing by pentatricopeptide repeat proteins. Plant Cell 21:17
MacIntosh GC, Castandet B (2020) Organellar and secretory ribonucleases: major players in plant RNA homeostasis. Plant Physiol 183:1438–1452
Majeran W, Friso G, Asakura Y, Qu X, Huang M, Ponnala L, Watkins KP, Barkan A, van Wijk KJ (2012) Nucleoid-enriched proteomes in developing plastids and chloroplasts from maize leaves: a new conceptual framework for nucleoid functions. Plant Physiol 158:156–189
Marchive C, Yehudai-Resheff S, Germain A, Fei Z, Jiang X, Judkins J, Wu H, Fernie AR, Fait A, Stern DB (2009) Abnormal physiological and molecular mutant phenotypes link chloroplast polynucleotide phosphorylase to the phosphorus deprivation response in Arabidopsis. Plant Physiol 151:905–924
Maris C, Dominguez C, Allain FH (2005) The RNA recognition motif, a plastic RNA-binding platform to regulate post-transcriptional gene expression. FEBS J 272:2118–2131
McCray TN, Azim MF, Burch-Smith TM (2024) The dicot homolog of maize PPR103 carries a C-terminal DYW domain and may have a role in C-to-U editing of some chloroplast RNA transcripts. Plant Mol Biol 114:28
McDermott JJ, Watkins KP, Williams-Carrier R, Barkan A (2019) Ribonucleoprotein capture by in vivo expression of a designer pentatricopeptide repeat protein in Arabidopsis. Plant Cell 31:1723–1733
McDowell R, Small I, Bond CS (2022) Synthetic PPR proteins as tools for sequence-specific targeting of RNA. Methods 208:19–26

Mermigka G, Helm JM, Vlatakis I, Schumacher HT, Vamvaka E, Kalantidis K (2016) ERIL1, the plant homologue of ERI-1, is involved in the processing of chloroplastic rRNAs. Plant J 88:839–853

Michel F, Umesono K, Ozeki H (1989) Comparative and functional anatomy of group II catalytic introns – a review. Gene 82:5–30

Mingam A, Toffano-Nioche C, Brunaud V, Boudet N, Kreis M, Lecharny A (2004) DEAD-box RNA helicases in Arabidopsis thaliana: establishing a link between quantitative expression, gene structure and evolution of a family of genes. Plant Biotechnol J 2:401–415

Miura MC, Nagata S, Tamaki S, Tomita M, Kanai A (2022) Distinct expansion of group II introns during evolution of prokaryotes and possible factors involved in its regulation. Front Microbiol 13:849080

Mower JP (2009) The PREP suite: predictive RNA editors for plant mitochondrial genes, chloroplast genes and user-defined alignments. Nucleic Acids Res 37:W253–W259

Nakamura T, Sugita M (2008) A conserved DYW domain of the pentatricopeptide repeat protein possesses a novel endoribonuclease activity. FEBS Lett 582:4163–4168

Nakamura T, Meierhoff K, Westhoff P, Schuster G (2003) RNA-binding properties of HCF152, an Arabidopsis PPR protein involved in the processing of chloroplast RNA. Eur J Biochem 270(20):4070–4081

Nawaz G, Kang H (2017) Chloroplast- or mitochondria-targeted DEAD-box RNA helicases play essential roles in organellar RNA metabolism and abiotic stress responses. Front Plant Sci 8:871

Nishimura K, Ashida H, Ogawa T, Yokota A (2010) A DEAD box protein is required for formation of a hidden break in Arabidopsis chloroplast 23S rRNA. Plant J 63:766–777

Okuda K, Myouga F, Motohashi R, Shinozaki K, Shikanai T (2007) Conserved domain structure of pentatricopeptide repeat proteins involved in chloroplast RNA editing. Proc Natl Acad Sci USA 104:8178–8183

Okuda K, Chateigner-Boutin AL, Nakamura T, Delannoy E, Sugita M, Myouga F, Motohashi R, Shinozaki K, Small I, Shikanai T (2009) Pentatricopeptide repeat proteins with the DYW motif have distinct molecular functions in RNA editing and RNA cleavage in Arabidopsis chloroplasts. Plant Cell 21:146–156

Oldenkott B, Yamaguchi K, Tsuji-Tsukinoki S, Knie N, Knoop V (2014) Chloroplast RNA editing going extreme: more than 3400 events of C-to-U editing in the chloroplast transcriptome of the lycophyte Selaginella uncinata. RNA 20:1499–1506

Oldenkott B, Yang Y, Lesch E, Knoop V, Schallenberg-Rudinger M (2019) Plant-type pentatricopeptide repeat proteins with a DYW domain drive C-to-U RNA editing in Escherichia coli. Commun Biol 2:85

Olinares PD, Ponnala L, van Wijk KJ (2010) Megadalton complexes in the chloroplast stroma of Arabidopsis thaliana characterized by size exclusion chromatography, mass spectrometry, and hierarchical clustering. Mol Cell Proteomics 9:1594–1615

Ostersetzer O, Cooke AM, Watkins KP, Barkan A (2005) CRS1, a chloroplast group II intron splicing factor, promotes intron folding through specific interactions with two intron domains. Plant Cell 17:241–255

Ostheimer GJ, Williams-Carrier R, Belcher S, Osborne E, Gierke J, Barkan A (2003) Group II intron splicing factors derived by diversification of an ancient RNA-binding domain. EMBO J 22:3919–3929

Ostheimer GJ, Rojas M, Hadjivassiliou H, Barkan A (2006) Formation of the CRS2-CAF2 group II intron splicing complex is mediated by a 22-amino acid motif in the COOH-terminal region of CAF2. J Biol Chem 281:4732–4738

Paieri F, Tadini L, Manavski N, Kleine T, Ferrari R, Morandini P, Pesaresi P, Meurer J, Leister D (2018) The DEAD-box RNA helicase RH50 Is a 23S-4.5S rRNA maturation factor that functionally overlaps with the plastid signaling factor GUN1. Plant Physiol 176:634–648

Palmer JD (1985) Chloroplast DNA and molecular phylogeny. BioEssays 2:263–267

Peeters NM, Hanson MR (2002) Transcript abundance supercedes editing efficiency as a factor in developmental variation of chloroplast gene expression. RNA 8:497–511

Pfalz J, Liebers M, Hirth M, Grubler B, Holtzegel U, Schroter Y, Dietzel L, Pfannschmidt T (2012) Environmental control of plant nuclear gene expression by chloroplast redox signals. Front Plant Sci 3:257

Podar M, Perlman PS, Padgett RA (1998) The two steps of group II intron self-splicing are mechanistically distinguishable. RNA 4:890–900

Prikryl J, Watkins KP, Friso G, van Wijk KJ, Barkan A (2008) A member of the Whirly family is a multifunctional RNA- and DNA-binding protein that is essential for chloroplast biogenesis. Nucleic Acids Res 36:5152–5165

Pyo YJ, Kwon KC, Kim A, Cho MH (2013) Seedling Lethal1, a pentatricopeptide repeat protein lacking an E/E+ or DYW domain in Arabidopsis, is involved in plastid gene expression and early chloroplast development. Plant Physiol 163:1844–1858

Qi Y, Armbruster U, Schmitz-Linneweber C, Delannoy E, de Longevialle AF, Ruhle T, Small I, Jahns P, Leister D (2012) Arabidopsis CSP41 proteins form multimeric complexes that bind and stabilize distinct plastid transcripts. J Exp Bot 63:1251–1270

Redko Y, Bechhofer DH, Condon C (2008) Mini-III, an unusual member of the RNase III family of enzymes, catalyses 23S ribosomal RNA maturation in B. subtilis. Mol Microbiol 68:1096–1106

Robbins JC, Heller WP, Hanson MR (2009) A comparative genomics approach identifies a PPR-DYW protein that is essential for C-to-U editing of the Arabidopsis chloroplast accD transcript. RNA 15:1142–1153

Roberti M, Polosa PL, Bruni F, Manzari C, Deceglie S, Gadaleta MN, Cantatore P (2009) The MTERF family proteins: mitochondrial transcription regulators and beyond. Biochim Biophys Acta 1787:303–311

Rovira AG, Smith AG (2019) PPR proteins – orchestrators of organelle RNA metabolism. Physiol Plant 166:451–459

Royan S, Gutmann B, Colas des Francs-Small C, Honkanen S, Schmidberger J, Soet A, Sun YK, Vincis Pereira Sanglard L, Bond CS, Small I (2021) A synthetic RNA editing factor edits its target site in chloroplasts and bacteria. Commun Biol 4:545

Rudinger M, Polsakiewicz M, Knoop V (2008) Organellar RNA editing and plant-specific extensions of pentatricopeptide repeat proteins in jungermanniid but not in marchantiid liverworts. Mol Biol Evol 25:1405–1414

Ruwe H, Castandet B, Schmitz-Linneweber C, Stern DB (2013) Arabidopsis chloroplast quantitative editotype. FEBS Lett 587:1429–1433

Ruwe H, Gutmann B, Schmitz-Linneweber C, Small I, Kindgren P (2019) The E domain of CRR2 participates in sequence-specific recognition of RNA in plastids. New Phytol 222:218–229

Sager R, Wang X, Hill K, Yoo BC, Caplan J, Nedo A, Tran T, Bennett MJ, Lee JY (2020) Auxin-dependent control of a plasmodesmal regulator creates a negative feedback loop modulating lateral root emergence. Nat Commun 11:364

Salone V, Rudinger M, Polsakiewicz M, Hoffmann B, Groth-Malonek M, Szurek B, Small I, Knoop V, Lurin C (2007) A hypothesis on the identification of the editing enzyme in plant organelles. FEBS Lett 581:4132–4138

Sandoval R, Boyd RD, Kiszter AN, Mirzakhanyan Y, Santibanez P, Gershon PD, Hayes ML (2019) Stable native RIP9 complexes associate with C-to-U RNA editing activity, PPRs, RIPs, OZ1, ORRM1 and ISE2. Plant J 99:1116–1126

Schallenberg-Rudinger M, Kindgren P, Zehrmann A, Small I, Knoop V (2013a) A DYW-protein knockout in Physcomitrella affects two closely spaced mitochondrial editing sites and causes a severe developmental phenotype. Plant J 76:420–432

Schallenberg-Rudinger M, Lenz H, Polsakiewicz M, Gott JM, Knoop V (2013b) A survey of PPR proteins identifies DYW domains like those of land plant RNA editing factors in diverse eukaryotes. RNA Biol 10:1549–1556

Schein A, Sheffy-Levin S, Glaser F, Schuster G (2008) The RNase E/G-type endoribonuclease of higher plants is located in the chloroplast and cleaves RNA similarly to the E. coli enzyme. RNA 14:1057–1068

Schmitz-Linneweber C, Williams-Carrier RE, Williams-Voelker PM, Kroeger TS, Vichas A, Barkan A (2006) A pentatricopeptide repeat protein facilitates the trans-splicing of the maize chloroplast rps12 pre-mRNA. Plant Cell 18:2650–2663
Schuster G, Stern D (2009) RNA polyadenylation and decay in mitochondria and chloroplasts. Prog Mol Biol Transl Sci 85:393–422
Sharwood RE, Halpert M, Luro S, Schuster G, Stern DB (2011) Chloroplast RNase J compensates for inefficient transcription termination by removal of antisense RNA. RNA 17:2165–2176
Shen C, Wang X, Liu Y, Li Q, Yang Z, Yan N, Zou T, Yin P (2015) Specific RNA recognition by designer pentatricopeptide repeat protein. Mol Plant 8:667–670
Shen L, Zhang Q, Wang Z, Wen H, Hu G, Ren D, Hu J, Zhu L, Gao Z, Zhang G, Guo L, Zeng D, Qian Q (2020) OsCAF2 contains two CRM domains and is necessary for chloroplast development in rice. BMC Plant Biol 20:381
Shikanai T (2006) RNA editing in plant organelles: machinery, physiological function and evolution. Cell Mol Life Sci 63:698–708
Shikanai T (2015) RNA editing in plants: machinery and flexibility of site recognition. Biochim Biophys Acta 1847:779–785
Silverman E, Edwalds-Gilbert G, Lin RJ (2003) DExD/H-box proteins and their partners: helping RNA helicases unwind. Gene 312:1–16
Simpson GG, Filipowicz W (1996) Splicing of precursors to mRNA in higher plants: mechanism, regulation and sub-nuclear organisation of the spliceosomal machinery. Plant Mol Biol 32:1–41
Singleton MR, Dillingham MS, Wigley DB (2007) Structure and mechanism of helicases and nucleic acid translocases. Annu Rev Biochem 76:23–50
Small ID, Schallenberg-Rudinger M, Takenaka M, Mireau H, Ostersetzer-Biran O (2020) Plant organellar RNA editing: what 30 years of research has revealed. Plant J 101:1040–1056
Small I, Melonek J, Bohne AV, Nickelsen J, Schmitz-Linneweber C (2023) Plant organellar RNA maturation. Plant Cell 35:1727–1751
Steinhauser S, Beckert S, Capesius I, Malek O, Knoop V (1999) Plant mitochondrial RNA editing. J Mol Evol 48:303–312
Stern DB, Goldschmidt-Clermont M, Hanson MR (2010) Chloroplast RNA metabolism. Annu Rev Plant Biol 61:125–155
Stoppel R, Meurer J (2012) The cutting crew – ribonucleases are key players in the control of plastid gene expression. J Exp Bot 63:1663–1673
Stoppel R, Manavski N, Schein A, Schuster G, Teubner M, Schmitz-Linneweber C, Meurer J (2012) RHON1 is a novel ribonucleic acid-binding protein that supports RNase E function in the Arabidopsis chloroplast. Nucleic Acids Res 40:8593–8606
Sugiura M (1992) The chloroplast genome. Plant Mol Biol 19:149–168
Sugita M, Miyata Y, Maruyama K, Sugiura C, Arikawa T, Higuchi M (2006) Extensive RNA editing in transcripts from the PsbB operon and RpoA gene of plastids from the enigmatic moss Takakia lepidozioides. Biosci Biotechnol Biochem 9:2268–2274
Sun T, Germain A, Giloteaux L, Hammani K, Barkan A, Hanson MR, Bentolila S (2013) An RNA recognition motif-containing protein is required for plastid RNA editing in Arabidopsis and maize. Proc Natl Acad Sci USA 110:E1169–E1178
Sun T, Shi X, Friso G, Van Wijk K, Bentolila S, Hanson MR (2015) A zinc finger motif-containing protein is essential for chloroplast RNA editing. PLoS Genet 11:e1005028
Sun T, Bentolila S, Hanson MR (2016) The unexpected diversity of plant organelle RNA editosomes. Trends Plant Sci 21:962–973
Symmons MF, Williams MG, Luisi BF, Jones GH, Carpousis AJ (2002) Running rings around RNA: a superfamily of phosphate-dependent RNases. Trends Biochem Sci 27:11–18
Takenaka M, Zehrmann A, Verbitskiy D, Kugelmann M, Hartel B, Brennicke A (2012) Multiple organellar RNA editing factor (MORF) family proteins are required for RNA editing in mitochondria and plastids of plants. Proc Natl Acad Sci USA 109:5104–5109
Takenaka M, Zehrmann A, Verbitskiy D, Hartel B, Brennicke A (2013) RNA editing in plants and its evolution. Annu Rev Genet 47:335–352

Takenaka M, Takenaka S, Barthel T, Frink B, Haag S, Verbitskiy D, Oldenkott B, Schallenberg-Rudinger M, Feiler CG, Weiss MS, Palm GJ, Weber G (2021) DYW domain structures imply an unusual regulation principle in plant organellar RNA editing catalysis. Nat Catal 4:510–522

Tanner NK, Linder P (2001) DExD/H box RNA helicases: from generic motors to specific dissociation functions. Mol Cell 8:251–262

Tavares-Carreon F, Camacho-Villasana Y, Zamudio-Ochoa A, Shingu-Vazquez M, Torres-Larios A, Perez-Martinez X (2008) The pentatricopeptide repeats present in Pet309 are necessary for translation but not for stability of the mitochondrial COX1 mRNA in yeast. J Biol Chem 283:1472–1479

Till B, Schmitz-Linneweber C, Williams-Carrier R, Barkan A (2001) CRS1 is a novel group II intron splicing factor that was derived from a domain of ancient origin. RNA 7:1227–1238

Tillich M, Krause K (2010) The ins and outs of editing and splicing of plastid RNAs: lessons from parasitic plants. New Biotechnol 27:256–266

Toor N, Hausner G, Zimmerly S (2001) Coevolution of group II intron RNA structures with their intron-encoded reverse transcriptases. RNA 7:1142–1152

Tseng CC, Sung TY, Li YC, Hsu SJ, Lin CL, Hsieh MH (2010) Editing of accD and ndhF chloroplast transcripts is partially affected in the Arabidopsis vanilla cream1 mutant. Plant Mol Biol 73:309–323

Tuteja N, Tuteja R (2004) Unraveling DNA helicases. Motif, structure, mechanism and function. Eur J Biochem/FEBS 271:1849–1863

Tzafrir I, Pena-Muralla R, Dickerman A, Berg M, Rogers R, Hutchens S, Sweeney TC, McElver J, Aux G, Patton D, Meinke D (2004) Identification of genes required for embryo development in Arabidopsis. Plant Physiol 135:1206–1220

Umate P, Tuteja R, Tuteja N (2010) Genome-wide analysis of helicase gene family from rice and Arabidopsis: a comparison with yeast and human. Plant Mol Biol 73:449–465

Wakasugi T, Hirose T, Horihata M, Tsudzuki T, Kössel H, Sugiura M (1996) Creation of a novel protein-coding region at the RNA level in black pine chloroplasts: the pattern of RNA editing in the gymnosperm chloroplast is different from that in angiosperms. Proc Natl Acad Sci U S A 93:8766–8770

Walter M, Kilian J, Kudla J (2002) PNPase activity determines the efficiency of mRNA 3′-end processing, the degradation of tRNA and the extent of polyadenylation in chloroplasts. EMBO J 21:6905–6914

Walter M, Piepenburg K, Schottler MA, Petersen K, Kahlau S, Tiller N, Drechsel O, Weingartner M, Kudla J, Bock R (2010) Knockout of the plastid RNase E leads to defective RNA processing and chloroplast ribosome deficiency. Plant J 64:851–863

Wang Y, Duby G, Purnelle B, Boutry M (2000) Tobacco VDL gene encodes a plastid DEAD box RNA helicase and is involved in chloroplast differentiation and plant morphogenesis. Plant Cell 12:2129–2142

Wang X, An Y, Qi Z, Xiao J (2021) PPR protein Early Chloroplast Development 2 is essential for chloroplast development at the early stage of Arabidopsis development. Plant Sci 308:110908

Watkins KP, Rojas M, Friso G, van Wijk KJ, Meurer J, Barkan A (2011) APO1 promotes the splicing of chloroplast group II introns and harbors a plant-specific zinc-dependent RNA binding domain. Plant Cell 23:1082–1092

Williams-Carrier R, Kroeger T, Barkan A (2008) Sequence-specific binding of a chloroplast pentatricopeptide repeat protein to its native group II intron ligand. RNA 14:1930–1941

Yagi Y, Hayashi S, Kobayashi K, Hirayama T, Nakamura T (2013) Elucidation of the RNA recognition code for pentatricopeptide repeat proteins involved in organelle RNA editing in plants. PLoS One 8:e57286

Yan J, Zhang Q, Guan Z, Wang Q, Li L, Ruan F, Lin R, Zou T, Yin P (2017) MORF9 increases the RNA-binding activity of PLS-type pentatricopeptide repeat protein in plastid RNA editing. Nat Plants 3:17037

Yan J, Zhang Q, Yin P (2018) RNA editing machinery in plant organelles. Sci China Life Sci 61:162–169

Yan J, Yao Y, Hong S, Yang Y, Shen C, Zhang Q, Zhang D, Zou T, Yin P (2019) Delineation of pentatricopeptide repeat codes for target RNA prediction. Nucleic Acids Res 47:3728–3738
Yang J, Stern DB (1997) The spinach chloroplast endoribonuclease CSP41 cleaves the 3′-untranslated region of petD mRNA primarily within its terminal stem-loop structure. J Biol Chem 272:12874–12880
Yang J, Schuster G, Stern DB (1996) CSP41, a sequence-specific chloroplast mRNA binding protein, is an endoribonuclease. Plant Cell 8:1409–1420
Yang J, Zhang M, Wang X (2018) Crystal structure of the chloroplast RNA editing factor MORF2. Biochem Biophys Res Commun 495:2038–2043
Yang Z, Li M, Sun Q (2020) RHON1 co-transcriptionally resolves R-loops for Arabidopsis chloroplast genome maintenance. Cell Rep 30:243–256.e245
Yehudai-Resheff S (2003) Domain analysis of the chloroplast polynucleotide phosphorylase reveals discrete functions in RNA degradation, polyadenylation, and sequence homology with exosome proteins. The Plant Cell Online 15:2003–2019
Yehudai-Resheff S, Hirsh M, Schuster G (2001) Polynucleotide phosphorylase functions as both an exonuclease and a poly(A) polymerase in spinach chloroplasts. Mol Cell Biol 21:5408–5416
Zhang F, Tang W, Hedtke B, Zhong L, Liu L, Peng L, Lu C, Grimm B, Lin R (2014) Tetrapyrrole biosynthetic enzyme protoporphyrinogen IX oxidase 1 is required for plastid RNA editing. Proc Natl Acad Sci USA 111:2023–2028
Zhang HD, Cui YL, Huang C, Yin QQ, Qin XM, Xu T, He XF, Zhang Y, Li ZR, Yang ZN (2015) PPR protein PDM1/SEL1 is involved in RNA editing and splicing of plastid genes in Arabidopsis thaliana. Photosynth Res 126:311–321
Zhang Q, Shen L, Wang Z, Hu G, Ren D, Hu J, Zhu L, Gao Z, Zhang G, Guo L, Zeng D, Qian Q (2019) OsCAF1, a CRM domain containing protein, influences chloroplast development. Int J Mol Sci 20:4386
Zhang Q, Shen L, Ren D, Hu J, Zhu L, Gao Z, Zhang G, Guo L, Zeng D, Qian Q (2020) Characterization of the CRM gene family and elucidating the function of OsCFM2 in rice. Biomol Ther 10:327
Zhang L, Chen J, Zhang L, Wei Y, Li Y, Xu X, Wu H, Yang ZN, Huang J, Hu F, Huang W, Cui YL (2021) The pentatricopeptide repeat protein EMB1270 interacts with CFM2 to splice specific group II introns in Arabidopsis chloroplasts. J Integr Plant Biol 63:1952–1966
Zhao Y, Cai M, Zhang X, Li Y, Zhang J, Zhao H, Kong F, Zheng Y, Qiu F (2014) Genome-wide identification, evolution and expression analysis of mTERF gene family in maize. PLoS One 9:e94126
Zhou W, Lu Q, Li Q, Wang L, Ding S, Zhang A, Wen X, Zhang L, Lu C (2017) PPR-SMR protein SOT1 has RNA endonuclease activity. Proc Natl Acad Sci USA 114:E1554–E1563
Zhu X, Mou C, Zhang F, Huang Y, Yang C, Ji J, Liu X, Cao P, Nguyen T, Lan J, Zhou C, Liu S, Jiang L, Wan J (2020) WSL9 encodes an HNH endonuclease domain-containing protein that is essential for early chloroplast development in rice. Rice (NY) 13:45
Zhu J, Li C, Peng X, Zhang X (2021) RNA architecture influences plant biology. J Exp Bot 72:4144–4160
Zoschke R, Nakamura M, Liere K, Sugiura M, Borner T, Schmitz-Linneweber C (2010) An organellar maturase associates with multiple group II introns. Proc Natl Acad Sci USA 107:3245–3250
Zoschke R, Watkins KP, Miranda RG, Barkan A (2016) The PPR-SMR protein PPR53 enhances the stability and translation of specific chloroplast RNAs in maize. Plant J 85:594–606

Chapter 3
The Roles of mTERF Proteins in Chloroplast Gene Expression

Kamel Hammani

Abstract The mitochondrial transcription termination factor (mTERF) proteins are tandem degenerate α-helical repeat proteins that are encoded by nuclear genomes of all eukaryotes except fungi. First described in animals in the late 1980s, mTERF proteins are nucleic acid binding proteins that are specifically targeted to mitochondria or chloroplasts where they regulate the expression of organellar genes. While metazoan genomes encode only 3 to 4 mitochondrial-localized mTERF proteins that influence mitochondrial gene transcription or ribosomal assembly and translation, the genomes of higher plants encode more than 30 mTERF proteins that have been associated with various abiotic stress responses. Reverse genetics in Arabidopsis combined with molecular phenotyping and *in vitro* biochemistry has recently brought considerable clarity into the mechanisms of action of some members of the plant mTERF family in gene expression. In this chapter, we will specifically discuss the results of recent investigations of mTERF functions in plants, focusing on their contribution to chloroplast gene expression.

Keywords mTERF · Chloroplast · Gene expression · Transcription · Translation · Intron splicing

3.1 Introduction

The chloroplast is the defining organelle of the plant cell and houses photosynthesis as well as many metabolic pathways that are essential for plant productivity, fertility and various developmental and stress response pathways (Van Dingenen et al. 2016; Schwenkert et al. 2022; Inaba and Ito-Inaba 2010; Jarvis and López-Juez 2013).

K. Hammani (✉)
Institut de Biologie Moléculaire des Plantes, Centre National de la Recherche Scientifique (CNRS), Université de Strasbourg, Strasbourg, France
e-mail: khammani@unistra.fr

T. M. Burch-Smith (ed.), *Chloroplast Gene Expression: Regulation, Stress Signaling and Biotechnology*, Nucleic Acids and Molecular Biology 37,
https://doi.org/10.1007/978-3-031-70098-9_3

Chloroplast genomes are typically ~150 kb and carry ~100 genes, the majority of which encode essential components of the photosynthetic apparatus and core components of the chloroplast gene expression machinery (Green 2011). Chloroplasts have acquired unique traits that differentiate them from their prokaryotic ancestor. In particular, chloroplasts employ distinct and complex RNA metabolism mechanisms that combine features inherited from their prokaryotic ancestors with novel features that evolved post-endosymbiosis in the eukaryotic host cell (Barkan 2011). For instance, gene transcription in chloroplasts is driven from multiple promoters by two types of RNA polymerases: a nuclear-encoded T3-T7 bacteriophage type and a multi-subunit plastid-encoded polymerase of prokaryotic origin; specific cytidine RNA bases are edited post-transcriptionally to uridines; some mRNAs and tRNAs harbor group I and group II introns that are spliced; and premature transcripts are further processed by endoribonucleolytic cleavage combined with $5'\rightarrow3'$ and $3'\rightarrow5'$ RNA trimming by endo- and exo-ribonucleases. Finally, mRNAs are translated by prokaryotic-like 70S chloroplast ribosomes.

Each step in chloroplast gene expression relies on nuclear-encoded nucleic acid binding proteins that are imported into organelles. The best-known group of chloroplast gene expression regulators is arguably the pentatricopeptide repeat (PPR) family (Small and Peeters 2000). In angiosperms, the PPR family encompasses more than 400 members that are targeted to mitochondria and chloroplasts where they control virtually all posttranscriptional steps in organellar gene expression (O'Toole et al. 2008; reviewed in Schmitz-Linneweber and Small 2008; Barkan and Small 2014). PPR proteins are composed of tandem repeats of a degenerate 35-amino acid motif (called the PPR motif) that folds into two antiparallel alpha helices that stack to form a solenoid structure which binds single-stranded RNA with sequence specificity (reviewed in Barkan and Small 2014; Shen et al. 2016). While the functions and mechanistic actions of plant PPR proteins have been extensively studied and described in the last two decades (reviewed in Schmitz-Linneweber and Small 2008; Barkan and Small 2014), another less-known organellar helical-repeat protein family called mTERF (for mitochondrial transcription termination factor) shares structural similarities with PPR proteins but their detailed molecular functions in plant organelles are just being understood. This chapter will discuss the latest findings on mTERF research in plants with a particular focus on their function in chloroplast gene expression.

3.2 mTERF Protein Structural Features and Family Organization

The mTERF family is found in metazoa, protozoa and plants and is characterized by the presence of a degenerate ~30 amino acids motif (Linder et al. 2005). The mTERF family was named after its founding member in mammals, MTERF1 which mediates mitochondrial transcription termination *in vitro* by binding to a specific mitochondrial DNA sequence between the 16S rRNA gene and the *tRNA-Leu* gene (Kruse et al. 1989). mTERF proteins have a modular architecture with variable numbers of mTERF repeats (Fig. 3.1a). In addition to mTERF repeats, most mTERF proteins harbor an N-terminal sequence predicted to target these proteins to mitochondria or chloroplasts. There are no atomic structures for plant mTERF proteins and our current structural knowledge of mTERF repeats comes from the resolution of crystal structures of three human mTERF proteins, namely MTERF1 (Yakubovskaya et al. 2010), MTERF3 (Spahr et al. 2010) and MTERF4 (Yakubovskaya et al. 2012; Spahr et al. 2012). An mTERF motif forms two antiparallel alpha-helices followed by a shorter 3.10 helix perpendicular to the first one (Fig. 3.1b) and the tandem mTERF repeats stack to form a super-helical protein (or solenoid) adopting a "croissant" shape (Fig. 3.1c).

The mTERF family is small in mammals and comprises 4 members that are localized to mitochondria and have been predominantly implicated in mitochondrial transcription, DNA replication and ribosomal biogenesis (reviewed in Kleine and Leister 2015; Roberti et al. 2009). DNA binding activity has been confirmed both *in vivo* and *in vitro* for three metazoan mTERF proteins (reviewed in Peralta et al. 2012). In addition, the resolution of *Hs*-MTERF1 structure in complex with its mitochondrial DNA target has provided mechanistic insights into its termination effect on mitochondrial transcription (Yakubovskaya et al. 2010) (Fig. 3.1c). *Hs*-MTERF1 binds double-stranded DNA as a monomer in its inner positively charged groove via numerous electrostatic interactions with the DNA sugar-phosphate backbone and makes base-specific contacts with only 5 bases of its 22-base pair DNA footprint. MTERF1 binding allows base flipping from the DNA duplex, which in turn interferes with transcriptional elongation. (Yakubovskaya et al. 2010). Computational methods (I-TASSER or AlphaFold) for predicting three-dimensional

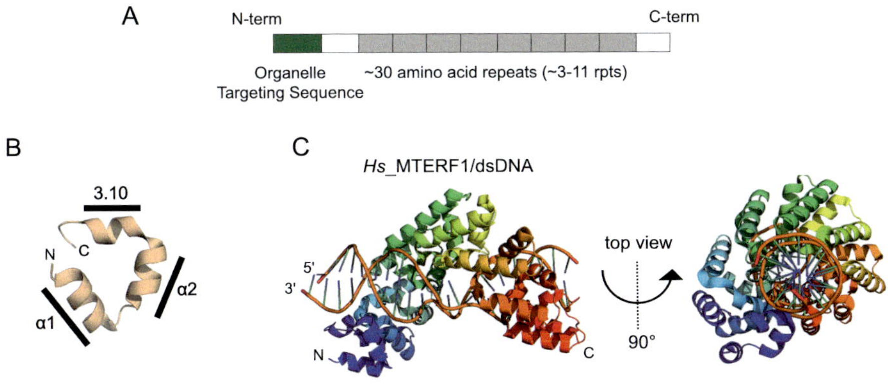

Fig. 3.1 Structural features of mTERF-repeat proteins. (**a**) Scheme of a typical mTERF protein architecture. mTERF proteins possess an N-terminal organellar transit peptide and tandem repeats (rpts) of a ~30 amino acid mTERF motif. (**b**) Structure of the first mTERF motif of *Hs*-MTERF1 (PDB: 3MVA). (**c**) Structure of *Hs*-MTERF1 bound to its double stranded (ds) 22-bp mitochondrial DNA ligand (PDB: 3MVA)

structures of proteins suggest that mTERF proteins in plants adopt similar alpha solenoid structures and conserved some aspects of the nucleic acid binding mechanism (Hammani and Barkan 2014; Zhao et al. 2014; Romani et al. 2015). Nevertheless, more recent investigations have reported RNA binding activity for two metazoan mTERF members, *Hs*-MTERF4 (Camara et al. 2011) and *Dm*-MTERF3 (Wredenberg et al. 2013) and three members of the plant mTERF family, *Zm*-mTERF4 (Hammani and Barkan 2014), *At*-mTERF6 (Romani et al. 2015) and *At*-mTERF9 (Méteignier et al. 2021) (note that none of the animal mTERF proteins are conserved in plants). In all these cases, the mTERF proteins interact with structured RNAs in organelles (rRNA, tRNA, introns) but how the mTERF repeats accommodate RNA in their solenoid structure is not clear. The analysis of the crystal structure of *Hs*-MTERF4 revealed the presence of a positively charged surface on the concave side of the solenoid and this is proposed to mediate RNA binding, but whether it binds double-stranded or single-stranded RNA remains unknown (Spahr et al. 2012).

Compared to the small size of the mTERF family in metazoans, the number of mTERF proteins has considerably increased during the evolution of land plants, due, on the one hand, to the acquisition of the chloroplast and, on the other hand, to the increased complexity of the steps and protein factors that are required to express organellar genes in plants. Non-vascular plant genomes encode fewer than 20 mTERF genes (https://phytozome-next.jgi.doe.gov/). For instance, the unicellular green alga Chlamydomonas genome (C.*reinhardtii* v5.6) encodes 14 putative mTERF genes while the terrestrial moss *Physcomitrium patens* genome (*P. patens* v3.3) carries 16 of these. In vascular plants, mTERF genes have undergone a large number of duplications to reach 35, 31 and 34 members in the genomes of the flowering plants Arabidopsis (Babiychuk et al. 2011; Kleine 2012), maize (*Zea mays*, (Zhao et al. 2014)) and rice (*Oryza sativa* v7.0), respectively. The number of mTERF motifs per protein in plants varies from three to eleven. In flowering plants, a third of the ~30 mTERF proteins are predicted to be targeted to chloroplasts (Babiychuk et al. 2011; Zhao et al. 2014) and their chloroplastic localization was further confirmed by GFP fusion and/or proteomic studies of the chloroplast nucleoid (Majeran et al. 2012; Babiychuk et al. 2011). Only 11 out of the ~30 mTERF proteins in angiosperms have been genetically studied (Table 3.1) and even so, the characterization of their molecular functions in organelles remains pretty superficial as only 6 of them have been assigned precise gene targets in organelles.

Mutations in plant mTERF genes can lead to embryo or developmental defects (i.e., loss of leaf pigmentation, slow-growth or dwarfism) due to defects in chloroplast or mitochondria biogenesis (Babiychuk et al. 2011; Tzafrir et al. 2004; Bryant et al. 2011). In addition, forward genetic screens in plants aimed at identifying nuclear genes that are important for plant responses to particular stresses or plant transcriptomic studies in response to these stresses have highlighted the importance of *mTERF* genes for the plant acclimation to abiotic stress (e.g. resistance to salt, heat or oxidative stress) (Xu et al. 2017; Robles et al. 2018; Zhou et al. 2016; Zhao et al. 2014; Nunez-Delegido et al. 2019). How *mTERF* genes fit into these stress-response networks is still a black box but is not very surprising considering their

Table 3.1 mTERF proteins whose molecular functions have been genetically characterized in angiosperms

Name	Other name	Gene	Organelle	Molecular function	References
mTERF1	SOLDAT10	*At2G03050*	C	Accumulation of 16S and 23S rRNAs	Meskauskiene et al. (2009)
mTERF2	EMB2219	*At2g21710*	C	Splicing of group IIb introns	Lee et al. (2021)
mTERF4	BSM RUG2	*At4g02990*	C	Splicing of group IIa introns	Babiychuk et al. (2011), Quesada et al. (2011)
mTERF5	MDA1	*At4g14605*	C	Transcription of *psbE* and *ndhA* genes and mRNA stabilization	Robles et al. (2012), Ding et al. (2019), Méteignier et al. (2020)
mTERF6		*At4g38160*	C	tRNA-Ile maturation*rpoA* gene transcription	Romani et al. (2015), Zhang et al. (2018)
mTERF8	PTAC15	*At5g54180*	C	Transcription termination of *psbJ* gene	Xiong et al. (2020)
mTERF9	TWIRT1	*At5g55580*	C	Accumulation of 16S and 23S rRNAs and ribosomal assembly	Méteignier et al. (2021)
mTERF15		*At1g74120*	M	Splicing of *nad2* intron 3	Hsu et al. (2014)
mTERF18	SHOT1	*At3g60400*	M	Accumulation of tRNA-Gly and *ccmFc* RNAs	Kim et al. (2021), Kim et al. (2012)
mTERF22		*At5g64950*	M	Gene transcription	Shevtsov et al. (2018)
Zm-mTERF4		*GRMZM2G029933*	C	Splicing of group II introns	Hammani and Barkan (2014)
Zm-SMK3		*GRMZM2G177019*	M	Splicing of *nad1* intron 4 and *nad4* intron 1	Pan et al. (2019)

C chloroplast, *M* mitochondria

roles as organellar gene regulators and the importance of organelles in sensing environmental perturbations. In the absence of mTERF functions, the alteration of organellar gene expression likely impacts chloroplast and mitochondrial functions, which in turn triggers a retrograde signal that influences the expression of nuclear genes in order to adjust the plant cell physiology. The physiological roles of mTERF proteins in the plant acclimation to stress have been reviewed by other colleagues (Kleine and Leister 2015; Robles and Quesada 2021; Wobbe 2021) and this chapter will specifically discuss chloroplastic mTERF proteins whose molecular functions have been resolved, with a particular emphasis on their mechanistic insights.

3.3 mTERF Functions in Chloroplast Gene Expression

3.3.1 Regulation of Chloroplast Gene Transcription

mTERF proteins were first studied in metazoans and as indicated by their name, were believed to control gene transcription termination in mitochondria. This paradigm was originally motivated by the ability of the founding member of the mTERF family in humans, *Hs*-MTERF1, to bind *in vivo* a mitochondrial DNA (mtDNA) sequence located in the intergenic region between the *16S* rRNA and *ND1* genes and the ability of the recombinantly-expressed protein to promote the termination of RNA synthesis in *in vitro* run-off transcription assays (Kruse et al. 1989; Asin-Cayuela et al. 2004, 2005). MTERF1 was believed to act as a DNA road-block to stop transcription by the mitochondrial RNA polymerase from the heavy strand promoter (HSP) at the *16S* rRNA gene. However, subsequent work that included the study of a *mterf1* knock-out mouse model did not confirm this but instead demonstrated that MTERF1 DNA binding terminates transcription of the light mtDNA strand opposite to the *16S* rRNA gene near its binding site to prevent the interference of transcription initiation between the HSP and LSP promoters which are on opposite strands but in close genomic proximity (Terzioglu et al. 2013).

In plants, chloroplast gene transcription is driven by at least two RNA polymerases, the multisubunit prokaryotic-type plastid-encoded RNA polymerase (PEP) (encoded by the plastid *rpoA*, *rpoB*, *rpoC1* and *rpoC2* genes) and one to two nuclear-encoded single subunit phage-type RNA polymerases of eukaryotic origin (NEP) (reviewed in Borner et al. 2015). The core PEP subunits associate with six sigma factors that are important for promoter recognition and additional PEP-associated proteins (PAPs) that are encoded in the nuclear genome (Yu et al. 2014). PEP and NEP recognize different promoters and PEP performs the major transcriptional activity in chloroplasts (Zhelyazkova et al. 2012b). RNA-seq studies of the chloroplast transcriptome revealed that the entire chloroplast genome is transcribed (Shi et al. 2016) and the mechanisms involved in the termination of chloroplast gene transcription are not well understood (reviewed in Ji et al. 2019). In fact, RNA 3′-ends in chloroplasts do not result from gene transcription termination but rather from post-transcriptional processing of longer RNAs. This processing includes RNA cleavage and degradation steps by endo- and exo-ribonucleases and the 3′-end RNA stabilization by intrinsic stem-loop RNA structures or RNA binding protein caps (reviewed in Stern et al. 2010) (See Chap. 1). In bacteria, 3′-end stem-loops (or hairpin) that are rich in GC and followed by a U tract act as transcription terminators by provoking RNA polymerase pausing and subsequent release from the DNA template. This sequence feature is less conserved in chloroplast genes and even when flanking inverted repeats are present at their 3′ end, these do not efficiently terminate transcription *in vitro* (Stern and Gruissem 1987; Rott et al. 1996). However, the 3′-end regions of only a few chloroplast genes have been tested in these transcription termination assays and the roles of 3′-end hairpin structures and nucleotide sequence context in terminating transcription cannot be completely ruled out. Many DNA binding proteins including the mTERF protein, mTERF8 (also named pTAC15), associate with the chloroplast transcriptional active chromosome (pTAC) in Arabidopsis and some of these protein factors could play a role in regulating of transcription (Pfalz et al. 2006). In addition, ten members of the mTERF family associate with nucleoids in maize chloroplasts in agreement with their hypothetical function in organellar gene regulation (Majeran et al. 2012). Nevertheless, so far among the 11 mTERF members that have been genetically studied, only three exhibit functions in chloroplast gene transcription.

The role of mTERF proteins in chloroplast gene transcription is certainly best characterized for Arabidopsis mTERF5 (also named MDA1, AT4G14605) since two recent and independent studies provided complementary molecular and mechanistic insights into the protein function in chloroplasts (Ding et al. 2019; Meteignier et al. 2020). Mutations in the *MDA1/mTERF5* gene in Arabidopsis causes a pale-yellow leaf and dwarf phenotype that is provoked by a reduction of photosystem II (PSII) and NDH dehydrogenase complex (NDH) in chloroplasts. In agreement with the reduction of PSII and NDH subunits, the analysis of the chloroplast transcriptome in *mda1* mutants revealed the loss of specific mRNA isoforms deriving from the *psbE* and *ndhH* operons compared to the wild type (Ding et al. 2019; Meteignier et al. 2020). That is, tetracistronic *psbE-F-L-J* mRNAs and *ndhA*-containing mRNAs are specifically reduced in *mda1/mterf5* mutants. These two physically-separated operons are transcribed by the PEP and *in vitro* transcription assays using chloroplast extracts (run-on assays) showed that the *psbE* and *ndhA* transcription rates are specifically reduced in *mda1*. This demonstrated that MDA1/mTERF5 positively regulates the transcription of these genes (Fig. 3.2a). To gain insights into the mechanism by which MDA1/mTERF5 could influence gene transcription, the protein and nucleic acid partners of MDA1/mTERF5 were identified by *in vivo* immunoprecipitation assays [immunoprecipitation followed by mass spectrometry (IP-MS) and chromatin immunoprecipitation (ChIP)] using complemented *mda1* Arabidopsis lines expressing a tagged version of MDA1/mTERF5. The results demonstrated that MDA1/mTERF5 interacts *in vivo* with the subunits of the PEP and many PAPs as well as DNA sequences near the *psbE* and *ndhA* promoters. Furthermore, electromobility gel shift and/or DNase footprinting assays using recombinant MDA1/mTERF5 demonstrated a 27-bp sequence-specific binding of the protein to its *in vivo psbE* and *ndhA* targets within the promoter region downstream of the transcription start site (TSS) for *psbE* and ~100 nt from the start codon for the *ndhA* gene. At first glance, these results supported a simple model for the activity of MDA1/mTERF5 as a transcription initiation factor where the protein binds near the promoters of *psbE* and *ndhA* to stimulate the local recruitment of the

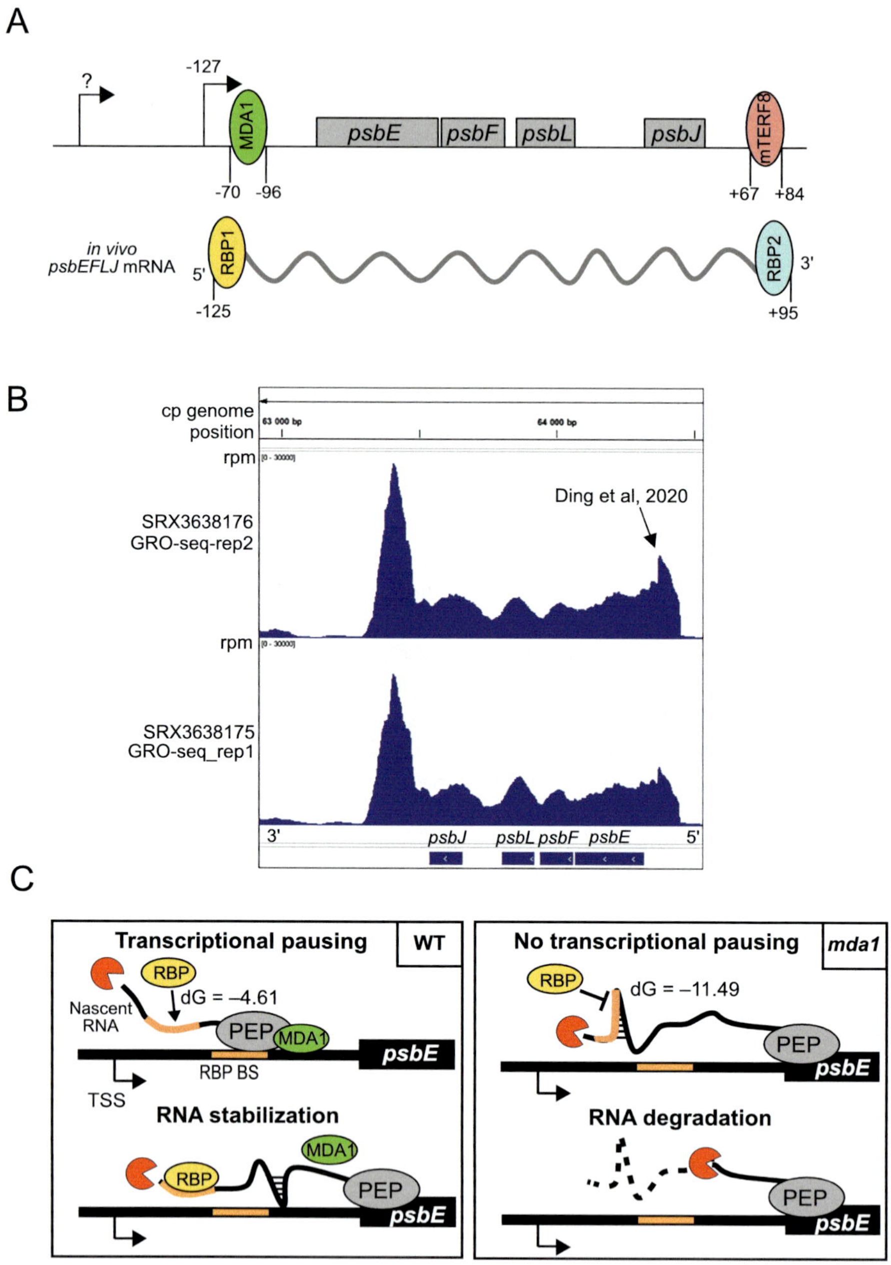

Fig. 3.2 (**a**) Genetic map of the chloroplast *psbEFLJ* gene operon and the protein factor controlling its gene expression. The positions of the transcription start sites (right squared arrows), mTERF protein DNA binding sites and the mapped transcript termini are indicated. The positions are specified relative to gene start or stop codon. The *in vivo psbEFLJ* mRNA termini positions coincide with the presence of an abundant sRNA in chloroplasts which are footprints of RNA bind-

(continued)

ing proteins (RBP) that protect the 5′- and 3′-ends of the processed mRNA after transcription. (**b**) GRO-seq profiles showing two transcription pausing sites in *psbEFLJ* operon represented as read peaks located in the 5′- and 3′-flanking region of *psbE* and *psbJ* genes, respectively. Sequencing reads from GRO-seq libraries (SRR6661079 and SRR6661080) generated from *A. thaliana* 6-day old seedling RNA were analyzed according to (Hetzel et al. 2016). The reads were aligned to the TAIR10 chloroplast genome using STAR (version 2.6.1d, with parameters alignIntronMin set to 60 and alignIntronMax set to 6000). Further analysis was performed with HOMER v4.10.4, using annotations from TAIR10. Only reads aligned to a single unique location with a mapping quality score > 10 were considered. All analysis or visualization data were normalized to a read depth of 10e7 total uniquely aligned read per experiment. The coverage files were produced by HOMER and visualized in IGV software. (**c**) Model for the dual-function of MDA1/mTERF5 in gene transcription and RNA stabilization in chloroplasts (Meteignier et al. 2020). MDA1/mTERF5 binds a DNA sequence near the promoter of *psbE* gene and stimulates gene transcription by recruiting the PEP transcription machinery. In addition, MDA1/mTERF5 halts PEP transcription at its binding site to prevent the formation of a secondary structure in the neosynthesized mRNA and facilitate the binding of an RNA binding protein (RBP) that is protecting the processed 5′-end of *psbE* mRNA against exoribonucleases. In *mda1/mterf5* mutants, transcription elongation is not paused and the neosynthesized *psbE* RNA folds into a secondary structure that prevents the RBP from binding its RNA target site (BS), leading to the degradation of the transcribed *psbE* mRNA

PEP/PAPs complex and enhance the transcription of these genes. However, there was more to the story. Additional mapping of the binding sites of the transcriptionally active PEP in chloroplasts (GRO-seq analysis) revealed the existence of a transcription pausing site at *psbE-F-L-J* in the promoter region that coincides with MDA1/mTERF5 DNA binding site and is dependent on the presence MDA1/mTERF5 in chloroplasts (Ding et al. 2019) (Fig. 3.2b). The function of this MDA1/mTERF5-dependent transcription pausing at *psbE* promoter was not clear but the end-mapping and quantification of the steady-state level of the mRNAs from *psbE* and *ndhH* operon and their *in vivo* transcription rates gave us a hint (Meteignier et al. 2020): (i) the mRNAs that are specifically lost in *mda1/mterf5* mutants result from post-transcriptional processing of their 5′-end and, (ii) the reduction in *psbE* and *ndhA* transcription rates in the mutants compared to wild type was not sufficient to explain by itself the severe reduction in the abundance of these processed mRNAs (Meteignier et al. 2020). Therefore, not only does MDA1/mTERF5 stimulate gene transcription, but it also promotes the post-transcriptional stabilization of the 5′-ends of the processed *psbE* and *ndhA* mRNAs. There is extensive evidence that the mature mRNAs in chloroplasts are stabilized by RNA-binding proteins like PPR proteins that specifically bind intergenic regions in RNAs post-transcriptionally to protect them from the degradation by 5′→3′ and 3′→5′ exoribonucleases (Ruwe and Schmitz-Linneweber 2012; Ruwe et al. 2016; Zhelyazkova et al. 2012a). The binding of these PPR-like proteins to RNA is tight enough to leave an RNA footprint of their binding site that accumulates *in vivo* as abundant ~20–30 nt sRNAs whose sequence boundaries coincide with the 5′- or 3′-end of the processed RNAs found *in vivo* (Ruwe and Schmitz-Linneweber 2012; Ruwe et al. 2016; Zhelyazkova et al. 2012a). Evidence that MDA1 plays a role in *psbE* and *ndhA* RNA stabilization came from the observation that sRNAs matching the 5′-end of the *psbE-F-L-J* and *ndhA* mRNAs, which normally accumulate in chloroplasts are specifically absent in absence of MDA1 (Meteignier et al. 2020) (Fig. 3.2a). These results suggested that MDA1/mTERF5 specifically promotes the binding of PPR-like proteins to the 5′-end of *psbE* and *ndhA* RNA to stabilize these transcripts, which raised the puzzling question of the mechanism involved.

The answer to this conundrum came from the analysis of the genomic location of the MDA1/mTERF5 binding site in *psbE*. It is placed 29 nucleotides downstream of the position of the 5′-end of processed *psbE* RNA and it coincides with a PEP transcriptional pausing site whose *in vivo* activity requires MDA1/mTERF5 (Ding et al. 2019). As previously explained, the stability of *psbE* and *ndhA* mRNAs depends on the capacity of an RNA binding protein to bind its RNA target and we know from biochemical studies that PPR proteins preferentially bind single-stranded RNA whereas their capacity to invade RNA secondary structures is very limited (McDermott et al. 2018; Williams-Carrier et al. 2008; Hammani et al. 2011). Therefore, *in vivo* mechanisms should be in place to facilitate PPR binding to their RNA target sequences when these are folded in secondary structures (Jiang et al. 2019). Many RNA helicases localize to chloroplasts and have been proposed to modulate RNA folding *in vivo* but evidence for this remains scarce (Nawaz and Kang 2017). On the contrary, a transcriptional pausing site strategically placed downstream of a PPR binding site offers an attractive mechanism by which to pause RNA synthesis and avoid the formation of secondary structures that would prevent PPR binding and RNA stabilization. The structural prediction of RNA structures of the neo-transcribed *psbE* RNA segments including or excluding pausing at the MDA1/mTERF5 binding site revealed that when transcription is paused at this site, the 5′-end of the *psbE* RNA and the PPR binding site do not fold into a stable RNA structure ($\Delta G > -5$ kcal/mol) and would allow PPR binding and *psbE* mRNA stabilization. On the contrary, farther RNA elongation to the MDA1/mTERF5 binding site is predicted to allow the formation of stable RNA-RNA interactions ($\Delta G < -10$ kcal/mol) that occlude the PPR binding site and presumably compete with the PPR binding. These predictions and the molecular defects caused by the loss of MDA1/mTERF5 function allowed us to draw an attractive mechanistic model that explains the function of this mTERF protein in coupling gene transcription with RNA stabilization in chloroplasts: MDA1/mTERF5 promotes *psbE* mRNA stability by pausing gene transcription to facilitate binding of a PPR protein to the RNA 5′-end which in turn stabilizes the mRNA *in vivo* (Meteignier et al. 2020) (Fig. 3.2c). No transcription pausing site has been identified in *ndhA* but the specific loss of the sRNA mapping to the 5′-end of processed *ndhA* mRNAs suggests that an analogous mechanism occurs at this site. This model challenged the established view of the functional repertoire of chloroplast DNA binding proteins: their functions are not only restricted to DNA-related processes (i.e. transcription, DNA replication and repair) but they can also cooperate with RNA binding proteins to regulate post-transcriptional processes in chloroplast gene expression.

Two additional mTERF proteins have been shown to act as gene transcription regulators in Arabidopsis chloroplasts: mTERF8 (AT5G54180, also named pTAC15) (Xiong et al. 2020) and mTERF6 (AT4G38160) (Zhang et al. 2018). mTERF8 is an interesting transcription regulator. It associates *in vivo* with the PEP and acts as a transcription terminator on the 3′-end of the *psbE-F-L-J* operon whose transcription is stimulated by MDA1/mTERF5 and whose processed 5′-end mRNA is stabilized by a PPR-like protein (Fig. 3.2a). The loss of mTERF8 function in Arabidopsis caused only a mild decrease in PSII and does not lead to a detrimental growth phenotype (Xiong et al. 2020). The analysis of the *mterf8* chloroplast transcriptome revealed an overaccumulation of the polycistronic *psbE-F-L-J* mRNA and in particular, the specific accumulation of a longer *psbE-F-L-J* mRNA in the mutant compared to the wild-type plants. The mapping of the *psbE-F-L-J* mRNA termini revealed that the loss of mTERF8 in chloroplasts conditioned the accumulation of a *psbE-F-L-J* mRNA possessing a longer 3′-end, indicating that mTERF8 promotes transcription termination at *psbJ*. ChIP analysis and electromobility gel shift assays using recombinant protein confirmed that mTERF8 has dsDNA binding activity and specifically binds to a DNA sequence located downstream of *psbJ* gene near the genomic position of the *psbE-F-L-J* mRNA 3′-end that it stabilizes *in vivo*. In addition, *in vitro* run-off transcription assays revealed that the recombinant mTERF8 has the capacity to terminate transcription at *psbJ* 3′-end and was proposed to act as a DNA roadblock that provokes transcription termination by the PEP. Intriguingly, the formation of the *psbE-F-L-J* 3′-end is not totally compromised in an *mterf8* loss-of-function mutant and we have demonstrated that this 3′-end is likely to be defined post-transcriptionally by a PPR-like protein (Meteignier et al. 2020). The re-analysis of the GRO-seq data revealed that not only is the PEP activity paused in the promoter region of *psbE-F-J-L* operon as proposed in (Ding et al. 2019), but also at a second site mapping to the *psbJ* 3′-end region downstream of mTERF8 binding site and the genomic location of the 3′-end processed *psbE-F-L-J* mRNA (Fig. 3.2b). The sequence of the 3′-end of *psbJ* transcript is predicted to form a GC-rich hairpin RNA structure that sequesters the putative binding site of a PPR-like protein that likely protects and stabilizes the 3′-end of the mature *psbE-F-L-J* mRNA. Thus, in light of the mechanistic results gained on MDA1/mTERF5, it is tempting to speculate that the binding of mTERF8 to the 3′-end region of *psbJ* slows down the PEP activity to enhance transcription termination near this site as well as to facilitate the processing and stabilization of the 3′-end of *psbE-F-L-J* mRNA by controlling the local RNA structure and the subsequent binding of a PPR-like protein.

Finally, another member of the mTERF family in Arabidopsis, mTERF6 has been suggested to be a transcription terminator in chloroplasts but the evidence for this activity is poor (Zhang et al. 2018). Plants lacking mTERF6 function developed an albino and seedling lethal phenotype when grown on soil. The transcriptomic analysis of the *mterf6* mutant revealed a reduction of the transcription of PEP-dependent genes and the accumulation in the mutant of a *rpoA* mRNA possessing a longer 3′-end than the wild-type. Additional ChIP and electromobility gel shift assays showed that mTERF6 binds to a DNA sequence located in the 3′-end of the *rpoA* gene and in light of these results, mTERF6 was proposed to promote the transcription termination of *rpoA* by a supposedly DNA roadblock activity (Zhang et al. 2018).

3.3.2 Regulation of Post-transcriptional Steps in Chloroplast Gene Expression

Three members of the plant mTERF family have been assigned functions in RNA-related processes in chloroplasts. The first one, Zm-mTERF4 (GRMZM2G029933) is a maize chloroplastic protein whose loss of function conditions chlorophyll-deficient seedlings that die after the development of three to four leaves (Hammani and Barkan 2014). *Zm-mTERF4* is orthologous to the Arabidopsis gene At4g02990 (also named *BSM*, *RUG2* and *mTERF4*) whose knock-out mutations in Arabidopsis led to embryo-lethality (Babiychuk et al. 2011; Quesada et al. 2011). *Zm-mterf4* mutants suffer from a general and severe loss of photosynthetic enzyme complexes as well as defects in chloroplast rRNA processing which are typical features of mutants lacking chloroplast ribosomes. Zm-mTERF4 is found in high-molecular weight ribonucleoprotein complexes in the stroma (Hammani and Barkan 2014). The profiling of chloroplast RNAs in *mterf4* mutants showed that the RNA splicing of group II introns is compromised for many genes including tRNAs and genes encoding ribosomal proteins. RNA and protein co-immunoprecipitation assays demonstrated the *in vivo* specific association of Zm-mTERF4 with a subset of these RNA intron sequences and many known chloroplast splicing factors. This confirmed a direct role of Zm-mTERF4 in promoting group II intron splicing. As such, the defect in chloroplast translation that was observed in *Zm-mterf4* could be explained as an indirect effect of the intron splicing defects in genes encoding essential components of the chloroplast ribosomes.

mTERF6, whose function was previously discussed, has also been reported to affect the maturation of tRNA-Ile (GAU) in chloroplasts in a study that employed different *mterf6* mutant alleles (Romani et al. 2015). The recombinant mTERF6 protein was shown to have the capacity to bind both DNA and RNA, and, as demonstrated by (Zhang et al. 2018), it bears an *in vitro* transcriptional termination activity, not at the *rpoA* site, but in the intron sequence of *tRNA-Ile* gene which belongs to the rRNA gene operon. *In vitro*, recombinant mTERF6 binds to the RNA sequence of the *tRNA-Ile* DNA binding site with specificity but the association of mTERF6 with chloroplast RNAs *in vivo* could not be confirmed. Although *mterf6* mutants displayed defects in the processing of the chloroplastic rRNAs and an over-accumulation of the tRNA-Ile and decrease of its aminoacylation, the mechanism underlying these effects could not clearly be established and whether mTERF6 DNA and/or RNA binding activity contributes to these effects on *tRNA-Ile* gene expression remains unclear.

Finally, Arabidopsis mTERF9 is the best functionally-characterized mTERF protein in a post-transcriptional process in chloroplasts. The molecular function of mTERF9 in chloroplast ribosomal assembly has been supported by genetic analysis combined with *in vivo* and *in vitro* studies of mTERF9's physical association with its RNA and protein targets (Méteignier et al. 2021). mTERF9 localizes to chloroplast and the loss of its activity in Arabidopsis conditioned a pale-yellow leaf and delayed growth phenotype that does not prevent growth in soil. *mterf9* mutants suffered from a reduction in chloroplast translation that leads to a ~50–75% decrease in chloroplast-encoded proteins compared to the wild-type. rRNA profiling revealed that *mterf9* is specifically affected in the accumulation of the 16S and 23S rRNAs with a more pronounced effect on the 16S rRNA. Sucrose gradient sedimentation analyses revealed that the assembly of the small 30S and large 50S ribosomal subunits into chloroplast 70S ribosomes was compromised in *mterf9*, and that mTERF9 predominantly co-sedimented with the small 30S subunit. In addition, mTERF9 associates with chloroplast polysomes. Protein and RNA co-immunoprecipitations revealed that mTERF9 associates *in vivo* with the 16S rRNA and many proteins that are involved in ribosomal biogenesis and chloroplast translation including ribosomal proteins and assembly factors. *In vitro*, the recombinant mTERF9 showed preferential binding to domain I and II of the 16S rRNA, and yeast-two hybrid assays demonstrated that the mTERF-repeats in mTERF9 have the capacity to directly interact with some of its *in vivo* protein partners. Therefore, mTERF9 is proposed to play a scaffolding function integrating protein-protein and protein-RNA interactions to stimulate chloroplast ribosomal assembly and translation in Arabidopsis (Méteignier et al. 2021).

3.4 Protein-Protein and Protein-Nucleic acid Binding Activities Underlay mTERF Effects on Gene Expression

The results described above have brought clarity to the mechanisms underlying the effects of mTERF proteins on chloroplast gene expression. mTERF repeats bind DNA or RNA molecules and some of them can be sequence-specific. In addition to their nucleic acid binding activity, mTERF repeats can directly interact with proteins. The capacity to interact with proteins and/or nucleic acids are two potent biochemical properties that are intrinsically linked to α-helices structures (reviewed in Grove et al. 2008; Rubinson and Eichman 2012), and the stabilization of nucleoprotein complexes by mTERF proteins via physical interactions certainly accounts for their function in gene transcription, RNA splicing or translation as all these processes require macromolecular assembly. In fact, most of the plant mTERF factors that have been studied in detail were found in high-molecular weight nucleoprotein complexes that contain constituents of the transcription, splicing or translation multicomponent machineries (Meteignier et al. 2020, 2021; Xiong et al. 2020; Babiychuk et al. 2011; Hammani and Barkan 2014). Some mTERF proteins with sequence-specific binding could play the role of guides for the recruitment of protein *trans*-effectors to particular DNA or RNA sites that are important for gene expression (e.g. promoters, transcription terminators, RNA *cis*-elements) or alternatively, their sequence-specific binding could trap particular DNA or RNA sites blocking their access by other macromolecules, which can be useful to halt some processes in gene expression. Finally, the capacity of some mTERF repeats to oligomerize (e.g. Arabidopsis mTERF9 (Méteignier et al. 2021) and human MTERF1

(Asin-Cayuela et al. 2004)) via intermolecular interactions likely confers the proteins new opportunities for ligand association by extending their binding surfaces. As discussed for the case of MDA1/mTERF5, it is worth keeping in mind that mTERF proteins involved in DNA-associated processes can have direct implications in RNA processes in organelles when these post-transcriptional processes are directly coupled to transcription.

3.5 Conclusions and Perspectives

It is rewarding to see that we have gained some insights into the functions of chloroplast mTERF proteins in the past decade. Nevertheless, in most of the studied cases, the mechanisms underlying mTERF effects on gene expression in organelles remain too superficially understood. This is mostly due to the fact that genetic analyses of plant mutants only are often not sufficient to determine the primary function of *mTERF* genes, as mutations in organellar gene regulators can cause both direct and indirect effects, and this is further confounded by the difficulty of confirming the physical association of mTERF proteins with their genetically-defined DNA or RNA targets *in vivo*. Thus, it is important to continue the characterization of the molecular functions of mTERF proteins, seeking evidence for their direct *in vivo* association with their nucleic acid and protein ligands in organelles. Such studies are likely to provide new insights into the intricate connections between DNA- and RNA-related processes in plant organelles. In addition, we currently lack structural data for plant mTERF proteins in complex with their *in vivo* nucleic acid and/or protein ligands, and we do not understand how these proteins integrate these two activities at the atomic level. Another knowledge gap resides in the understanding of the mechanisms that regulate mTERF intrinsic activities *in vivo* in response to environmental or developmental cues that demand the adjustment of chloroplast gene expression. Therefore, we should push the envelope of mTERF research in these directions and seek to unravel the molecular and structural determinants of mTERF activities in plants, and how these are regulated *in vivo*.

Acknowledgments This work was supported by the Centre National de la Recherche Scientifique (CNRS), a EU research funding grant [CIG-618492-plantMTERF] and Agence Nationale de la Recherche grant [ANR-16-CE20-0007]. The author thanks Valérie Cognat for the GRO-seq analysis.

References

Asin-Cayuela J, Helm M, Attardi G (2004) A monomer-to-trimer transition of the human mitochondrial transcription termination factor (mTERF) is associated with a loss of in vitro activity. J Biol Chem 279(15):15670–15677

Asin-Cayuela J, Schwend T, Farge G, Gustafsson CM (2005) The human mitochondrial transcription termination factor (mTERF) is fully active in vitro in the non-phosphorylated form. J Biol Chem 280(27):25499–25505

Babiychuk E, Vandepoele K, Wissing J, Garcia-Diaz M, De Rycke R, Akbari H, Joubes J, Beeckman T, Jansch L, Frentzen M, Van Montagu MC, Kushnir S (2011) Plastid gene expression and plant development require a plastidic protein of the mitochondrial transcription termination factor family. Proc Natl Acad Sci USA 108(16):6674–6679

Barkan A (2011) Expression of plastid genes: organelle-specific elaborations on a prokaryotic scaffold. Plant Physiol 155(4):1520–1532

Barkan A, Small I (2014) Pentatricopeptide repeat proteins in plants. Annu Rev Plant Biol 65:415–442

Borner T, Aleynikova AY, Zubo YO, Kusnetsov VV (2015) Chloroplast RNA polymerases: role in chloroplast biogenesis. Biochim Biophys Acta 1847(9):761–769

Bryant N, Lloyd J, Sweeney C, Myouga F, Meinke D (2011) Identification of nuclear genes encoding chloroplast-localized proteins required for embryo development in Arabidopsis. Plant Physiol 155(4):1678–1689

Camara Y, Asin-Cayuela J, Park CB, Metodiev MD, Shi Y, Ruzzenente B, Kukat C, Habermann B, Wibom R, Hultenby K, Franz T, Erdjument-Bromage H, Tempst P, Hallberg BM, Gustafsson CM, Larsson NG (2011) MTERF4 regulates translation by targeting the methyltransferase NSUN4 to the mammalian mitochondrial ribosome. Cell Metab 13(5):527–539

Ding S, Zhang Y, Hu Z, Huang X, Zhang B, Lu Q, Wen X, Wang Y, Lu C (2019) mTERF5 acts as a transcriptional pausing factor to positively regulate transcription of chloroplast *psbEFLJ*. Mol Plant 12(9):1259–1277

Green BR (2011) Chloroplast genomes of photosynthetic eukaryotes. Plant J 66(1):34–44

Grove TZ, Cortajarena AL, Regan L (2008) Ligand binding by repeat proteins: natural and designed. Curr Opin Struct Biol 18(4):507–515

Hammani K, Barkan A (2014) An mTERF domain protein functions in group II intron splicing in maize chloroplasts. Nucleic Acids Res 42(8):5033–5042

Hammani K, des Francs-Small CC, Takenaka M, Tanz SK, Okuda K, Shikanai T, Brennicke A, Small I (2011) The pentatricopeptide repeat protein OTP87 is essential for RNA editing of *nad7* and *atp1* transcripts in Arabidopsis mitochondria. J Biol Chem 286(24):21361–21371

Hetzel J, Duttke SH, Benner C, Chory J (2016) Nascent RNA sequencing reveals distinct features in plant transcription. Proc Natl Acad Sci USA 113(43):12316–12321

Hsu YW, Wang HJ, Hsieh MH, Hsieh HL, Jauh GY (2014) Arabidopsis mTERF15 is required for mitochondrial *nad2* intron 3 splicing and functional complex I activity. PLoS One 9(11):e112360

Inaba T, Ito-Inaba Y (2010) Versatile roles of plastids in plant growth and development. Plant Cell Physiol 51(11):1847–1853

Jarvis P, López-Juez E (2013) Biogenesis and homeostasis of chloroplasts and other plastids. Nat Rev Mol Cell Biol 14(12):787–802

Ji D, Manavski N, Meurer J, Zhang L, Chi W (2019) Regulated chloroplast transcription termination. Biochim Biophys Acta Bioenerg 1860(1):69–77

Jiang J, Chai X, Manavski N, Williams-Carrier R, He B, Brachmann A, Ji D, Ouyang M, Liu Y, Barkan A, Meurer J, Zhang L, Chi W (2019) An RNA chaperone-like protein plays critical roles in chloroplast mRNA stability and translation in Arabidopsis and Maize. Plant Cell 31(6):1308–1327

Kim M, Lee U, Small I, des Francs-Small CC, Vierling E (2012) Mutations in an Arabidopsis mitochondrial transcription termination factor-related protein enhance thermotolerance in the absence of the major molecular chaperone HSP101. Plant Cell 24(8):3349–3365

Kim M, Schulz V, Brings L, Schoeller T, Kuhn K, Vierling E (2021) mTERF18 and ATAD3 are required for mitochondrial nucleoid structure and their disruption confers heat tolerance in Arabidopsis thaliana. New Phytol 232(5):2026–2042

Kleine T (2012) Arabidopsis thaliana mTERF proteins: evolution and functional classification. Front Plant Sci 3:233

Kleine T, Leister D (2015) Emerging functions of mammalian and plant mTERFs. Biochim Biophys Acta 1847(9):786–797

Kruse B, Narasimhan N, Attardi G (1989) Termination of transcription in human mitochondria: identification and purification of a DNA binding protein factor that promotes termination. Cell 58(2):391–397

Lee K, Leister D, Kleine T (2021) Arabidopsis mitochondrial transcription termination factor mTERF2 promotes splicing of group IIB introns. Cells 10(2)

Linder T, Park CB, Asin-Cayuela J, Pellegrini M, Larsson NG, Falkenberg M, Samuelsson T, Gustafsson CM (2005) A family of putative transcription termination factors shared amongst metazoans and plants. Curr Genet 48(4):265–269

Majeran W, Friso G, Asakura Y, Qu X, Huang M, Ponnala L, Watkins KP, Barkan A, van Wijk KJ (2012) Nucleoid-enriched proteomes in developing plastids and chloroplasts from maize leaves: a new conceptual framework for nucleoid functions. Plant Physiol 158(1):156–189

McDermott JJ, Civic B, Barkan A (2018) Effects of RNA structure and salt concentration on the affinity and kinetics of interactions between pentatricopeptide repeat proteins and their RNA ligands. PLoS One 13(12):e0209713

Meskauskiene R, Wursch M, Laloi C, Vidi PA, Coll NS, Kessler F, Baruah A, Kim C, Apel K (2009) A mutation in the Arabidopsis mTERF-related plastid protein SOLDAT10 activates retrograde signaling and suppresses (1)O(2)-induced cell death. Plant J 60(3):399–410

Méteignier LV, Ghandour R, Meierhoff K, Zimmerman A, Chicher J, Baumberger N, Alioua A, Meurer J, Zoschke R, Hammani K (2020) The Arabidopsis mTERF-repeat MDA1 protein plays a dual function in transcription and stabilization of specific chloroplast transcripts within the *psbE* and *ndhH* operons. New Phytol 227(5):1376–1391

Méteignier L-V, Ghandour R, Zimmerman A, Kuhn L, Meurer J, Zoschke R, Hammani K (2021) Arabidopsis mTERF9 protein promotes chloroplast ribosomal assembly and translation by establishing ribonucleoprotein interactions in vivo. Nucleic Acids Res 49(2):1114–1132

Nawaz G, Kang H (2017) Chloroplast- or mitochondria-targeted DEAD-box RNA helicases play essential roles in organellar RNA metabolism and abiotic stress responses. Front Plant Sci 8:871

Nunez-Delegido E, Robles P, Ferrandez-Ayela A, Quesada V (2019) Functional analysis of mTERF5 and mTERF9 contribution to salt tolerance, plastid gene expression and retrograde signalling in Arabidopsis thaliana. Plant Biol 22(3):459–471

O'Toole N, Hattori M, Andres C, Iida K, Lurin C, Schmitz-Linneweber C, Sugita M, Small I (2008) On the expansion of the pentatricopeptide repeat gene family in plants. Mol Biol Evol 25(6):1120–1128

Pan Z, Ren X, Zhao H, Liu L, Tan Z, Qiu F (2019) A mitochondrial transcription termination factor, ZmSmk3, is required for *nad1* Intron4 and *nad4* Intron1 splicing and kernel development in maize. G3 Genes|Genomes|Genetics 9(8):2677–2686

Peralta S, Wang X, Moraes CT (2012) Mitochondrial transcription: lessons from mouse models. Biochim Biophys Acta 1819(9–10):961–969

Pfalz J, Liere K, Kandlbinder A, Dietz KJ, Oelmuller R (2006) pTAC2, -6, and -12 are components of the transcriptionally active plastid chromosome that are required for plastid gene expression. Plant Cell 18(1):176–197

Quesada V, Sarmiento-Manus R, Gonzalez-Bayon R, Hricova A, Perez-Marcos R, Gracia-Martinez E, Medina-Ruiz L, Leyva-Diaz E, Ponce MR, Micol JL (2011) Arabidopsis RUGOSA2 encodes an mTERF family member required for mitochondrion, chloroplast and leaf development. Plant J 68(4):738–753

Roberti M, Polosa PL, Bruni F, Manzari C, Deceglie S, Gadaleta MN, Cantatore P (2009) The MTERF family proteins: mitochondrial transcription regulators and beyond. Biochim Biophys Acta 1787(5):303–311

Robles P, Micol JL, Quesada V (2012) Arabidopsis MDA1, a nuclear-encoded protein, functions in chloroplast development and abiotic stress responses. PLoS One 7(8):e42924

Robles P, Quesada V (2021) Research progress in the molecular functions of plant mTERF proteins. Cells 10(2)

Robles P, Navarro-Cartagena S, Ferrandez-Ayela A, Nunez-Delegido E, Quesada V (2018) The characterization of Arabidopsis mterf6 mutants reveals a new role for mTERF6 in tolerance to abiotic stress. Int J Mol Sci 19(8)

Romani I, Manavski N, Morosetti A, Tadini L, Maier S, Kuhn K, Ruwe H, Schmitz-Linneweber C, Wanner G, Leister D, Kleine T (2015) A member of the Arabidopsis mitochondrial transcription termination factor family is required for maturation of chloroplast transfer RNAIle(GAU). Plant Physiol 169(1):627–646

Rott R, Drager RG, Stern DB, Schuster G (1996) The 3′ untranslated regions of chloroplast genes in Chlamydomonas reinhardtii do not serve as efficient transcriptional terminators. Mol Gen Genet 252(6):676–683

Rubinson EH, Eichman BF (2012) Nucleic acid recognition by tandem helical repeats. Curr Opin Struct Biol 22(1):101–109

Ruwe H, Schmitz-Linneweber C (2012) Short non-coding RNA fragments accumulating in chloroplasts: footprints of RNA binding proteins? Nucleic Acids Res 40(7):3106–3116

Ruwe H, Wang G, Gusewski S, Schmitz-Linneweber C (2016) Systematic analysis of plant mitochondrial and chloroplast small RNAs suggests organelle-specific mRNA stabilization mechanisms. Nucleic Acids Res 44(15):7406–7417

Schmitz-Linneweber C, Small I (2008) Pentatricopeptide repeat proteins: a socket set for organelle gene expression. Trends Plant Sci 13(12):663–670

Schwenkert S, Fernie AR, Geigenberger P, Leister D, Mohlmann T, Naranjo B, Neuhaus HE (2022) Chloroplasts are key players to cope with light and temperature stress. Trends Plant Sci 27(6):577–587

Shen C, Zhang D, Guan Z, Liu Y, Yang Z, Yang Y, Wang X, Wang Q, Zhang Q, Fan S, Zou T, Yin P (2016) Structural basis for specific single-stranded RNA recognition by designer pentatricopeptide repeat proteins. Nat Commun 7:11285

Shevtsov S, Nevo-Dinur K, Faigon L, Sultan LD, Zmudjak M, Markovits M, Ostersetzer-Biran O (2018) Control of organelle gene expression by the mitochondrial transcription termination factor mTERF22 in Arabidopsis thaliana plants. PLoS One 13(7):e0201631

Shi C, Wang S, Xia EH, Jiang JJ, Zeng FC, Gao LZ (2016) Full transcription of the chloroplast genome in photosynthetic eukaryotes. Sci Rep 6:30135

Small ID, Peeters N (2000) The PPR motif – a TPR-related motif prevalent in plant organellar proteins. Trends Biochem Sci 25(2):46–47

Spahr H, Samuelsson T, Hallberg BM, Gustafsson CM (2010) Structure of mitochondrial transcription termination factor 3 reveals a novel nucleic acid-binding domain. Biochem Biophys Res Commun 397(3):386–390

Spahr H, Habermann B, Gustafsson CM, Larsson NG, Hallberg BM (2012) Structure of the human MTERF4-NSUN4 protein complex that regulates mitochondrial ribosome biogenesis. Proc Natl Acad Sci USA 109(38):15253–15258

Stern DB, Gruissem W (1987) Control of plastid gene expression: 3′ inverted repeats act as mRNA processing and stabilizing elements, but do not terminate transcription. Cell 51(6):1145–1157

Stern DB, Goldschmidt-Clermont M, Hanson MR (2010) Chloroplast RNA metabolism. Annu Rev Plant Biol 61:125–155

Terzioglu M, Ruzzenente B, Harmel J, Mourier A, Jemt E, Lopez MD, Kukat C, Stewart JB, Wibom R, Meharg C, Habermann B, Falkenberg M, Gustafsson CM, Park CB, Larsson NG (2013) MTERF1 binds mtDNA to prevent transcriptional interference at the light-strand promoter but is dispensable for rRNA gene transcription regulation. Cell Metab 17(4):618–626

Tzafrir I, Pena-Muralla R, Dickerman A, Berg M, Rogers R, Hutchens S, Sweeney TC, McElver J, Aux G, Patton D, Meinke D (2004) Identification of genes required for embryo development in Arabidopsis. Plant Physiol 135(3):1206–1220

Van Dingenen J, Blomme J, Gonzalez N, Inzé D (2016) Plants grow with a little help from their organelle friends. J Exp Bot 67(22):6267–6281

Williams-Carrier R, Kroeger T, Barkan A (2008) Sequence-specific binding of a chloroplast pentatricopeptide repeat protein to its native group II intron ligand. RNA 14(9):1930–1941

Wobbe L (2021) The molecular function of plant mTERFs as key regulators of organellar gene expression. Plant Cell Physiol 61(12):2004–2017

Wredenberg A, Lagouge M, Bratic A, Metodiev MD, Spahr H, Mourier A, Freyer C, Ruzzenente B, Tain L, Gronke S, Baggio F, Kukat C, Kremmer E, Wibom R, Polosa PL, Habermann B, Partridge L, Park CB, Larsson NG (2013) MTERF3 regulates mitochondrial ribosome biogenesis in invertebrates and mammals. PLoS Genet 9(1):e1003178

Xiong HB, Wang J, Huang C, Rochaix JD, Lin FM, Zhang JX, Ye LS, Shi XH, Yu QB, Yang ZN (2020) mTERF8, a member of the mitochondrial transcription termination factor family, is involved in the transcription termination of chloroplast gene *psbJ*. Plant Physiol 182(1):408–423

Xu D, Leister D, Kleine T (2017) Arabidopsis thaliana mTERF10 and mTERF11, but Not mTERF12, are involved in the response to salt stress. Front Plant Sci 8:1213

Yakubovskaya E, Mejia E, Byrnes J, Hambardjieva E, Garcia-Diaz M (2010) Helix unwinding and base flipping enable human MTERF1 to terminate mitochondrial transcription. Cell 141(6):982–993

Yakubovskaya E, Guja KE, Mejia E, Castano S, Hambardjieva E, Choi WS, Garcia-Diaz M (2012) Structure of the essential MTERF4:NSUN4 protein complex reveals how an MTERF protein collaborates to facilitate rRNA modification. Structure 20(11):1940–1947

Yu QB, Huang C, Yang ZN (2014) Nuclear-encoded factors associated with the chloroplast transcription machinery of higher plants. Front Plant Sci 5:316

Zhang Y, Cui YL, Zhang XL, Yu QB, Wang X, Yuan XB, Qin XM, He XF, Huang C, Yang ZN (2018) A nuclear-encoded protein, mTERF6, mediates transcription termination of *rpoA* polycistron for plastid-encoded RNA polymerase-dependent chloroplast gene expression and chloroplast development. Sci Rep 8(1):11929

Zhao Y, Cai M, Zhang X, Li Y, Zhang J, Zhao H, Kong F, Zheng Y, Qiu F (2014) Genome-wide identification, evolution and expression analysis of mTERF gene family in maize. PLoS One 9(4):e94126

Zhelyazkova P, Hammani K, Rojas M, Voelker R, Vargas-Suarez M, Borner T, Barkan A (2012a) Protein-mediated protection as the predominant mechanism for defining processed mRNA termini in land plant chloroplasts. Nucleic Acids Res 40(7):3092–3105

Zhelyazkova P, Sharma CM, Forstner KU, Liere K, Vogel J, Borner T (2012b) The primary transcriptome of barley chloroplasts: numerous noncoding RNAs and the dominating role of the plastid-encoded RNA polymerase. Plant Cell 24(1):123–136

Zhou B, Zhang L, Ullah A, Jin X, Yang X, Zhang X (2016) Identification of multiple stress responsive genes by sequencing a normalized cDNA library from sea-land cotton (Gossypium barbadense L.). PLoS One 11(3):e0152927

Chapter 4
Chloroplast Translation Machinery

Ansul Lokdarshi

Abstract Chloroplasts are the only organelles capable of harnessing light energy and converting it into chemical energy through photosynthesis. These specialized organelles, found predominantly in photosynthetic organisms like plants, cyanobacteria, and some protists, possess their own translation machinery that is essential for synthesizing proteins encoded by the chloroplast genome. Originating from endosymbiotic events with ancient photosynthetic prokaryotes, chloroplasts exhibit distinct features and regulatory mechanisms reminiscent of their ancestors. This chapter delves into the intricate world of chloroplast translation, exploring its unique characteristics, regulation under stress conditions, and techniques used for studying chloroplast translation. By examining the process of translation within chloroplasts and discussing methodologies for studying this system, we gain valuable insights into the fundamental mechanisms that sustain plant life and contribute to our understanding of cellular evolution and adaptation.

Keywords Protein synthesis · Translation · Chloroplast ribosome · Retro-Anterograde signaling · Pentatricopeptide Repeat (PPR) Proteins

4.1 Introduction

Chloroplasts, essential to the photosynthetic machinery of plants and algae, are organelles with a sophisticated translation apparatus crucial for synthesizing proteins encoded within their own genomes (Zhang et al. 2023b; Zoschke and Bock 2018). These organelles originated from an ancient endosymbiotic event more than a billion years ago, when a eukaryotic host cell engulfed a photosynthetic cyanobacterium, giving rise to the plastid lineage (Martin et al. 2002; Timmis et al. 2004).

A. Lokdarshi (✉)
Department of Biology, Valdosta State University, Valdosta, GA, USA
e-mail: alokdarshi@valdosta.edu

T. M. Burch-Smith (ed.), *Chloroplast Gene Expression: Regulation, Stress Signaling and Biotechnology*, Nucleic Acids and Molecular Biology 37,
https://doi.org/10.1007/978-3-031-70098-9_4

This evolutionary partnership led to a reduction in the cyanobacterial genome, with many genes being transferred to the host nucleus over time. A notable outcome of this coevolution is the formation of multimeric chloroplast protein complexes, where subunits are encoded in both the plastid and the nucleus, necessitating coordinated gene expression between these compartments (Jarvis and López-Juez 2013; Kleine and Leister 2016). This intricate regulatory network ensures the proper assembly and function of essential photosynthetic complexes. Today, chloroplasts maintain a small genome encoding approximately 120 genes in green plants, including crucial components such as the large subunit of Rubisco and subunits of photosystem I and II (PSI and PSII), as well as RNA polymerases, ribosomal RNAs (rRNAs), transfer RNAs (tRNAs), and ribosomal proteins necessary for chloroplast gene expression and photosynthetic function (Allen et al. 2011; Green 2011).

Chloroplast translation encompasses precisely regulated events from initiation to elongation and termination akin to those of the bacterial ancestor, mediated by tRNAs and translation factors (Marín-Navarro et al. 2007; Sadhu et al. 2023; Sun and Zerges 2015). These processes are finely tuned by regulatory proteins that modulate mRNA stability, translational efficiency, and protein folding, ensuring optimal protein synthesis in response to developmental and environmental cues (Ahlert et al. 2003; Zoschke and Bock 2018). At the heart of chloroplast translation are chloroplast ribosomes (chloro-ribosomes), composed of rRNAs and associated proteins, responsible for translating chloroplast-encoded mRNAs into functional proteins (Beligni et al. 2004a; Sugiura 2014; Tiller and Bock 2014; Tiller et al. 2012). The chloro-ribosomes play critical roles in mRNA recognition and peptide bond formation, ensuring the accurate synthesis of proteins required for chloroplast function (Bieri et al. 2017; Graf et al. 2017). The biogenesis of chloro-ribosomes involves a complex series of steps, starting with transcription and processing of rRNA precursors in the chloroplast (Schmid et al. 2023). This process is facilitated by various proteins, including endonucleases, RNA helicases, GTPases, RNA chaperones, and Pentatricopeptide Repeat (PPR) proteins, which collaborate to cleave, trim, and assemble rRNA transcripts into mature ribosomal subunits, crucial for efficient translation (Schmid et al. 2023; Zhang et al. 2023b).

The advancements in genomic and structural analyses have provided detailed insights into the spatial organization and regulatory networks governing chloroplast translation (Ahmed et al. 2016; Anand et al. 2010; Chotewutmontri and Barkan 2020; Chotewutmontri et al. 2020; Graf et al. 2017; Sharma et al. 2007; Zoschke and Barkan 2015). Techniques such as ribosome profiling, RNA co-immunoprecipitation assays, and high-resolution structural analyses have enhanced our understanding of translation dynamics within chloroplasts (Kwasniak-Owczarek and Janska 2024; Pesaresi 2011; Zoschke and Bock 2018).

Despite their prokaryotic ancestry, chloro-ribosomes and their operational mechanisms have evolved distinct features, including regulatory responses to light and integration with the thylakoid membrane system to optimize photosynthetic efficiency (Zhang et al. 2023b). This chapter explores the intriguing world of chloroplast translation, emphasizing its unique characteristics, regulatory mechanisms, and the innovative methodologies that have expanded our understanding. The chapter also discusses a relatively new concept of translation control, "retrograde control of translation" in chloroplast biology that may offer fresh insights into the plant's rapid adjustments towards wide variety of environmental conditions (Dietz 2015; Lokdarshi et al. 2020a; Lokdarshi and von Arnim 2022; Urquidi Camacho et al. 2020). By unraveling these intricate processes, we gain deeper insights into fundamental aspects of chloroplast biology, evolutionary adaptations in photosynthetic organisms, and potential applications in agricultural biotechnology. Ultimately, these insights pave the way for enhancing crop productivity and developing sustainable solutions to global challenges in food security and environmental sustainability.

4.2 Chloroplast Translation Machinery: Basic Overview

The chloroplast translation machinery comprises ribosomes, transfer RNAs (tRNAs), messenger RNAs (mRNAs), and various protein factors that facilitate the translation of mRNA transcripts originating from the chloroplast genome.

4.2.1 Proteins of Chloroplast Ribosomes

Chloro-ribosomes, similar in structure to the *Escherichia coli* 70S ribosomes, consist of a large (50S) and small (30S) subunit (Beligni et al. 2004b; Manuell et al. 2007; Yamaguchi and Subramanian 2000, 2003; Yamaguchi et al. 2000) (Fig. 4.1; Table 4.1). Chloro-ribosomes possess a total of 58 plastid ribosomal proteins (PRPs), with 25 proteins in the 30S subunit and 33 proteins in the 50S subunit (Stoppel and Meurer 2012; Yamaguchi and Subramanian 2000; Yamaguchi et al. 2000) with the exception that most plant species do not possess Rpl25 and Rpl30 (Yamaguchi and Subramanian 2000). While one-third of all PRPs are encoded by the chloroplast genome, the remaining proteins are nuclear-encoded and post-translationally imported from the cytosol (Cline and Henry 1996). A feature distinguishing PRPs from their *E. coli* counterparts is their larger sizes due to extensions at their N and C termini (e.g., *Chlamydomonas reinhardtii* small subunit S2, S3) and in some cases, insertions within the protein (e.g., Chlamydomonas S5 protein) as evidenced by proteomic studies (Yamaguchi et al. 2000).

In addition to PRPs, the chloro-ribosome in higher plants possess six non-orthologous proteins termed plastid-specific ribosomal proteins (PSRPs). Four of these proteins (PSRP1–4) are associated with the 30S subunit, and two (PSRP5 and PSRP6) are associated with the 50S subunit (Tiller et al. 2012; Yamaguchi and Subramanian 2000; Yamaguchi et al. 2000). These proteins are encoded by the chloroplast genome and are essential for the translation of chloroplast-encoded mRNAs. The cyanobacterium *Synechocystis* sp. PCC6803 contains genes for only two of the six PSRPs present in the higher plants (PSRP1 and PSRP3), and the genome of

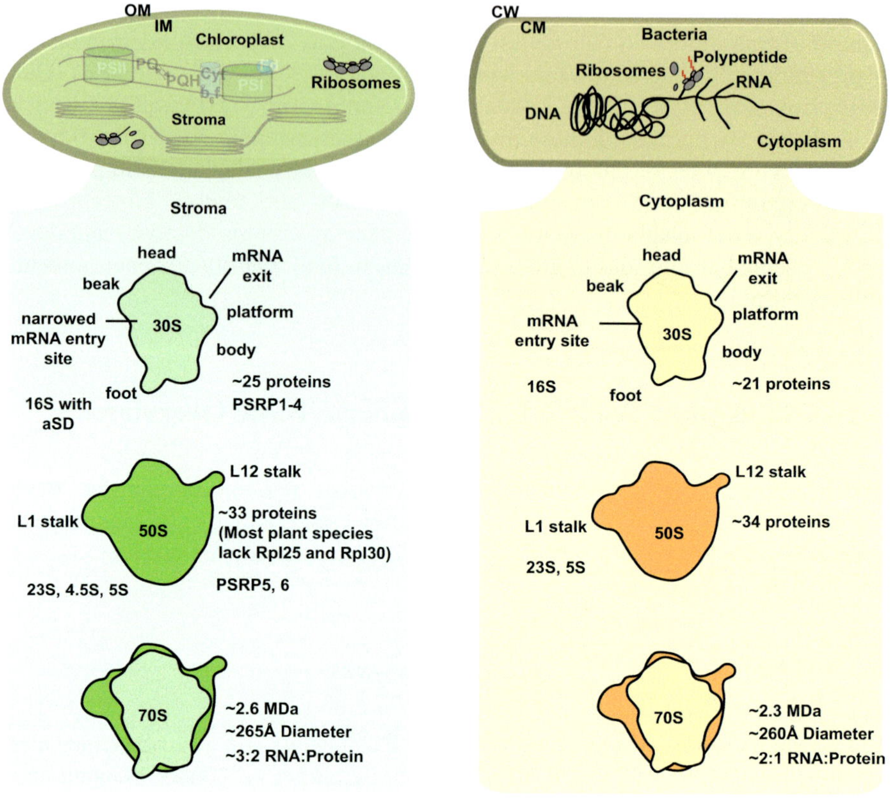

Fig. 4.1 Chloro-ribosome and bacterial ribosome. The three-dimensional cryo-EM map of the spinach 70S chloro-ribosome revealed structural similarities and differences compared to bacterial ribosomes. Key features such as the body, head, and platform of the 30S subunit, and the central protuberance, and L1 and L7/L12 stalks of the 50S subunit are conserved, resembling those of the bacterial ribosome. Chloroplast 16S rRNA (with an anti-Shine-Dalgarno segment) in the 30S subunit is slightly smaller with 1491 nucleotides and a distinct foot or "spur" feature, whereas the 50S subunit contains three rRNAs, totaling 3033 nucleotides, slightly larger than their bacterial counterparts. Chloro-ribosomes harbor 58 plastid ribosomal proteins (PRPs), with 25 and 33 PRPs in the 30S and 50S subunits, respectively. These PRPs are generally larger than *E. coli* counterparts, containing extensions at N and C-termini. Additionally, six plastid-specific ribosomal proteins (PSRPs) are present, enhancing the protein to RNA mass ratio to 2/3 compared to 1/2 for *E. coli*. Combined, the 70S chloro-ribosome is slightly larger in size (approximately 10 Å larger along its longest diameter) and weight versus the *E. coli* 70S ribosome. For more details, please refer to (Zoschke and Bock 2018; Schmid et al. 2023; Agarwal et al. 2011). Abbreviations: *OM* Outer membrane, *IM* Inner membrane, *CW* Cell wall, *CM* Cell membrane, *aSD* anti-Shine-Dalgarno

Chlamydomonas lacks genes for PSRP2 and PSRP5, possibly indicating that at least some PSRPs were acquired rather late in evolution (Yamaguchi and Subramanian 2003). While the function of PSRPs is largely unknown, some PSRPs have been associated with regulation under stress conditions. For example, the *Arabidopsis thaliana* PSRP2, harboring RNA chaperone activity, plays a role as a

Table 4.1 Similarities and differences between bacterial and plastid translational machinery

Feature	Bacteria	Plastid	References
Origin	Prokaryotic	Endosymbiotic origin, derived from cyanobacteria	Barbrook et al. (2006), Reuveni et al. (2017)
Location	Cytoplasm	Stroma of plastids	Maier et al. (2013), Shajani et al. (2011)
Ribosome Size (Svedberg units)	70S	70S	Maier et al. (2013), Shajani et al. (2011)
Subunit Composition	50S large, 30S small	50S large, 30S small	Maier et al. (2013), Shajani et al. (2011)
rRNA Components	16S (30S), 23S and 5S (50S)	16S (30S), 23S and 5S (50S); 23S split into 23S and 4.5S rRNA	Bieri et al. (2017), Kössel et al. (1985), Whitfeld et al. (1978)
Protein Components	~55 proteins (e.g., RplA, RplB, RplC, RplD, RplE, RpsA, RpsB, RpsC, RpsD, RpsE)	~60–70 proteins, including plastid-specific ribosomal proteins (PSRPs such as PSRP1, PSRP2)	Bieri et al. (2017), Sharma et al. (2010), Sharma et al. (2007), Yamaguchi and Subramanian (2000)
Sensitivity to Antibiotics	Sensitive to many antibiotics (e.g., tetracycline, streptomycin)	Sensitive to some antibiotics but generally more resistant	Ahmed et al. (2016), Yamaguchi and Subramanian (2000)
Evolutionary Origin	Ancestral prokaryotes	Endosymbiotic theory suggests origin from cyanobacteria	Barbrook et al. (2006), Reuveni et al. (2017)
Functional Similarity	Protein synthesis	Protein synthesis, adapted for plastid-specific functions	Maier et al. (2013), Shajani et al. (2011)
Genomic Encoding of Proteins	Ribosomal proteins encoded by bacterial genome	Ribosomal proteins encoded by plastid (e.g., rpl16) and nuclear genome (e.g., RPL11, RPL24, RPS17, RPS21)	Barbrook et al. (2006), Reuveni et al. (2017)
Translation Initiation	Shine-Dalgarno sequence in mRNA	Shine-Dalgarno-like sequences in mRNA; anti-Shine-Dalgarno sequence	Scharff et al. (2017)
Structural Differences	Typical bacterial ribosome structure	Additional secondary structures in 23S rRNA	Bieri et al. (2017), Kössel et al. (1985)

(continued)

Table 4.1 (continued)

Feature	Bacteria	Plastid	References
Translational Regulation	Regulated by bacterial factors	Regulated by plastid and nuclear-encoded factors (e.g., PRORP1)	Ahmed et al. (2016), Bieri et al. (2017), Graf et al. (2017), Manuell et al. (2007), Tiller et al. (2012), Yamaguchi and Subramanian (2003)
Loss and Replacement of Proteins	Retains original set of ribosomal proteins	Some bacterial proteins lost; some replaced by cytosolic counterparts (e.g., Rpl33 replaced by nuclear-encoded protein); presence of PSRPs	Bubunenko et al. (1994), Yamaguchi et al. (2000)
Protein Extensions	Typically lack extensive modifications	Some proteins exhibit extensions (e.g., L23, L29); mediate new interactions	Ahmed et al. (2016), Bieri et al. (2017), Graf et al. (2017), Yamaguchi and Subramanian (2000)
mRNA Entry Site	Standard prokaryotic structure	Narrowed mRNA entry site due to extensions in ribosomal proteins	Bieri et al. (2017)
Polypeptide Exit Tunnel	Standard prokaryotic structure	Structural changes support cotranslational binding of SRP (Signal Recognition Particle)	Ahmed et al. (2016), Bieri et al. (2017), Graf et al. (2017)
Protein to RNA Ratio	Lower protein to RNA ratio (1:3)	Higher protein to RNA ratio (2:3)	Barbrook et al. (2006), Reuveni et al. (2017)
Evolutionary Trends	Stable ribosome composition	Tendency to lose RNA functions, replaced by protein elements	Barbrook et al. (2006), Pinker et al. (2013), Ziehe et al. (2017)
Presence in Eukaryotic Cells	Not present	Present in photosynthetic eukaryotes	Barbrook et al. (2006), Reuveni et al. (2017)
Essentiality of Ribosomal Proteins	Many ribosomal proteins essential	Some ribosomal proteins nonessential, varies by species and conditions (e.g., RPL11, RPL24, RPS17, RPS21, PSRP3, PSRP6, Rps15, Rpl33, Rpl36)	Ahlert et al. (2003), Pesaresi et al. (2001), Sosso et al. (2012), Tiller and Bock (2014), Yamaguchi and Subramanian (2003)
Ribosome Heterogeneity	Exists in response to different growth conditions	Unknown if nonessential proteins modulate ribosome affinity to specific mRNAs	Sauert et al. (2015), Shi et al. (2017)
tRNA Completeness	Typically contains full set of tRNAs	Reduced set of tRNAs, relies on superwobble base pairing for translation (e.g., trnG-GCC, trnL-CAA, trnS-GGA, trnT-GGU, trnV-GAC)	Alkatib et al. (2012)
Essential Plastid Genes	Not applicable	Some plastid genes (e.g., accD, clpP, ycf1, ycf2) essential for viability	Bock (2007a), Tiller and Bock (2014)

negative regulator in seed germination under different abiotic stress conditions and in seedling growth under salt stress but not under cold or dehydration stress (Xu et al. 2013).

With advancements in structure prediction and analysis tools, cryo-electron microscopy (cryo-EM) structures of the spinach chloroplast 50S subunit (Ahmed et al. 2016) and the 70S ribosome (Bieri et al. 2017; Graf et al. 2017) have provided detailed insights into the uniqueness of the chloro-ribosome and associated proteins (Fig. 4.1). The cryo-EM structure of the spinach chloro-ribosome shows where specific proteins (PSRP1, PSRP4, PSRP5, and PSRP6) bind and how certain protein extensions interact with the ribosome (Graf et al. 2017). These interactions suggest that the PSRP proteins and extensions have evolved together with the chloro-ribosome. Another cryo-EM study of the 70S chloro-ribosome reveals the organization of the 4.5S rRNA and five plastid-specific ribosomal proteins (Bieri et al. 2017). These proteins stabilize unique chloroplast rRNA regions and remodel the mRNA entry/exit and polypeptide tunnel sites. Specifically, the translation factor pY binds to the 30S subunit's mRNA channel, inhibiting translation by preventing tRNA binding and stabilizing the ribosome in a non-rotated state. Additionally, a large protein mass near the ribosome's exit site likely helps it interact with the thylakoid membrane and protein-targeting machinery. Overall, the chloro-ribosome differs from the *E. coli* 70S ribosome mainly in peripheral regions, and not the core functional areas. A key difference is in the ribosomal tunnel, with modifications that may aid in the formation of α-helical structures and the targeting of transmembrane proteins to the thylakoid membrane. These changes are linked to the shorter β-hairpin of chloroplast L23 and the presence of the N- and C-terminal extensions in the chloroplast L24 (Graf et al. 2017).

Generally, the larger size of the PRPs and the presence of PSRPs significantly alter the protein/RNA mass ratio to 2/3 in chloro-ribosomes, compared with a 1/3 ratio in *E. coli* (Yamaguchi et al. 2000). The higher protein content of chloro-ribosomes and the presence of PSRPs suggest that chloroplasts have evolved specific proteins that have unique functional roles during translation, such as light-dependent stimulation of protein synthesis (Sharma et al. 2007). Combined, these unique features are predicted to play crucial roles in the translation of chloroplast-encoded mRNAs and contribute to the organelle's autonomy in protein synthesis.

4.2.2 rRNA Processing and Ribosome Biogenesis

Chloro-ribosome assembly is an intricate process that necessitates a coordinated interplay of several maturation factors to ensure the proper processing, modification, and assembly of rRNAs and PRP proteins (Germain et al. 2013; Majeran et al. 2011; Stoppel and Meurer 2012) (Fig. 4.2a). The biogenesis of chloro-ribosomes mirrors some aspects of bacterial ribosome assembly but involves unique plant-specific proteins and mechanisms (Pablo Pulido et al. 2018; Schmid et al. 2023;

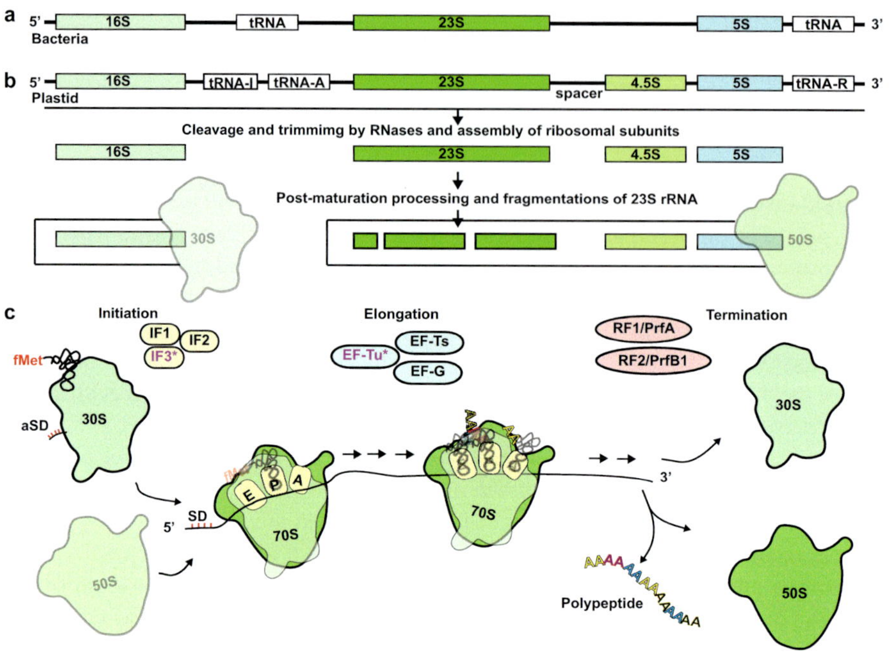

Fig. 4.2 Schematic of bacterial and plastid rRNA and the process of translation in chloroplast. (**a**) Primary transcript of bacterial rRNA operon showing the clustering of rRNA genes. The primary transcript is processed by a combination of RNase III and RNases E, G, and T. Based on Davies et al. (2010). (**b**) Primary transcript of the plastid rRNA operon (~4524 nucleotides in spinach) is similar in length as the bacterial rRNA operon (~4566 nucleotides in *Escherichia coli*) with an additional 4.5S rRNA encoding segment that shares sequence homology to the 3′ tail of the bacterial 23S rRNA. The 4.5S rRNA segment is separated from the 23S rRNA by a short spacer (~115 nucleotides). The plastid rRNA processing events involve action of RNases and fragmentation of the 23S rRNA into three smaller segments based on "hidden breaks" within the 23S sequence. Although information about some of the RNases are known, the intricate details about the maturation of the primary transcript in chloroplasts remain unknown. Adapted and modified from Bieri et al. (2017). (**c**) Chloroplast translation: (1) Initiation: The process begins with the formation of a preinitiation complex comprising the 30S subunit and initiator tRNA (fMet), facilitated by initiation factors (IF1, 2, 3—many plants contain multiple paralogous genes for plastid IF3). The preinitiation complex scans mRNA in a 5′-to-3′ direction, preferentially selecting start codons and aligning the fMe anticodon to the start codon in the peptidyl (P) site. Also shown are the charged tRNA acceptor (A) and the uncharged tRNA exit (E) sites in the 70S ribosome. Shine-Dalgarno (SD) sequences, present before approximately two-thirds of chloroplast reading frames, interact with the anti-Shine-Dalgarno sequence (aSD) in 16S rRNA of the 30S ribosome, aiding proper positioning. Some genes initiate translation independently of SD sequences, favored by low mRNA secondary structure near the start codon and interactions upstream of the start codon with initiator tRNA-fMet and aSD. (2) Elongation: Upon binding of the 50S subunit, the 70S ribosome moves along mRNA with the aid of elongation factors (EF-Tu, EF-G, EF-Ts), synthesizing the polypeptide chain. (3) Termination: Stop codons trigger release factors RF1/PrfA and RF2/PrfB1 to hydrolyze the ester bond, releasing the synthesized protein. PrfB1 also stabilizes mRNAs with UGA codons, while PrfB3 aids in petB mRNA stabilization. RF3 facilitates release factor dissociation. Ribosome recycling is facilitated by EF-G and IF3, leading to the disassembly of ribosomal subunits and the release of mRNA and tRNA for the next round of translation initiation. For more details, please refer to Zoschke and Bock (2018)

Shajani et al. 2011). The complexity of this process is underscored by the necessity for precise enzymatic actions that facilitate the maturation of rRNAs from large precursor molecules into their functional forms, which can then be accurately assembled into ribosomes.

4.2.2.1 Chloroplast Ribosomal RNAs

Plastids, notably chloroplasts, feature a distinctive genetic architecture where the plastid genome (ptDNA) is organized into a circular structure comprising a large single-copy region (LSC), a small single-copy region (SSC), and a pair of large, inverted repeats (IRs) (Bock 2007b; Sugiura 1992). These IRs, constituting approximately one-third of the genome size in seed plant plastid genomes, harbor the ribosomal RNA (rRNA) operon, a vital component encoding the 16S rRNA of the small ribosomal subunit and the 23S, 5S, and 4.5S rRNAs of the large ribosomal subunit (Schwarz and Kössel 1979, 1980) (Fig. 4.2a). Despite the absence of a 4.5S rRNA in bacteria, the plastid 4.5S rRNA shares homology with the 3′-terminal segment of the bacterial 23S rRNA, indicating its evolution through fragmentation (Edwards and Kössel 1981). However, the functional significance of this fragmentation remains uncertain. The plastid rRNA gene cluster encompasses four distinct rRNA species interspersed with intron-encoded *trnI* and *trnA* tRNA genes, followed by the *trnR* gene downstream (Manavski et al. 2018; Stoppel and Meurer 2012). Processing of the large precursor RNA molecule into mature monocistronic rRNA and tRNA isoforms involves multiple steps facilitated by various RNA-binding proteins as well as exo- and endonucleases, akin to observed processes in eubacteria (Shajani et al. 2011). The presence of the rRNA operon within the IR suggests a potential role in augmenting rRNA gene dosage, crucial for ribosome biogenesis, considering the high cellular RNA content in both prokaryotes and eukaryotes. Removal of one IR copy in tobacco chloroplasts resulted in reduced chloro-ribosome abundance, emphasizing the significance of doubled rRNA gene dosage associated with the IR (Krämer et al. 2024). Moreover, the IRs undergo homologous recombination, termed "flip-flop" recombination (Day and Madesis 2007; Stein et al. 1986), which alters the orientation of the single-copy regions. Studies indicate that the slow nucleotide substitution rates within the IRs, attributed to copy correction mechanisms, contribute to chloroplast genome stability (Birky and Walsh 1992; Khakhlova and Bock 2006; Maier et al. 1995; Yamane et al. 2006; Zhu et al. 2016). These findings underscore the multifaceted roles of the IR and the rRNA operon in plastid genome organization and function, implicating them in maintaining genome stability and regulating ribosome biogenesis (Krämer et al. 2024).

4.2.2.2 rRNA Processing Proteins

Through their precise actions, specialized maturation factors encoded by the nucleus enable the correct incorporation of rRNAs and PRPs into functional chloro-ribosomal complexes, ensuring the efficiency and fidelity of the chloroplast translation machinery. Here we describe several of these factors with crucial roles in rRNA processing and ribosome biogenesis.

4.2.2.3 Endonucleases and rRNA Stability

One critical factor for rRNA processing is the endonuclease YbeY in Arabidopsis chloroplasts (Liu et al. 2015). YbeY's role is pivotal in the maturation of the 5′ and 3′ ends of 16S, 23S, and 4.5S rRNAs, ensuring proper cleavage and preparation for ribosomal subunit assembly. Additionally, a PPR protein, the SUPPRESSOR OF THYLAKOID FORMATION1 (SOT1) aids in maturation of the dicistronic 23S-4.5S rRNA and safeguards its 5′ overhang from exonucleolytic degradation by RNase J (Wu et al. 2016; Zhou et al. 2017). This protection enables necessary base pairing between the 5′ and 3′ ends, crucial for further processing and maturation, while the mini-endoribonuclease III (mini III) cleaves double-stranded RNA loops formed during the maturation of 23S and 4.5S rRNAs, producing their mature ends (Hotto et al. 2015, 2020). A similar rRNA processing program is executed by the Chloroplast Stem-loop-binding Protein of 41 kDa (CSP41), which trims the 16S rRNA by approximately 180 nucleotides downstream of its mature 3′ end (Beligni and Mayfield 2008) and associates with pre-ribosomal particles, suggesting a role in the final steps of 23S rRNA maturation. Additionally, the RNA-binding protein RAP, with an octotricopeptide repeat (OPR) domain, binds the 5′ region of the 16S rRNA precursor, specifying the trimming site for proper maturation and stability (Kleinknecht et al. 2014; Stoppel and Meurer 2012).

In bacteria, the endoribonuclease RNase E is crucial for RNA processing, stability, and degradation, playing a key role in rRNA maturation, a function conserved in Arabidopsis, as its RNase E can rescue *E. coli rne* mutants (Bandyra and Luisi 2018; Mudd et al. 2008). In Arabidopsis, *rne* mutants show impaired chloro-ribosome production partly due to defective maturation of essential ribosomal protein transcripts (Stoppel and Meurer 2012; Walter et al. 2010). Additionally, chloroplast RNase J enzymes prevent antisense RNA duplex formation by eliminating dysfunctional RNAs (Sharwood et al. 2011). The chloroplast exoribonucleases RNase R and PNPase are vital for trimming 3′ ends of precursor rRNA and ensuring proper 16S, 23S, and 5S rRNA maturation (Bollenbach et al. 2005). In Arabidopsis, knockouts of *cpRNR* and *cpPNPase* lead to embryo lethality, highlighting their cooperative function (Germain et al. 2012). These enzymes' roles are crucial for producing and processing chloroplast rRNA (Stoppel and Meurer 2012). For example, the Arabidopsis CHLOROPLAST RIBOSOME ASSOCIATED (CRASS) co-expresses with nucleus-encoded chloro-ribosomal proteins and plays a crucial role in ribosome biogenesis and stability under stress conditions (P. Pulido et al. 2018).

4.2.2.4 RNA Helicases

As described above, chloroplast rRNAs undergo maturation steps through proper processing and possess a compact but elaborate structure with many potential base-pairing configurations, requiring partial unwinding to fit into the ribosomal subunit structure and interact with ribosomal proteins. Several RNA helicases (RHs) are involved in these additional processing and restructuring tasks within chloroplasts (Jankowsky and Fairman 2007; Jarmoskaite and Russell 2014). These enzymes represent a highly conserved clade of RNA-binding proteins found across all domains of life and many viruses that adjust RNA secondary structures and rearrange RNA–protein complexes in an ATP-dependent manner (Byrd and Raney 2012). Most RHs belong to the helicase superfamily 2 (SF2), with DEAD-box helicases being the largest family characterized by the D-E-A-D (Asp-Glu-Asp-Glu) motif in their amino acid sequence (Linder et al. 1989). DEAD-box RHs play crucial roles in transcription, mRNA processing, translation initiation, RNA decay, and ribosomal biogenesis (Jarmoskaite and Russell 2011; Nawaz and Kang 2017; Robles and Quesada 2019). In Arabidopsis, 10 of the 58 identified DEAD-box proteins are involved in chloroplast RNA metabolism (Aubourg et al. 1999; Zybailov et al. 2008). These proteins are essential for ribosomal RNA biogenesis and the assembly of ribosomal subunits, particularly under abiotic stress conditions (Kant et al. 2007; Nawaz and Kang 2017; Owttrim 2006, 2013; Vashisht and Tuteja 2006).

Chloroplast RHs, such as RH22, RH39, RH50, and RH3, facilitate the maturation and processing of rRNAs, implicating them in ribosomal biogenesis (Chi et al. 2012; Nishimura et al. 2010). For instance, RH22 interacts with RPL24 and aids in the assembly of the 50S ribosomal subunit, while RH39 processes the 23S rRNA's hidden break B site (Chi et al. 2012; Nishimura et al. 2010). RH50 is crucial for the maturation of 23S and 4.5S rRNAs, and RH3 is involved in splicing several group II introns (Asakura et al. 2012; Gu et al. 2014; Lee et al. 2013; Paieri et al. 2018). Another essential nucleus-encoded RNA helicase is the chloroplast INCREASED SIZE EXCLUSION LIMIT2 (ISE2), which is required for rRNA processing and ribosome assembly (Bobik et al. 2017, 2019). ISE2 interacts with plastid ribosomal protein L15 (uL15c/RPL15), essential for autotrophic growth and the knockdown of Arabidopsis *RPL15* disrupts rRNA processing, ribosomal subunit assembly, and chloroplast translation, indicating its vital role in chloroplast function (Bobik et al. 2019). Additionally, ISE2 contributes to chloroplast translation by facilitating ribosome assembly, maturation, and the assembly of functional ribosomal complexes (Bobik et al. 2019). The importance of these helicases in maintaining chloroplast function under stress conditions, such as cold temperatures, underscores their role in forming cold-adapted ribosomes and RNA degradosomes, ensuring proper RNA metabolism and overall chloroplast health (Aguirre et al. 2017; Prud'homme-Généreux et al. 2004; Stoppel and Meurer 2012).

4.2.2.5 GTPases

The importance of GTPases in ribosome assembly has been extensively studied in bacterial systems (Britton 2009). These enzymes support the energetically demanding assembly process by providing extra energy through GTP hydrolysis for intramolecular rearrangements or disassembling auxiliary factors after their tasks are completed (Verstraeten et al. 2011). Additionally, GTPases connect the translation machinery with the cell cycle and other metabolic processes, linking protein production with various cellular activities (Karbstein 2007). In the context of chloroplasts, several GTPases are implicated in the processing of the 16S rRNA and serve as factors for 30S assembly. For instance, the Arabidopsis ribosome small subunit-dependent GTPase A (AtRsgA), plays a crucial role in the maturation of 16S rRNA, like its bacterial counterpart RsgA, although it only partially rescues the Δrsga *E. coli* strain, indicating slightly different and/or additional functions (Janowski et al. 2018). Like in *E. coli*, AtRsgA interacts with a homolog of the eubacterial ribosome-binding factor A (RbfA) called ribosome-binding factor (RBF1), which is essential for the 5′ end maturation of 16S rRNA and also implicated to have an additional role in the 3′ end maturation of 16S rRNA in plants (Fristedt et al. 2014).

Another class of GTPases involved in ribosome assembly is the YqeH group, which contains circularly permuted G (CPG) motif domains prevalent in bacteria. In chloroplasts, YqeH proteins have rearranged CPGs, allowing the formation of the three-dimensional fold known for canonical GTPases (Sudhamsu et al. 2008). For example, the essential YqeH protein in *Bacillus subtilis* binds to the 30S subunit in a GTP/GDP-dependent manner during initial assembly stages (Anand et al. 2010). In Arabidopsis, *NITRIC OXIDE-ASSOCIATED PROTEIN 1,* AtNOA1 (initially thought to function as a nitric-oxide synthase) is a YqeH-type GTPase participating in ribosome assembly, with studies showing that its mutation affects 16S rRNA levels and translation of chloroplast-encoded proteins (H. Liu et al. 2010a). Furthermore, the maize EMB14 protein, another YqeH class GTPase, plays a crucial role in the formation of the 30S ribosomal subunit, and its Arabidopsis homolog, NOA1, can restore defects in the *emb14* mutant (Li et al. 2015). As another example, the BRZ-INSENSITIVE PALE GREEN 2 (BPG2) protein, plays pivotal roles in ribosomal RNA (rRNA) processing and ribosome biogenesis. In maize, it is implicated in assembling the 30S ribosomal subunit by binding specifically to both 16S and 23S ribosomal RNAs (Komatsu et al. 2010). In Arabidopsis, BPG2 is essential for the greening process, with its expression regulated by light through phytochrome mediation (Kim et al. 2012). Recent investigations reveal BPG2's association with chloroplast nucleoids and its interaction with uS10c, a crucial plastid ribosomal protein, independent of rRNA presence (Sun et al. 2024). This interaction modulates ribosome biogenesis, affecting the processing of 16S and 23S-4.5S rRNAs, plastid protein accumulation, and plastid signaling response, indicating a coordinated role of uS10c-BPG2 module in chloroplast nucleoids for ribosome biogenesis (Sun et al. 2024). Finally, Obg proteins, such as ObgC, are involved in the assembly of the 50S ribosome subunit, highlighting their essential function in ribosomal biogenesis across varied species (Bang et al. 2012).

4.2.2.6 RNA Chaperones

RNA chaperones play versatile roles in RNA folding, stabilization, and remodeling, ensuring the proper structure and function of RNA molecules during translation. Some RNA chaperones are specific to chloroplasts and assist in the folding of chloroplast-encoded mRNAs or ribosomal RNAs, thereby promoting efficient translation. A notable RNA chaperone in chloroplasts is the petB/petD Stabilizing Factor (BSF), a protein enriched with S1 domains, pivotal for stabilizing and translating specific chloroplast mRNAs in Arabidopsis and maize. BSF's engagement with the intergenic region of petB-petD stabilizes the 3′ processed petB transcripts and enhances petD translation, while also activating petA translation by binding to its 5′ untranslated region (Jiang et al. 2019). The nucleic-acid-melting activity exhibited by BSF underscores its RNA chaperone role, as it remodels RNA structures to ensure optimal mRNA function, a task bolstered by its interaction with other proteins like AtCRP1 and PrfB3, thereby aiding mRNA stability and translation (Jiang et al. 2019). Lastly, as previously mentioned under RNA helicases, the AtRH3 protein is intricately involved in chloroplast intron splicing and RNA stability (Asakura et al. 2012).

In addition to these, the plastid RNA-binding protein (PRBP) in *Nicotiana benthamiana* and Rho-like RNA helicase (RHON1) in Arabidopsis contribute significantly to chloroplast RNA metabolism. Specifically, the NtPRBP, a DEAD-box RNA helicase, orchestrates the maturation of the 4.5S rRNA, crucial for ribosome function, while AtRHON1 guides ribonuclease E (RNase E) to specific RNA targets, thus facilitating rRNA processing and overall chloroplast RNA stability. These RNA helicases exemplify the diverse roles RNA chaperones play in maintaining the proper function and stability of chloroplast RNAs (Park et al. 2011; Stoppel et al. 2012). Furthermore, the Pale Cress (PAC) protein epitomizes the plant-specific RNA chaperones' significance in chloroplasts. The Arabidopsis *pac* mutations led to defects in chloroplast development and RNA metabolism, underlining its crucial role in chloroplast function. PAC's interaction with the 23S rRNA, where it remodels a specific region, facilitates early ribosome assembly and function, underscoring its importance in ribosome stability and regulation of translation (Meurer et al. 1998, 2017; Reiter et al. 1994).

The Arabidopsis S1 domain-containing proteins, SDP and SRRP1, are vital for chloroplast RNA metabolism. SDP is essential for chloroplast biogenesis and photosynthetic activity, facilitating the processing of rRNA transcripts. Similarly, SRRP1's role in splicing and RNA cleavage underscores its RNA chaperone function. Both proteins' ability to complement RNA chaperone-deficient *E. coli* strains further confirms their chaperone activity (Gu et al. 2015; Han et al. 2015). The Arabidopsis chloroplast RNA splicing and ribosome maturation (CRM)-domain-containing protein family member, CRM4 is crucial for the correct processing of 16S and 4.5S rRNAs and plays a role in growth and stress responses (Lee et al. 2014). Its RNA chaperone activity, demonstrated through cross-species complementation and transcription anti-termination assays, highlights its importance in RNA processing and stability (Lee et al. 2014).

Interestingly, around one-third of all ribosomal proteins (RPs) in the 50S subunit of *E. coli* possess RNA chaperone activity, with the L1 subunit displaying the highest activity (Semrad et al. 2004). This capability may explain the ribosomes' ability to self-assemble *in vitro* under certain conditions. Given the conservation of many ribosomal subunits between eubacteria and chloroplasts, it is plausible that similar chaperone functions exist in plastid RPs (Ameres et al. 2007; Semrad et al. 2004). Therefore, understanding the roles and interactions of RNA chaperones provides crucial insights into the post-transcriptional regulation of the chloroplast genome at the level of translation and highlights the complexity and specificity of chloroplast RNA metabolism.

4.2.2.7 Pentatricopeptide Repeat (PPR) Proteins

The PPR proteins represent a large family of RNA-binding proteins found in semi-autonomous organelles, including chloroplasts (Barkan and Small 2014). These proteins are encoded in the nucleus and imported into the chloroplasts, where certain PPR proteins regulate ribosome biogenesis, assembly, stability, interaction with ribosomal RNAs (rRNAs) and are essential for the proper functioning of chloroplast ribosomes (Barkan and Small 2014; Lee et al. 2019a, b; Sugita 2022; Wang et al. 2021). The PPR proteins, such as the PPR4 and EMBRYO DEFECTIVE (EMB)2654, chloroplast RNA splicing and ribosome maturation protein CAF2, the peptidyl-tRNA hydrolase (PTH) protein CRS2, and RH3 are required for the removal of intron 1 from *rps12* pre-mRNA (Ding et al. 2006; Jenkins and Barkan 2001; Lee et al. 2019a, b; Ostheimer et al. 2003; Schmitz-Linneweber et al. 2006; Tadini et al. 2018). Additionally, PPR4 and EMB2654 facilitate the splicing of the *rps12* intron 1 in maize and Arabidopsis, respectively, while the rice PPR protein, WHITE STRIPE LEAF (WSL) is involved in the splicing of chloroplast *rpl2* (Tan et al. 2014). Failures in these essential post-transcriptional processing events are associated with a broad spectrum of growth and developmental abnormalities. For example, the rice chloroplast *wsl* mutants display early chlorotic striations and enhanced sensitivity to abscisic acid (ABA), salt, and sugar (Tan et al. 2014) while the rice *wsl5* mutant develops white-striped leaves and exhibits albino phenotype at low temperatures (Lee et al. 2019a, b; Liu et al. 2018). Another example is the Arabidopsis PPR protein LOW PHOTOSYNTHETIC EFFECIENCY 1 (LPE1) and its maize ortholog TPJ1 (Translation of psbJ1). Mutations in these PPR proteins result in a significant reduction in psbJ translation, leading to a PSII deficiency. In Arabidopsis, *lpe1* mutants also show decreased translation of psbN, suggesting that LPE1 has multiple roles in the translation of chloroplast mRNAs critical for PSII assembly and function (Rosalind Williams-Carrier et al. 2019). These findings highlight the requirement of a reliable mRNA editing and rRNA processing programs for the transcripts encoding chloroplast ribosomal proteins under both normal and different stress conditions.

4.2.3 Process of Translation in Chloroplasts

Like bacterial translation, chloroplast translation involves initiation, elongation, termination, and ribosome recycling. However, chloroplast translation exhibits unique features, including the involvement of specific initiation factors and the utilization of non-canonical start codons (Fig. 4.2).

4.2.3.1 Initiation

The initiation of translation commences with the formation of a preinitiation complex, involving the 30S ribosomal subunit and the initiator tRNA for N-formylmethionine (fMet) at the mRNA's initiation site. In bacterial systems, this initiation relies heavily on initiation factors IF1, IF2, and IF3, facilitating the binding of the initiator tRNA and ribosomal subunit assembly. Given the conservation in fundamental aspect of the process of translation, It is no surprise that chloroplast orthologs of these initiation factors have been identified (Campos et al. 2001; Miura et al. 2007; Sijben-Müller et al. 1986; Zheng et al. 2016). Interestingly, many plants contain multiple paralogous genes for plastid IF3, and their expression levels have been proposed to regulate chloroplast translation initiation (Nesbit et al. 2015).

Like bacterial translation initiation, approximately two-thirds of chloroplast open reading frames are preceded by Shine-Dalgarno (SD) sequences, purine-rich regions on the mRNA (Scharff et al. 2011; Shine and Dalgarno 1974). These SD sequences interact via base pairing with a pyrimidine-rich sequence in the 16S rRNA, termed the anti-Shine-Dalgarno sequence (aSD), ensuring accurate positioning of the initiation complex at the start codon. The conservation of the aSD sequence in plant plastids underscores its functional importance, facilitating translation initiation for genes harboring SD sequences (Hirose and Sugiura 2004; Kim and Mullet 1994). Nonetheless, the significance of SD-dependent initiation in chloroplasts has been debated (Fargo et al. 1998). This was resolved by studies introducing mutations into the aSD sequence in tobacco chloroplasts that provided conclusive evidence for the functionality of SD-aSD interactions (Scharff et al. 2017). Conversely, roughly a third of chloroplast genes, as well as many bacterial genes, lack SD sequences or exhibit SD-independent translation initiation. While the chloroplast ribosomal protein CS1 prefers adenine or uracil-rich sequences, its specific contribution to SD-independent translation initiation has not been fully deduced (Franzetti et al. 1992; Shteiman-Kotler and Schuster 2000). One scenario is that the low mRNA secondary structure around the start codon facilitates SD-independent translation initiation, with CS1 likely acting as an RNA chaperone that unfolds structured mRNA regions (Nakagawa et al. 2017; Scharff et al. 2011, 2017).

Studies in both bacterial and chloroplast systems have implicated a 5′-to-3′ ribosome scanning mechanism and the preferential utilization of the 5′-most start codon. Additionally, interactions between mRNA sequences upstream of the start codon and adjacent to the SD with bases downstream of the initiator tRNA-fMet anticodon and adjacent to the aSD have been proposed to facilitate chloroplast translation initiation (Esposito et al. 2003; Kuroda et al. 2007). Nonetheless, the precise mechanism and quantitative contribution of these interactions to translation efficiency warrant further investigation. Moreover, like bacterial systems, chloroplasts exhibit versatility in start codon recognition, with AUG, GUG, and UUG serving as start codons. However, the recognition efficiency of non-AUG start codons presumably depends on the sequence context (Boeck and Kolakofsky 1994; Hirose et al. 1999; Kuroda et al. 2007). Stable transplastomic tobacco plants expressing the *β-glucuronidase* (*GUS*) reporter gene under the *psbA* and *psbC* promoters, differing in start codon usage (AUG and GUG, respectively), revealed significant differences in expression levels of the transgene (Sadhu et al. 2023). Specifically, a comparative analysis between *psbC* and *psbA* promoters demonstrated differential expression patterns, with *psbC:GUG* yielding the highest GUS activity (Sadhu et al. 2023). Further comparison in *E. coli* indicated similar expression profiles under both *psbC* and *psbA* promoters, contrasting with the significant differences observed in chloroplasts. RNA sequencing elucidated that the non-AUG start codons were not subjected to RNA editing, and proteomic analysis revealed multiple translation initiation sites and post-translational modifications, highlighting their utility in regulating protein expression (Sadhu et al. 2023). These findings underscore chloroplasts' use of non-AUG start codons as an additional layer of gene regulation, particularly for proteins with high turnover rates.

Chloroplast protein translation is predominantly controlled at the initiation step. For example, ATP4, PPR10, HIGH CHLOROPHYLL FLUORESCENCE107 (HCF107), HCF173, and HCF244 promote translation of specific chloroplast genes primarily at the initiation step (Felder et al. 2001; Link et al. 2012; Prikryl et al. 2011; Schult et al. 2007; Zoschke et al. 2012). They ensure efficient assembly of the translational machinery and accurate initiation of translation at the appropriate start codon. For instance, the 5′-untranslated region (UTR) of the *psbA* mRNA, which encodes the PSII reaction center polypeptide D1, contains essential elements for its translation. These include sequences complementary to the 3′-terminus of the chloroplast 16S rRNA, termed RBS1 and RBS2, and an AU-rich sequence known as the AU box. The cooperative binding of RBS1 and RBS2 to the 16S rRNA results in the looping out of the AU box, facilitating the interaction of regulatory proteins with the mRNA (Hirose and Sugiura 1996). The stability and translation of chloroplast mRNAs are further enhanced by interactions between their 5′- and 3′-UTRs and regulatory proteins. The 3′-UTR of the *psbA* mRNA, for instance, stabilizes the binding of regulatory proteins to the 5′-UTR, enhancing the stability of the translation initiation complex. This interaction prevents premature degradation and ensures efficient translation. High-affinity binding of regulatory proteins to the 5′-UTR is stabilized by the presence of the 3′-UTR, which may provide a means for translational discrimination against incorrectly processed mRNAs. These complex interactions between trans-acting factors and UTRs highlight the multifaceted mechanisms by which these proteins regulate chloroplast mRNA translation, ensuring precise and efficient expression of essential chloroplast proteins (Katz and Danon 2002; Pfalz et al. 2009). Moreover, studies on chloroplast development underscore the critical roles of translation initiation factors. The SVR9 protein, a chloroplast-localized prokaryotic type translation initiation factor 3 (IF3), is essential for chloroplast and normal leaf development in Arabidopsis. *SVR9* and its homolog *SVR9L1* play crucial roles in chloroplast and leaf development, highlighting the multifaceted functions of translation initiation factors in plant growth (Zheng et al. 2016).

4.2.3.2 Elongation

The translation elongation process in plastid and bacteria begins with the association of the 50S subunit and the preinitiation complex leading to the formation of a 70S ribosome. The functional 70S ribosome then commences decoding of the mRNA into a polypeptide chain. This process in bacteria relies on elongation factors EF-Tu, EF-G, and EF-Ts, for which conserved chloroplast orthologs have been identified (Breitenberger et al. 1979; Fox et al. 1980; Sreedharan et al. 1985). Numerous stimuli are known to affect chloroplast translation via the modulation of expression of the elongation factors (Akkaya and Breitenberger 1992; Albrecht et al. 2006; Bhadula et al. 2001; X. Liu et al. 2010b; Schröter et al. 2010; Singh et al. 2004). In addition, mutations within different elongation factor genes have been shown to adversely affect plastid development and function. For instance, mutations in the Arabidopsis *SUPPRESSOR OF OVEREXPRESSION OF CONSTANS1 (SCO1)*, which encodes a plastid-targeted elongation factor G, lead to defects in plastid development during embryogenesis (Ruppel and Hangarter 2007). T-DNA insertion mutations in *SCO1* result in embryo lethality, while the point mutation allele *sco1-1* and a promoter insertion allele *sco1-4* cause conditional seedling lethality. These seedlings exhibit albinism in cotyledons and hypocotyls due to improper chloroplast development but can produce green leaves when supplemented with sucrose (Ruppel and Hangarter 2007).

Similarly, Arabidopsis *SUPPRESSOR OF VARIEGATION3 (SVR3)*, encoding a TypA-like translation elongation factor, is essential for chloroplast development, especially under cold conditions (X. Liu et al. 2010b). Disruption of *SVR3* leads to pale green leaves at 22 °C and pronounced chlorosis at 8 °C, highlighting its role in chloroplast function under stress. Double mutant analysis suggests a potential interaction between SVR3 and another chloroplast-targeted protein, SVR7, involved in RNA processing. Further illustrating the complexity of chloroplast development, Arabidopsis *SVR11*, which encodes a putative chloroplast-localized prokaryotic-type translation elongation factor EF-Tu (Liu et al. 2019), has been shown to be crucial for chloroplast development and plant survival, with high expression in juvenile leaves, lateral roots, and root tips. Additionally, *SVR11* and *SVR9* together regulate leaf development, including leaf margin development and cotyledon venation patterns. Moreover, EF-Tu (delivers aminoacyl-tRNA to the ribosome), is susceptible to inactivation by reactive oxygen species (ROS) in the cyanobacterium Synechocystis sp. PCC 6803 (Toriu et al. 2023). Interestingly, a recombinant cpEF-Tu protein of Arabidopsis showed that Cys149 and Cys451 were sensitive to oxidation by H_2O_2. Thioredoxin f1, a major thioredoxin in the Arabidopsis chloroplast, reduced oxidized cpEF-Tu, suggesting that the oxidation of cpEF-Tu might be a reversible regulatory mechanism that suppresses the chloroplast translation system in a redox-dependent manner (Toriu et al. 2023).

4.2.3.3 Termination

Termination of translation is orchestrated by the action of peptide chain release factors, which specifically execute the following two tasks: recognition of the stop codon and catalysis of peptidyl-tRNA hydrolysis. Despite significant progress in understanding translation termination in bacterial systems, the mechanisms underlying this process in chloroplasts remain largely elusive (Zoschke and Bock 2018). Characterization of peptide chain release factors, such as RF1/PrfA and RF2/PrfB1, has provided insights into their roles in chloroplast translation termination (Buckingham et al. 1997; Meurer et al. 2002; Motohashi et al. 2007). RF1/PrfA and RF2/PrfB1 play essential roles in liberating synthesized proteins through the hydrolysis of the ester bond, ensuring the accurate termination of translation. Notably, PrfB1 also contributes to the stabilization of plastid mRNAs harboring reading frames terminated by UGA stop codons, highlighting the multifunctional nature of peptide chain release factors within chloroplasts (Meurer et al. 2002). Mutational analyses have demonstrated the critical importance of these factors in maintaining chloroplast function, as mutations can lead to aberrant phenotypes and dysfunctional chloroplasts in model organisms like Arabidopsis (Meurer et al. 2002; Stoppel et al. 2011). Moreover, the emergence of nonfunctional paralogs like PrfB3, which gain novel functions in transcript stabilization, adds layers of complexity to the termination process, emphasizing the need for further exploration into the regulatory mechanisms governing chloroplast translation termination (Stoppel et al. 2011). Lastly, investigations into the interplay between peptide chain release factors such as DIG8 and regulatory elements, such as ROS, offer promising avenues for unraveling the complexities of chloroplast translation termination and its implications for plant growth and stress responses (X. Zhang et al. 2023a).

4.2.4 Retrograde and Anterograde Control of Translation

Plant survival depends on both fast regulatory switches and slower responses to numerous stresses. For example, fluctuating light intensities may demand a nimble response compared to long-term acclimation to slowly changing stressors like drought. Chloroplasts play a key role in coordinating both rapid and slow responses, using retrograde and anterograde signaling (Hwang et al. 2022; Jan et al. 2022). Retrograde signaling pathways relay information from the chloroplast to the nucleus, allowing the plant cell to adjust nuclear gene expression in accordance with chloroplast function and stress conditions. Conversely, anterograde signaling pathways regulate chloroplast gene expression and translation in response to signals from the nucleus, ensuring coordinated gene expression between the two compartments.

Although primarily associated with transcriptional management, anterograde and retrograde signaling also affect the faster regulatory mode of chloroplast/cytosolic translation, but this remains poorly understood. Translational control is conditionally superior to transcriptional regulation due to faster response time and potential subcellular heterogeneity (Urquidi Camacho et al. 2020). For example, within 15 min after transferring Chlamydomonas from low growth light to high light, the synthesis of the Rubisco large subunit is suppressed, and the synthesis of D1 protein is enhanced to facilitate the D1 repair cycle (Shapira et al. 1997). Light also regulates the translation of psbA mRNA in Chlamydomonas by modulating the binding of activator proteins to its 5′ untranslated region, influenced by the photosynthesis-generated redox state of these proteins (Danon and Mayfield 1994). Additionally, the translation of light-harvesting complex proteins of Chlamydomonas PSII (LHCII) is under the control of a translational repressor (Wobbe et al. 2009). In higher plants, signals derived from the chloroplast control the association of the cytosolic ferredoxin mRNA with the polysomal (actively translating) fraction (Petracek et al. 1997). This preferential ribosome loading hypothesis suggests that by exploiting the preexisting mRNA population, cells are able to synthesize the most appropriate proteins for corresponding environmental conditions.

Redox changes in photosynthetic electron transport components occur immediately after altering incident light intensities (Dietz 2015; Dietz et al. 1985, 2002). High light conditions link overreduction of the photosynthetic electron transport chain to increased ROS release (Tripathy and Oelmüller 2012). Combined redox signals from photosynthetic electron transport regulate both short-term and long-term acclimation (Escoubas et al. 1995; Fey et al. 2005). Among the list of redox cues involved in retrograde signaling, singlet oxygen ($^1O_2$) and hydrogen peroxide (H_2O_2) have attracted significant attention for mediating chloroplastic and cytosolic translation. For example, $^1O_2$ inhibits the PSII reaction center repair cycle in Synechocystis by inhibiting D1-protein translational elongation (Nishiyama et al. 2004). Kojima and co-workers observed redox-dependent elongation via thioredoxin-linked redox changes in elongation factor G (elF-G) in *Synechocystis* (Kojima et al. 2009).

Significant evidence indicates that eukaryotic translational initiation factors (eIFs) are under redox and phosphorylation/dephosphorylation control (Boex-Fontvieille et al. 2013). Specifically, phosphoproteome analysis revealed a correlation between the phosphorylation state of eukaryotic cytosolic translation initiation factors eIF3, eIF4A, eIF4B, eIF4G, and eIF5 and chloroplast photosynthetic activity (Boex-Fontvieille et al. 2013). Further evidence for the chloroplast's role in the retrograde control of translation in Arabidopsis is provided by Lokdarshi et al. (2020a, b), showing the rapid phosphorylation of the cytosolic protein eIF2 in response to various stresses. Specifically, conditions that trigger ROS accumulation in the chloroplast, such as excess light, cold, salt, and herbicides, induce the phosphorylation of the alpha (α) subunit of eIF2α by the pan-eukaryotic serine/threonine protein kinase, General Control of Non-depressible (GCN)2 within 10–30 min

(Lokdarshi et al. 2020a, b, 2022). This phosphorylation can be ameliorated by inhibitors of photosynthetic electron transport and antioxidants, suggesting that under numerous stresses, the cytosolic GCN2-eIF2α module operates under the command of chloroplastic ROS to rapidly manage the status of the cytosolic protein synthesis apparatus. This process represents a novel signaling mechanism for the chloroplast to communicate directly with the cytosolic protein synthesis machinery (Lokdarshi and von Arnim 2022). In addition to the plant stress sentinel kinase, GCN2, the serine/threonine protein kinases, TARGET OF RAPAMYCIN (TOR) and Sucrose nonfermenting 1-related kinase (SnRK) have been extensively studied for their influence on photosynthesis and chloroplast function. For example, there is a general correlation between the chloroplast originating factors known to stimulate the TOR pathway, such as light (Chen et al. 2018; Pfeiffer et al. 2016), glucose and sucrose (Dobrenel et al. 2016; Enganti et al. 2017; Xiong et al. 2013), and the absence of abiotic stressors that signal through the ABA pathway (Wang et al. 2018). Conversely, under sugar and energy deprivation conditions SnRK is activated, while TOR is inhibited (Robaglia et al. 2012; Urquidi Camacho et al. 2020) (Fig. 4.3). Although direct evidence of retrograde control of translation linking TOR and SnRKs to the chloroplast remains to be defined more fully, potential mechanisms to execute these functions are in fact emerging (Merchante et al. 2017; Rodriguez et al. 2019; Song et al. 2021; Urquidi Camacho et al. 2020; Wurzinger et al. 2018).

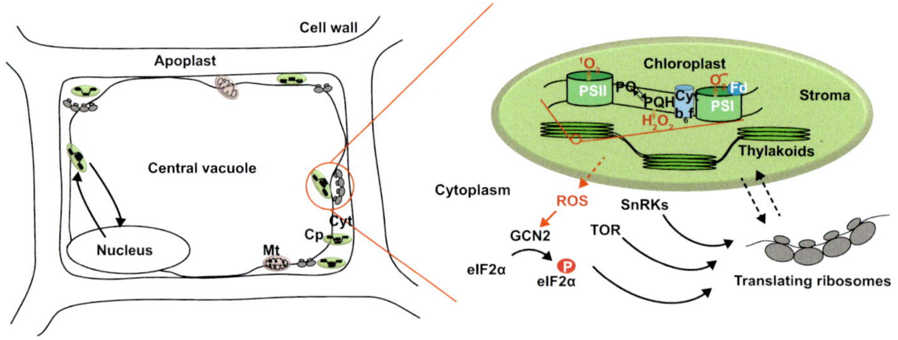

Fig. 4.3 Proposed model of retrograde control of translation. Photosynthetic cells exhibit spatial heterogeneity in nuclear gene regulation due to the variable distance between chloroplasts and the nucleus, alongside the presence of translation-competent ribosomes in the surrounding cytosol. Stressors such as high light, cold, salt, and herbicides induce over-accumulation of reactive oxygen species (ROS) by affecting the photosynthesis process and its components, triggering retrograde control of translation for rapid adjustment of the chloroplast proteome. ROS triggers GCN2 activation which then phosphorylates eIF2α within 10–30 min of stress onset, establishing a direct signaling pathway from chloroplasts to the cytosolic protein synthesis machinery. Additionally, the TOR and SnRK kinases are implicated in chloroplast function regulation, responding to factors such as light, glucose, sucrose, and abiotic stress signals. Based on Dietz (2015) and Urquidi Camacho et al. (2020). Abbreviations: *Cp* Chloroplast, *Cyt* Cytoplasm, *Mt* Mitochondrion. Dotted arrows indicate signaling events that remain to be verified

Finally, photosynthetic plant cells usually have multiple chloroplasts but a single nucleus, and the distance between individual chloroplasts and the nucleus often spans several tens of micrometers, leading to heterogeneity in nuclear gene regulation between chloroplasts depending on their proximity to the nucleus. However, each chloroplast is surrounded by cytosol containing translation-competent ribosomes. Given that ROS accumulation is localized in Arabidopsis leaves exposed to high light or herbicide stress (Lokdarshi et al. 2020a, b), it is logical that retrograde control of translation may be a potent mechanism for immediate spatial and temporal adjustment of the chloroplast proteome. Furthermore, it is hypothesized that retrograde control of translation could allow for spatially heterogeneous protein synthesis. This molecular arrangement would mean preferential loading of transcripts that directly code for repair and photosynthesis related proteins thereby supporting a faster translational retrograde circuit, a plausible idea that will surely attract the attention of researchers in the future (Fig. 4.3). Innovative approaches to dissect the spatiotemporal communication between the chloroplast and cytosolic translation machinery will open new frontiers in understanding chloroplast biology and plant responses to diverse environmental conditions.

4.2.5 Tools for Studying Chloroplast Translation

Studying chloroplast translation entails a multifaceted approach employing various techniques (Kwasniak-Owczarek and Janska 2024; Pesaresi 2011) and a major challenge is the need to isolate chloroplasts quickly and gently to preserve their protein synthesis potential. As the prolonged storage of isolated chloroplasts compromises their bioenergetic capacity, translation-related studies are recommended immediately after isolation. Protein synthesis in isolated chloroplasts requires no external chemical energy source since the triphosphonucleosides needed can be regenerated via light-dependent processes (Gnanam et al. 1988). Interestingly, chloroplast protein synthesis can also occur in the dark if ATP is supplied, though with lower efficiency compared to light-driven synthesis (Siddell and Ellis 1975). These observations provide unique opportunities to formulate future studies with isolated chloroplasts under specific conditions, for example, low vs high light, or testing the effect of different metabolites.

Traditional methods, such as pulse-chase labeling allow tracking of newly synthesized proteins over time, while translation inhibitors like chloramphenicol or spectinomycin halt protein synthesis for assessment. Immunoblotting enables the detection and quantification of chloroplast-encoded proteins, providing insights into translation dynamics. Quantitative PCR measures mRNA levels, indirectly reflecting translation activity. Ribosome profiling is a powerful technique used to globally assess translation activity at the transcriptome-wide level, while mass spectrometry quantifies protein abundance. Mutant analyses illuminate the importance of specific translation components while cross-linking and immunoprecipitation elucidate RNA-binding protein involvement. Integrating these techniques facilitates a comprehensive understanding of chloroplast translation regulation and dynamics. This section discusses some of these techniques with their advantages and limitations.

4.2.5.1 Antibiotics

While chloroplast translation is a crucial process in plants, it remains untargeted by commercial herbicides (Sukhoverkov et al. 2022). This void provides an opportunity to investigate antimicrobial inhibitors of protein synthesis as potential herbicides. This is an attractive strategy considering the conservation of bacterial and chloroplast translational machinery and the differences between eukaryotic and prokaryotic translation, reducing the likelihood of off-target effects (Ellis 1970; Zoschke and Bock 2018). Aminoglycoside antibiotics, such as streptomycin, kanamycin, neomycin, and gentamicin, have been shown to inhibit chloroplast translation by targeting the 16S and 23S rRNAs (Conte et al. 2009; Kasai et al. 2004). High concentrations of tetracyclines (500 mM) rapidly inhibit chloroplast translation by binding to 16S rRNA and blocking the entry of amino-acylated tRNA into the A site of the 70S ribosome (Kasai et al. 2004). Among these notable antibiotics, chloramphenicol (which inhibits peptidyl transferase activity on the 50S ribosomal subunit) and puromycin (which causes premature chain termination during translation by mimicking the structure of aminoacyl-tRNA) have been widely used in chloroplast translation studies by several laboratories (Sabine Kahlau and Ralph Bock 2008; Uniacke and Zerges 2009; Yamamoto et al. 1981). Linezolid 1, a known inhibitor of bacterial protein translation, efficiently inhibits chloroplast translation (Sukhoverkov et al. 2022). Notably, linezolid 1 does not disrupt electron transfer in thylakoid membranes, distinguishing it from photosynthesis inhibitors like atrazine (Sabine Kahlau & Ralph Bock 2008; Uniacke and Zerges 2009; Yamamoto et al. 1981).

Despite the high similarity between chloroplast and prokaryotic translational machinery, the exact targets of some antibiotics in chloroplasts remain unknown. Additionally, other studies have demonstrated that aminoglycosides (e.g., spectinomycin), lincosamides (e.g., lincomycin), and macrolides (e.g., erythromycin) inhibit chloroplast translation without affecting cytoplasmic protein synthesis (Mulo et al. 2003). Tetracyclines and amphenicols (e.g., chloramphenicol) also repress photosynthesis (Li et al. 2011; Margulies 1966; Wen et al. 2012; Xie et al. 2011). Furthermore, several studies have reported additional effects of antibiotic treatments on plants, such as inhibition of PSII function and light-harvesting complexes, and changes in the transcription rate of certain nuclear-encoded photosynthesis-related genes, such as *Lhcb* and *RbcS* (Krupka et al. 2022; Mulo et al. 2003). Future research is needed towards the development of cleaner tools that are specific for chloroplast translation and mitigate the confounding effects stemming from off-site targets and organellar mistranslation (e.g., mitochondria).

4.2.5.2 Polysome Profiling and Ribo-seq

Polysome profiling is a key method for indirectly measuring translational activity by analyzing the association of mRNAs with ribosomes. Polysomes, which are assemblies of actively translating ribosomes held together by mRNA strands, can be separated from free mRNAs and ribosomes through ultracentrifugation across sucrose density gradients. RNA extracted from these fractions and analyzed by RNA gel blotting reveals the distribution of specific mRNAs across the gradient, providing qualitative insights into their translational activity (Barkan 1998). For a genome-wide perspective, mRNAs from different density fractions can be examined via microarray hybridization (Kahlau and Bock 2008). Polysome analysis in plastids is complex due to the operon-like organization of genes and the varied RNA species resulting from polycistronic transcription, which complicates the resolution of individual translation rates. Furthermore, this method can only detect translational regulation at the initiation level since mRNAs loaded with actively elongating ribosomes are indistinguishable from those with paused or stalled ribosomes. This limitation prevents a comprehensive understanding of the overall translational dynamics and regulation within the cell.

A more advanced technique called ribosome profiling, also known as ribosome footprinting or Ribo-seq, is a powerful technique used to globally assess *in vivo* translation activity at the transcriptome-wide level (Chotewutmontri et al. 2018; Ingolia et al. 2009; Wu et al. 2023). Ribo-seq involves isolating ribosomes bound to mRNA transcripts and fragmentation using nucleases such as RNase I. This process generates ribosome-protected fragments (RPFs), which encompass the mRNA region shielded by ribosomes. High-throughput sequencing of the RPFs enables the precise mapping of ribosome occupancy along transcripts, providing insights into translation dynamics, translational efficiency, and ribosome pausing events.

Ribo-seq has revolutionized our understanding of translation in chloroplasts, particularly under varying light conditions (Chotewutmontri and Barkan 2018, 2020; Chotewutmontri et al. 2020; Gawroński et al. 2018; Hristou et al. 2019; McDermott et al. 2019; Swift et al. 2020; Watkins et al. 2020; R. Williams-Carrier et al. 2019; Zoschke and Barkan 2015). Chotewutmontri and Barkan (2018) utilized this method to show that light significantly enhances the translation elongation rate of the *psbA* mRNA, which encodes a core protein of Photosystem II (PSII). This increase in translation is crucial for the rapid synthesis of PsbA protein needed for PSII repair, demonstrating a unique regulatory mechanism that responds dynamically to light (Chotewutmontri and Barkan 2018). Further insights into the chloroplast translation machinery were provided by Hristou et al. (2019), who investigated the role of the chloroplast signal recognition particle subunit cpSRP54. Their ribosome profiling studies revealed that cpSRP54 is essential for the co-translational membrane insertion of key photosynthetic proteins including PsbA. This process is initiated by a novel interaction between cpSRP54 and ribosomal protein uL4, which is crucial for early membrane targeting of nascent polypeptide chains. The absence of cpSRP54 significantly disrupted the integration of these proteins, ultimately affecting photosynthetic efficiency (Hristou et al. 2019). These findings emphasize the critical role of cpSRP54 in maintaining the efficiency of the photosynthetic machinery.

Ribosome footprinting has also uncovered the regulatory role of ribosome pausing in chloroplasts. Gawroński et al. (2018) found that specific sequence motifs and secondary structures induce ribosome pausing, which is closely linked to the assembly of photosynthetic complexes. This pausing ensures the synchronization of protein synthesis with the assembly process, optimizing photosynthetic efficiency (Gawroński et al. 2018). Additionally, Zoschke and Barkan (2015) provided a genome-wide analysis of thylakoid-bound ribosomes in maize, revealing principles of co-translational targeting essential for the proper assembly and function of thylakoid membrane proteins. Their findings highlighted that specific pauses in translation are associated with the insertion of newly synthesized proteins into the thylakoid membrane, emphasizing the complex regulatory mechanisms in chloroplast translation (Zoschke and Barkan 2015). Moreover, Swift et al. (2020) analyzed the function of PSRP1, a chloroplast homolog of a cyanobacterial ribosome hibernation factor, demonstrating its role in modulating ribosome activity during stress conditions to maintain photosynthetic efficiency (Swift et al. 2020). Chotewutmontri et al. (2020) explored the link between Photosystem II assembly and the translation of the chloroplast *psbA* mRNA, revealing that disruptions in PSII assembly feedback regulate psbA translation (Chotewutmontri et al. 2020). These studies collectively illustrate the intricate regulatory networks controlling chloroplast translation and underscore the importance of ribosome profiling in uncovering these mechanisms.

4.2.5.3 Traditional and Non-traditional Approaches

Pulse labeling, involving ^{35}S-radiolabeled amino acids like methionine or cysteine, offers a direct measure of translational activity *in vivo* (Barkan 1998). This technique, first performed by John Ellis and colleagues, involves the incorporation of radioisotopes into newly synthesized proteins, reflecting their synthesis rate (Blair and Ellis 1973; Eaglesham and John Ellis 1974). These proteins can be separated by gel electrophoresis for visualization and quantification. Pulse labeling can independently measure protein synthesis rates from ribosome dynamics, particularly for large plastid-encoded subunits of the photosynthesis machinery. However, the method faces challenges: small or similarly sized subunits require selective purification, low expression levels of certain proteins complicate measurement, and protein turnover rates can obscure results. Additionally, in multicellular organisms, pulsing and chasing may not occur uniformly across all cells, complicating quantitative comparisons.

Mass spectrometry (MS)-based methods, like bio-orthogonal non-canonical amino acid tagging (BONCAT), are emerging for detecting and quantifying newly synthesized proteins. BONCAT involves tagging nascent polypeptides with a non-canonical amino acid, followed by affinity chromatography and MS analysis. Despite its potential, only a few chloroplast-encoded proteins have been identified using this technique (Glenn et al. 2017). Furthermore, cryo-EM studies of ribosomes, such as the specialized functions of chloroplast-specific ribosomal proteins, provide new insights into organellar translation mechanics. The unique localization of proteins like cS23 in the chloroplast ribosomes suggests different mechanisms of translation initiation in chloroplasts, which could be further elucidated with improved cryo-EM resolution and biochemical analyses (Ahmed et al. 2017).

4.2.5.4 Genetic Manipulation Techniques

Genetic manipulation techniques, such as chloroplast transformation and gene editing, are indispensable tools for studying the roles of specific genes and regulatory elements in chloroplast translation. Chloroplast transformation allows for the targeted introduction of transgenes or mutations into the chloroplast genome, enabling researchers to manipulate gene expression and assess the functional consequences on translation and chloroplast function. By analyzing the phenotypic effects of genetic alterations on chloroplast translation and plant growth, researchers can infer the functions of targeted genes and unravel their roles in chloroplast biogenesis and protein synthesis. This topic is discussed in detail in Chap. 6 in this volume.

4.3 Concluding Statement

The study of chloroplast translation has revealed both similarities and distinct features compared to bacterial translation, yet significant gaps remain in our understanding. Current research has elucidated basic mechanisms, such as the conservation of core ribosomal components and the translation initiation process mediated by specific RNA cis-elements and trans-factors. However, fundamental questions persist regarding the regulation and coordination of chloroplast translation.

One pressing issue is the role of regulatory RNA elements and their interaction with protein factors in modulating translation efficiency and specificity within chloroplasts. While progress has been made in identifying key elements involved in translation initiation and elongation, the molecular details governing these processes remain poorly understood. Future research efforts should aim to uncover how these elements function in concert to ensure precise control over protein synthesis in response to environmental cues and developmental signals. Another critical area for exploration is the interplay between chloroplast translation and broader cellular signaling pathways. Understanding how chloroplasts coordinate their translation with nuclear-encoded proteins synthesized in the cytosol and mitochondria could provide insights into integrated cellular responses to environmental changes. This includes investigating potential communication channels and regulatory checkpoints that facilitate crosstalk between organelles during stress adaptation and growth phases. In addition to exploring the fundamental mechanisms of chloroplast translation, a burgeoning area of research focuses on retrograde signaling and its influence on translation regulation. Specifically, how retrograde signals such as ROS originating from chloroplasts modulate translation in the cytosol and how this management feeds back into the chloroplast function remain critical questions. Investigating these rapid response mechanisms will shed light on how plants adapt to dynamic environmental conditions.

Finally, technological advancements such as ribosome profiling and structural biology techniques offer promising avenues to dissect the dynamic interactions between ribosomes, nascent polypeptides, and regulatory factors in chloroplast translation. These approaches hold potential for mapping translation landscapes at high resolution and uncovering novel mechanisms underlying organelle-specific protein synthesis.

Looking forward, addressing these questions will not only deepen our understanding of chloroplast biology but also pave the way for applications in agriculture and biotechnology. Enhancing chloroplast translation efficiency is needed to improve crop yields and resilience to environmental stresses, contributing to sustainable food production in a changing climate. As research continues to unravel the intricacies of chloroplast translation, the field stands poised to unlock new insights into organelle function and its broader implications for cellular physiology and plant biology.

Acknowledgments This work was supported by the grant to Ansul Lokdarshi (AL) from the National Science Foundation (Award # 2334869). AL sincerely thanks Tessa Burch-Smith (editor) for initial discussions about the different topics in this chapter and ideas related to the formulation of figures and Table 4.1. AL also thanks Daniel Rincon Diaz and Morgan E Wynn (Graduate students in AL lab, Department of Biology, Valdosta State University) for reading the text and providing valuable comments.

References

Agarwal D, Gregory ST, O'Connor M (2011) Error-prone and error-restrictive mutations affecting ribosomal protein S12. J Mol Biol 410(1):1–9. 10.1016/j.jmb.2011.04.068. Epub 2011 May 7. PMID: 21575643

Aguirre AA, Vicente AM, Hardwick SW, Alvelos DM, Mazzon RR, Luisi BF, Marques MV (2017) Association of the cold shock DEAD-box RNA helicase RhlE to the RNA degradosome in caulobacter crescentus. J Bacteriol 199(13). https://doi.org/10.1128/jb.00135-17

Ahlert D, Ruf S, Bock R (2003) Plastid protein synthesis is required for plant development in tobacco. Proc Natl Acad Sci USA 100(26):15730–15735. https://doi.org/10.1073/pnas.2533668100

Ahmed T, Yin Z, Bhushan S (2016) Cryo-EM structure of the large subunit of the spinach chloroplast ribosome. Sci Rep 6:35793. https://doi.org/10.1038/srep35793

Ahmed T, Shi J, Bhushan S (2017) Unique localization of the plastid-specific ribosomal proteins in the chloroplast ribosome small subunit provides mechanistic insights into the chloroplastic translation. Nucleic Acids Res 45(14):8581–8595. https://doi.org/10.1093/nar/gkx499

Akkaya MS, Breitenberger CA (1992) Light regulation of protein synthesis factor EF-G in pea chloroplasts. Plant Mol Biol 20(5):791–800. https://doi.org/10.1007/bf00027150

Albrecht V, Ingenfeld A, Apel K (2006) Characterization of the snowy cotyledon 1 mutant of Arabidopsis thaliana: the impact of chloroplast elongation factor G on chloroplast development and plant vitality. Plant Mol Biol 60(4):507–518. https://doi.org/10.1007/s11103-005-4921-0

Alkatib S, Scharff LB, Rogalski M, Fleischmann TT, Matthes A, Seeger S, Schöttler MA, Ruf S, Bock R (2012) The contributions of wobbling and superwobbling to the reading of the genetic code. PLoS Genet 8(11):e1003076. https://doi.org/10.1371/journal.pgen.1003076

Allen JF, de Paula WB, Puthiyaveetil S, Nield J (2011) A structural phylogenetic map for chloroplast photosynthesis. Trends Plant Sci 16(12):645–655. https://doi.org/10.1016/j.tplants.2011.10.004

Ameres SL, Shcherbakov D, Nikonova E, Piendl W, Schroeder R, Semrad K (2007) RNA chaperone activity of L1 ribosomal proteins: phylogenetic conservation and splicing inhibition. Nucleic Acids Res 35(11):3752–3763. https://doi.org/10.1093/nar/gkm318

Anand B, Surana P, Prakash B (2010) Deciphering the catalytic machinery in 30S ribosome assembly GTPase YqeH. PLoS One 5(4):e9944. https://doi.org/10.1371/journal.pone.0009944

Asakura Y, Galarneau E, Watkins KP, Barkan A, van Wijk KJ (2012) Chloroplast RH3 DEAD box RNA helicases in maize and Arabidopsis function in splicing of specific group II introns and affect chloroplast ribosome biogenesis. Plant Physiol 159(3):961–974. https://doi.org/10.1104/pp.112.197525

Aubourg S, Kreis M, Lecharny A (1999) The DEAD box RNA helicase family in Arabidopsis thaliana. Nucleic Acids Res 27(2):628–636. https://doi.org/10.1093/nar/27.2.628

Bandyra KJ, Luisi BF (2018) RNase E and the high-fidelity orchestration of RNA metabolism. Microbiol Spectr 6(2). https://doi.org/10.1128/microbiolspec.RWR-0008-2017

Bang WY, Chen J, Jeong IS, Kim SW, Kim CW, Jung HS, Lee KH, Kweon HS, Yoko I, Shiina T, Bahk JD (2012) Functional characterization of ObgC in ribosome biogenesis during chloroplast development. Plant J 71(1):122–134. https://doi.org/10.1111/j.1365-313X.2012.04976.x

Barbrook AC, Howe CJ, Purton S (2006) Why are plastid genomes retained in non-photosynthetic organisms? Trends Plant Sci 11(2):101–108. https://doi.org/10.1016/j.tplants.2005.12.004

Barkan A (1998) [4] Approaches to investigating nuclear genes that function in chloroplast biogenesis in land plants. In: Methods in enzymology, vol 297. Academic, pp 38–57. https://doi.org/10.1016/S0076-6879(98)97006-9

Barkan A, Small I (2014) Pentatricopeptide repeat proteins in plants. Ann Rev Plant Biol 65:415–442. https://doi.org/10.1146/annurev-arplant-050213-040159

Beligni MV, Mayfield SP (2008) Arabidopsis thaliana mutants reveal a role for CSP41a and CSP41b, two ribosome-associated endonucleases, in chloroplast ribosomal RNA metabolism. Plant Mol Biol 67(4):389–401. https://doi.org/10.1007/s11103-008-9328-2

Beligni MV, Yamaguchi K, Mayfield SP (2004a) The translational apparatus of Chlamydomonas reinhardtii chloroplast. Photosynth Res 82(3):315–325. https://doi.org/10.1007/s11120-004-2440-5

Beligni MV, Yamaguchi K, Mayfield SP (2004b) Chloroplast elongation factor ts pro-protein is an evolutionarily conserved fusion with the s1 domain-containing plastid-specific ribosomal protein-7. Plant Cell 16(12):3357–3369. https://doi.org/10.1105/tpc.104.026708

Bhadula SK, Elthon TE, Habben JE, Helentjaris TG, Jiao S, Ristic Z (2001) Heat-stress induced synthesis of chloroplast protein synthesis elongation factor (EF-Tu) in a heat-tolerant maize line. Planta 212(3):359–366. https://doi.org/10.1007/s004250000416

Bieri P, Leibundgut M, Saurer M, Boehringer D, Ban N (2017) The complete structure of the chloroplast 70S ribosome in complex with translation factor pY. EMBO J 36(4):475–486. https://doi.org/10.15252/embj.201695959

Birky CW Jr, Walsh JB (1992) Biased gene conversion, copy number, and apparent mutation rate differences within chloroplast and bacterial genomes. Genetics 130(3):677–683. https://doi.org/10.1093/genetics/130.3.677

Blair GE, Ellis RJ (1973) Protein synthesis in chloroplasts. I. Light-driven synthesis of the large subunit of fraction I protein by isolated pea chloroplasts. Biochim Biophys Acta 319(2):223–234. https://doi.org/10.1016/0005-2787(73)90013-0

Bobik K, McCray TN, Ernest B, Fernandez JC, Howell KA, Lane T, Staton M, Burch-Smith TM (2017) The chloroplast RNA helicase ISE2 is required for multiple chloroplast RNA processing steps in Arabidopsis thaliana. Plant J 91(1):114–131. https://doi.org/10.1111/tpj.13550

Bobik K, Fernandez JC, Hardin SR, Ernest B, Ganusova EE, Staton ME, Burch-Smith TM (2019) The essential chloroplast ribosomal protein uL15c interacts with the chloroplast RNA helicase ISE2 and affects intercellular trafficking through plasmodesmata. New Phytol 221(2):850–865. https://doi.org/10.1111/nph.15427

Bock R (2007a) Cell and molecular biology of plastids. Springer. Publisher description http://www.loc.gov/catdir/enhancements/fy0825/2007936025-d.html Table of contents only http://www.loc.gov/catdir/enhancements/fy0825/2007936025-t.html

Bock R (2007b) Structure, function, and inheritance of plastid genomes. In: Bock R (ed) Cell and molecular biology of plastids. Springer, Berlin, Heidelberg, pp 29–63. https://doi.org/10.1007/4735_2007_0223

Boeck R, Kolakofsky D (1994) Positions +5 and +6 can be major determinants of the efficiency of non-AUG initiation codons for protein synthesis. EMBO J 13(15):3608–3617. https://doi.org/10.1002/j.1460-2075.1994.tb06668.x

Boex-Fontvieille E, Daventure M, Jossier M, Zivy M, Hodges M, Tcherkez G (2013) Photosynthetic control of arabidopsis leaf cytoplasmic translation initiation by protein phosphorylation. PLoS One 8(7):e70692. https://doi.org/10.1371/journal.pone.0070692

Bollenbach TJ, Lange H, Gutierrez R, Erhardt M, Stern DB, Gagliardi D (2005) RNR1, a 3′–5′ exoribonuclease belonging to the RNR superfamily, catalyzes 3′ maturation of chloroplast ribosomal RNAs in Arabidopsis thaliana. Nucleic Acids Res 33(8):2751–2763. https://doi.org/10.1093/nar/gki576

Breitenberger CA, Graves MC, Spremulli LL (1979) Evidence for the nuclear location of the gene for chloroplast elongation factor G. Arch Biochem Biophys 194(1):265–270. https://doi.org/10.1016/0003-9861(79)90617-9

Britton RA (2009) Role of GTPases in bacterial ribosome assembly. Ann Rev Microbiol 63:155–176. https://doi.org/10.1146/annurev.micro.091208.073225

Bubunenko MG, Schmidt J, Subramanian AR (1994) Protein substitution in chloroplast ribosome evolution. A eukaryotic cytosolic protein has replaced its organelle homologue (L23) in spinach. J Mol Biol 240(1):28–41. https://doi.org/10.1006/jmbi.1994.1415

Buckingham RH, Grentzmann G, Kisselev L (1997) Polypeptide chain release factors. Mol Microbiol 24(3):449–456. https://doi.org/10.1046/j.1365-2958.1997.3711734.x

Byrd AK, Raney KD (2012) Superfamily 2 helicases. Front Biosci (Landmark Ed) 17(6):2070–2088. https://doi.org/10.2741/4038

Campos F, García-Gómez BI, Solórzano RM, Salazar E, Estevez J, León P, Alvarez-Buylla ER, Covarrubias AA (2001) A cDNA for nuclear-encoded chloroplast translational initiation factor 2 from a higher plant is able to complement an infB Escherichia coli null mutant. J Biol Chem 276(30):28388–28394. https://doi.org/10.1074/jbc.M100605200

Chen GH, Liu MJ, Xiong Y, Sheen J, Wu SH (2018) TOR and RPS6 transmit light signals to enhance protein translation in deetiolating Arabidopsis seedlings. Proc Natl Acad Sci USA 115(50):12823–12828. https://doi.org/10.1073/pnas.1809526115

Chi W, He B, Mao J, Li Q, Ma J, Ji D, Zou M, Zhang L (2012) The function of RH22, a DEAD RNA helicase, in the biogenesis of the 50S ribosomal subunits of Arabidopsis chloroplasts. Plant Physiol 158(2):693–707. https://doi.org/10.1104/pp.111.186775

Chotewutmontri P, Barkan A (2018) Multilevel effects of light on ribosome dynamics in chloroplasts program genome-wide and psbA-specific changes in translation. PLoS Genet 14(8):e1007555. https://doi.org/10.1371/journal.pgen.1007555

Chotewutmontri P, Barkan A (2020) Light-induced psbA translation in plants is triggered by photosystem II damage via an assembly-linked autoregulatory circuit. Proc Natl Acad Sci USA 117(35):21775–21784. https://doi.org/10.1073/pnas.2007833117

Chotewutmontri P, Stiffler N, Watkins KP, Barkan A (2018) Ribosome profiling in maize. Methods Mol Biol 1676:165–183. https://doi.org/10.1007/978-1-4939-7315-6_10

Chotewutmontri P, Williams-Carrier R, Barkan A (2020) Exploring the link between photosystem II assembly and translation of the chloroplast psbA mRNA. Plants (Basel) 9(2). https://doi.org/10.3390/plants9020152

Cline K, Henry R (1996) Import and routing of nucleus-encoded chloroplast proteins. Annu Rev Cell Dev Biol 12:1–26. https://doi.org/10.1146/annurev.cellbio.12.1.1

Conte S, Stevenson D, Furner I, Lloyd A (2009) Multiple antibiotic resistance in Arabidopsis is conferred by mutations in a chloroplast-localized transport protein. Plant Physiol 151(2):559–573. https://doi.org/10.1104/pp.109.143487

Danon A, Mayfield SP (1994) Light-regulated translation of chloroplast messenger RNAs through redox potential. Science 266(5191):1717–1719. https://doi.org/10.1126/science.7992056

Davies BW, Köhrer C, Jacob AI, Simmons LA, Zhu J, Aleman LM, Rajbhandary UL, Walker GC (2010) Role of Escherichia coli YbeY, a highly conserved protein, in rRNA processing. Mol Microbiol 78(2):506–518. https://doi.org/10.1111/j.1365-2958.2010.07351.x. Epub 2010 Sep 16. PMID: 20807199; PMCID: PMC2959132

Day A, Madesis P (2007) DNA replication, recombination, and repair in plastids. In: Bock R (ed) Cell and molecular biology of plastids. Springer, Berlin, Heidelberg, pp 65–119. https://doi.org/10.1007/4735_2007_0231

Dietz K-J (2015) Efficient high light acclimation involves rapid processes at multiple mechanistic levels. J Exp Bot 66(9):2401–2414. https://doi.org/10.1093/jxb/eru505

Dietz KJ, Schreiber U, Heber U (1985) The relationship between the redox state of Q A and photosynthesis in leaves at various carbon-dioxide, oxygen and light regimes. Planta 166(2):219–226. https://doi.org/10.1007/bf00397352

Dietz K-J, Link G, Pistorius EK, Scheibe R (2002) Redox regulation in oxigenic photosynthesis. In: Esser K, Lüttge U, Beyschlag W, Hellwig F (eds) Progress in botany: genetics, physiology, ecology. Springer, Berlin, Heidelberg, pp 207–245. https://doi.org/10.1007/978-3-642-56276-1_8

Ding YH, Liu NY, Tang ZS, Liu J, Yang WC (2006) Arabidopsis GLUTAMINE-RICH PROTEIN23 is essential for early embryogenesis and encodes a novel nuclear PPR motif protein that interacts with RNA polymerase II subunit III. Plant Cell 18(4):815–830. https://doi.org/10.1105/tpc.105.039495

Dobrenel T, Caldana C, Hanson J, Robaglia C, Vincentz M, Veit B, Meyer C (2016) TOR signaling and nutrient sensing. Annu Rev Plant Biol 67:261–285. https://doi.org/10.1146/annurev-arplant-043014-114648

Eaglesham ARJ, John Ellis R (1974) Protein synthesis in chloroplasts: II. Light-driven synthesis of membrane proteins by isolated pea chloroplasts. Biochim Biophys Acta (BBA) – Nucl Acids Protein Synt 335(3):396–407. https://doi.org/10.1016/0005-2787(74)90162-2

Edwards K, Kössel H (1981) The rRNA operon from Zea mays chloroplasts: nucleotide sequence of 23S rDNA and its homology with E. coli 23S rDNA. Nucleic Acids Res 9(12):2853–2869. https://doi.org/10.1093/nar/9.12.2853

Ellis RJ (1970) Further similarities between chloroplast and bacterial ribosomes. Planta 91(4):329–335. https://doi.org/10.1007/bf00387506

Enganti R, Cho SK, Toperzer JD, Urquidi-Camacho RA, Cakir OS, Ray AP, Abraham PE, Hettich RL, von Arnim AG (2017) Phosphorylation of ribosomal protein RPS6 integrates light signals and circadian clock signals. Front Plant Sci 8:2210. https://doi.org/10.3389/fpls.2017.02210

Escoubas JM, Lomas M, LaRoche J, Falkowski PG (1995) Light intensity regulation of cab gene transcription is signaled by the redox state of the plastoquinone pool. Proc Natl Acad Sci 92(22):10237–10241. https://doi.org/10.1073/pnas.92.22.10237

Esposito D, Fey JP, Eberhard S, Hicks AJ, Stern DB (2003) In vivo evidence for the prokaryotic model of extended codon-anticodon interaction in translation initiation. EMBO J 22(3):651–656. https://doi.org/10.1093/emboj/cdg072

Fargo DC, Zhang M, Gillham NW, Boynton JE (1998) Shine-Dalgarno-like sequences are not required for translation of chloroplast mRNAs in Chlamydomonas reinhardtii chloroplasts or in Escherichia coli. Mol Gen Genet 257(3):271–282. https://doi.org/10.1007/s004380050648

Felder S, Meierhoff K, Sane AP, Meurer JR, Driemel C, Plücken H, Klaff P, Stein B, Bechtold N, Westhoff P (2001) The nucleus-encoded HCF107 gene of Arabidopsis provides a link between intercistronic rna processing and the accumulation of translation-competent psbH transcripts in chloroplasts. Plant Cell 13(9):2127–2141. https://doi.org/10.1105/tpc.010090

Fey V, Wagner R, Braütigam K, Wirtz M, Hell R, Dietzmann A, Leister D, Oelmüller R, Pfannschmidt T (2005) Retrograde plastid redox signals in the expression of nuclear genes for chloroplast proteins of Arabidopsis thaliana. J Biol Chem 280(7):5318–5328. https://doi.org/10.1074/jbc.M406358200

Fox L, Erion J, Tarnowski J, Spremulli L, Brot N, Weissbach H (1980) Euglena gracilis chloroplast EF-Ts. Evidence that it is a nuclear-coded gene product. J Biol Chem 255(13):6018–6019

Franzetti B, Carol P, Mache R (1992) Characterization and RNA-binding properties of a chloroplast S1-like ribosomal protein. J Biol Chem 267(27):19075–19081

Fristedt R, Scharff LB, Clarke CA, Wang Q, Lin C, Merchant SS, Bock R (2014) RBF1, a plant homolog of the bacterial ribosome-binding factor RbfA, acts in processing of the chloroplast 16S ribosomal RNA. Plant Physiol 164(1):201–215. https://doi.org/10.1104/pp.113.228338

Gawroński P, Jensen PE, Karpiński S, Leister D, Scharff LB (2018) Pausing of chloroplast ribosomes is induced by multiple features and is linked to the assembly of photosynthetic complexes. Plant Physiol 176(3):2557–2569. https://doi.org/10.1104/pp.17.01564

Germain A, Kim SH, Gutierrez R, Stern DB (2012) Ribonuclease II preserves chloroplast RNA homeostasis by increasing mRNA decay rates, and cooperates with polynucleotide phosphorylase in 3′ end maturation. Plant J 72(6):960–971. https://doi.org/10.1111/tpj.12006

Germain A, Hotto AM, Barkan A, Stern DB (2013) RNA processing and decay in plastids. WIREs RNA 4(3):295–316. https://doi.org/10.1002/wrna.1161

Glenn WS, Stone SE, Ho SH, Sweredoski MJ, Moradian A, Hess S, Bailey-Serres J, Tirrell DA (2017) Bioorthogonal noncanonical amino acid tagging (BONCAT) enables time-resolved analysis of protein synthesis in native plant tissue. Plant Physiol 173(3):1543–1553. https://doi.org/10.1104/pp.16.01762

Gnanam A, Subbaiah CC, Mannan RM (1988) Protein synthesis by isolated chloroplasts. Photosynth Res 19(1-2):129–152. https://doi.org/10.1007/bf00114572

Graf M, Arenz S, Huter P, Dönhöfer A, Novácek J, Wilson DN (2017) Cryo-EM structure of the spinach chloroplast ribosome reveals the location of plastid-specific ribosomal proteins and extensions. Nucleic Acids Res 45(5):2887–2896. https://doi.org/10.1093/nar/gkw1272

Green BR (2011) Chloroplast genomes of photosynthetic eukaryotes. Plant J 66(1):34–44. https://doi.org/10.1111/j.1365-313X.2011.04541.x

Gu L, Xu T, Lee K, Lee KH, Kang H (2014) A chloroplast-localized DEAD-box RNA helicaseAtRH3 is essential for intron splicing and plays an important role in the growth and stress response in Arabidopsis thaliana. Plant Physiol Biochem 82:309–318. https://doi.org/10.1016/j.plaphy.2014.07.006

Gu L, Jung HJ, Kim BM, Xu T, Lee K, Kim YO, Kang H (2015) A chloroplast-localized S1 domain-containing protein SRRP1 plays a role in Arabidopsis seedling growth in the presence of ABA. J Plant Physiol 189:34–41. https://doi.org/10.1016/j.jplph.2015.10.003

Han JH, Lee K, Lee KH, Jung S, Jeon Y, Pai HS, Kang H (2015) A nuclear-encoded chloroplast-targeted S1 RNA-binding domain protein affects chloroplast rRNA processing and is crucial for the normal growth of Arabidopsis thaliana. Plant J 83(2):277–289. https://doi.org/10.1111/tpj.12889

Hirose T, Sugiura M (1996) Cis-acting elements and trans-acting factors for accurate translation of chloroplast psbA mRNAs: development of an in vitro translation system from tobacco chloroplasts. EMBO J 15(7):1687–1695

Hirose T, Sugiura M (2004) Functional Shine-Dalgarno-like sequences for translational initiation of chloroplast mRNAs. Plant Cell Physiol 45(1):114–117. https://doi.org/10.1093/pcp/pch002

Hirose T, Ideue T, Wakasugi T, Sugiura M (1999) The chloroplast infA gene with a functional UUG initiation codon. FEBS Lett 445(1):169–172. https://doi.org/10.1016/s0014-5793(99)00123-4

Hotto AM, Castandet B, Gilet L, Higdon A, Condon C, Stern DB (2015) Arabidopsis chloroplast mini-ribonuclease III participates in rRNA maturation and intron recycling. Plant Cell 27(3):724–740. https://doi.org/10.1105/tpc.114.134452

Hotto AM, Stern DB, Schuster G (2020) Plant Ribonuclease J: an essential player in maintaining chloroplast RNA quality control for gene expression. Plan Theory 9(3):334. https://www.mdpi.com/2223-7747/9/3/334

Hristou A, Gerlach I, Stolle DS, Neumann J, Bischoff A, Dünschede B, Nowaczyk MM, Zoschke R, Schünemann D (2019) Ribosome-associated chloroplast SRP54 enables efficient cotranslational membrane insertion of key photosynthetic proteins. Plant Cell 31(11):2734–2750. https://doi.org/10.1105/tpc.19.00169

Hwang Y, Han S, Yoo CY, Hong L, You C, Le BH, Shi H, Zhong S, Hoecker U, Chen X, Chen M (2022) Anterograde signaling controls plastid transcription via sigma factors separately from nuclear photosynthesis genes. Nat Commun 13(1):7440. https://doi.org/10.1038/s41467-022-35080-0

Ingolia NT, Ghaemmaghami S, Newman JR, Weissman JS (2009) Genome-wide analysis in vivo of translation with nucleotide resolution using ribosome profiling. Science 324(5924):218–223. https://doi.org/10.1126/science.1168978

Jan M, Liu Z, Rochaix JD, Sun X (2022) Retrograde and anterograde signaling in the cross-talk between chloroplast and nucleus. Front Plant Sci 13:980237. https://doi.org/10.3389/fpls.2022.980237

Jankowsky E, Fairman ME (2007) RNA helicases – one fold for many functions. Curr Opin Struct Biol 17(3):316–324. https://doi.org/10.1016/j.sbi.2007.05.007

Janowski M, Zoschke R, Scharff LB, Martinez Jaime S, Ferrari C, Proost S, Ng Wei Xiong J, Omranian N, Musialak-Lange M, Nikoloski Z, Graf A, Schöttler MA, Sampathkumar A, Vaid N, Mutwil M (2018) AtRsgA from Arabidopsis thaliana is important for maturation of the small subunit of the chloroplast ribosome. Plant J 96(2):404–420. https://doi.org/10.1111/tpj.14040

Jarmoskaite I, Russell R (2011) DEAD-box proteins as RNA helicases and chaperones. Wiley Interdiscip Rev RNA 2(1):135–152. https://doi.org/10.1002/wrna.50

Jarmoskaite I, Russell R (2014) RNA helicase proteins as chaperones and remodelers. Annu Rev Biochem 83:697–725. https://doi.org/10.1146/annurev-biochem-060713-035546

Jarvis P, López-Juez E (2013) Biogenesis and homeostasis of chloroplasts and other plastids. Nat Rev Mol Cell Biol 14(12):787–802. https://doi.org/10.1038/nrm3702

Jenkins BD, Barkan A (2001) Recruitment of a peptidyl-tRNA hydrolase as a facilitator of group II intron splicing in chloroplasts. EMBO J 20(4):872–879. https://doi.org/10.1093/emboj/20.4.872

Jiang J, Chai X, Manavski N, Williams-Carrier R, He B, Brachmann A, Ji D, Ouyang M, Liu Y, Barkan A, Meurer J, Zhang L, Chi W (2019) An RNA chaperone-like protein plays critical roles in chloroplast mRNA stability and translation in Arabidopsis and Maize. Plant Cell 31(6):1308–1327. https://doi.org/10.1105/tpc.18.00946

Kahlau S, Bock R (2008) Plastid transcriptomics and translatomics of tomato fruit development and chloroplast-to-chromoplast differentiation: chromoplast gene expression largely serves the production of a single protein. Plant Cell 20(4):856–874. https://doi.org/10.1105/tpc.107.055202

Kant P, Kant S, Gordon M, Shaked R, Barak S (2007) STRESS RESPONSE SUPPRESSOR1 and STRESS RESPONSE SUPPRESSOR2, two DEAD-box RNA helicases that attenuate Arabidopsis responses to multiple abiotic stresses. Plant Physiol 145(3):814–830. https://doi.org/10.1104/pp.107.099895

Karbstein K (2007) Role of GTPases in ribosome assembly. Biopolymers 87(1):1–11. https://doi.org/10.1002/bip.20762

Kasai K, Kanno T, Endo Y, Wakasa K, Tozawa Y (2004) Guanosine tetra- and pentaphosphate synthase activity in chloroplasts of a higher plant: association with 70S ribosomes and inhibition by tetracycline. Nucleic Acids Res 32(19):5732–5741. https://doi.org/10.1093/nar/gkh916

Katz YS, Danon A (2002) The 3′-untranslated region of chloroplast psbA mRNA stabilizes binding of regulatory proteins to the leader of the message. J Biol Chem 277(21):18665–18669. https://doi.org/10.1074/jbc.M201033200

Khakhlova O, Bock R (2006) Elimination of deleterious mutations in plastid genomes by gene conversion. Plant J 46(1):85–94. https://doi.org/10.1111/j.1365-313X.2006.02673.x

Kim J, Mullet JE (1994) Ribosome-binding sites on chloroplast rbcL and psbA mRNAs and light-induced initiation of D1 translation. Plant Mol Biol 25(3):437–448. https://doi.org/10.1007/bf00043872

Kim BH, Malec P, Waloszek A, von Arnim AG (2012) Arabidopsis BPG2: a phytochrome-regulated gene whose protein product binds to plastid ribosomal RNAs. Planta 236(2):677–690. https://doi.org/10.1007/s00425-012-1638-6

Kleine T, Leister D (2016) Retrograde signaling: organelles go networking. Biochim Biophys Acta 1857(8):1313–1325. https://doi.org/10.1016/j.bbabio.2016.03.017

Kleinknecht L, Wang F, Stübe R, Philippar K, Nickelsen J, Bohne AV (2014) RAP, the sole octotricopeptide repeat protein in Arabidopsis, is required for chloroplast 16S rRNA maturation. Plant Cell 26(2):777–787. https://doi.org/10.1105/tpc.114.122853

Kojima K, Motohashi K, Morota T, Oshita M, Hisabori T, Hayashi H, Nishiyama Y (2009) Regulation of translation by the redox state of elongation factor G in the cyanobacterium Synechocystis sp. PCC 6803. J Biol Chem 284(28):18685–18691. https://doi.org/10.1074/jbc.M109.015131

Komatsu T, Kawaide H, Saito C, Yamagami A, Shimada S, Nakazawa M, Matsui M, Nakano A, Tsujimoto M, Natsume M, Abe H, Asami T, Nakano T (2010) The chloroplast protein BPG2 functions in brassinosteroid-mediated post-transcriptional accumulation of chloroplast rRNA. Plant J 61(3):409–422. https://doi.org/10.1111/j.1365-313X.2009.04077.x

Kössel H, Natt E, Strittmatter G, Fritzche E, Gozdzicka-Jozefiak A, Przybyl D (1985) Structure and expression of rRNA operons from plastids of higher plants. In: van Vloten-Doting L, Groot G, Hall T (eds) Molecular form and function of the plant genome, pp 183–198

Krämer C, Boehm CR, Liu J, Ting MKY, Hertle AP, Forner J, Ruf S, Schöttler MA, Zoschke R, Bock R (2024) Removal of the large inverted repeat from the plastid genome reveals gene dosage effects and leads to increased genome copy number. Nature Plants. https://doi.org/10.1038/s41477-024-01709-9

Krupka M, Piotrowicz-Cieślak AI, Michalczyk DJ (2022) Effects of antibiotics on the photosynthetic apparatus of plants. J Plant Interact 17(1):96–104. https://doi.org/10.1080/17429145.2021.2014579

Kuroda H, Suzuki H, Kusumegi T, Hirose T, Yukawa Y, Sugiura M (2007) Translation of psbC mRNAs starts from the downstream GUG, not the upstream AUG, and requires the extended Shine-Dalgarno sequence in tobacco chloroplasts. Plant Cell Physiol 48(9):1374–1378. https://doi.org/10.1093/pcp/pcm097

Kwasniak-Owczarek M, Janska H (2024) Experimental approaches to studying translation in plant semi-autonomous organelles. J Exp Bot. https://doi.org/10.1093/jxb/erae151

Lee KH, Park J, Williams DS, Xiong Y, Hwang I, Kang BH (2013) Defective chloroplast development inhibits maintenance of normal levels of abscisic acid in a mutant of the Arabidopsis RH3 DEAD-box protein during early post-germination growth. Plant J 73(5):720–732. https://doi.org/10.1111/tpj.12055

Lee K, Lee HJ, Kim DH, Jeon Y, Pai H-S, Kang H (2014) A nuclear-encoded chloroplast protein harboring a single CRM domain plays an important role in the Arabidopsis growth and stress response. BMC Plant Biol 14(1):98. https://doi.org/10.1186/1471-2229-14-98

Lee K, Park SJ, Colas des Francs-Small C, Whitby M, Small I, Kang H (2019a) The coordinated action of PPR4 and EMB2654 on each intron half mediates trans-splicing of rps12 transcripts in plant chloroplasts. Plant J 100(6):1193–1207. https://doi.org/10.1111/tpj.14509

Lee K, Park SJ, Han JH, Jeon Y, Pai H-S, Kang H (2019b) A chloroplast-targeted pentatricopeptide repeat protein PPR287 is crucial for chloroplast function and Arabidopsis development. BMC Plant Biol 19(1):244. https://doi.org/10.1186/s12870-019-1857-0

Li Z-J, Xie X-Y, Zhang S-Q, Liang Y-C (2011) Wheat growth and photosynthesis as affected by oxytetracycline as a soil contaminant. Pedosphere 21(2):244–250. https://doi.org/10.1016/S1002-0160(11)60124-0

Li C, Shen Y, Meeley R, McCarty DR, Tan BC (2015) Embryo defective 14 encodes a plastid-targeted cGTPase essential for embryogenesis in maize. Plant J 84(4):785–799. https://doi.org/10.1111/tpj.13045

Linder P, Lasko PF, Ashburner M, Leroy P, Nielsen PJ, Nishi K, Schnier J, Slonimski PP (1989) Birth of the D-E-A-D box. Nature 337(6203):121–122. https://doi.org/10.1038/337121a0

Link S, Engelmann K, Meierhoff K, Westhoff P (2012) The atypical short-chain dehydrogenases HCF173 and HCF244 are jointly involved in translational initiation of the psbA mRNA of Arabidopsis. Plant Physiol 160(4):2202–2218. https://doi.org/10.1104/pp.112.205104

Liu H, Lau E, Lam MPY, Chu H, Li S, Huang G, Guo P, Wang J, Jiang L, Chu IK, Lo C, Tao Y (2010a) OsNOA1/RIF1 is a functional homolog of AtNOA1/RIF1: implication for a highly conserved plant cGTPase essential for chloroplast function. New Phytol 187(1):83–105. https://doi.org/10.1111/j.1469-8137.2010.03264.x

Liu X, Rodermel SR, Yu F (2010b) A var2 leaf variegation suppressor locus, SUPPRESSOR OF VARIEGATION3, encodes a putative chloroplast translation elongation factor that is important for chloroplast development in the cold. BMC Plant Biol 10(1):287. https://doi.org/10.1186/1471-2229-10-287

Liu J, Zhou W, Liu G, Yang C, Sun Y, Wu W, Cao S, Wang C, Hai G, Wang Z, Bock R, Huang J, Cheng Y (2015) The conserved endoribonuclease YbeY is required for chloroplast ribosomal RNA processing in Arabidopsis. Plant Physiol 168(1):205–221. https://doi.org/10.1104/pp.114.255000

Liu X, Lan J, Huang Y, Cao P, Zhou C, Ren Y, He N, Liu S, Tian Y, Nguyen T, Jiang L, Wan J (2018) WSL5, a pentatricopeptide repeat protein, is essential for chloroplast biogenesis in rice under cold stress. J Exp Bot 69(16):3949–3961. https://doi.org/10.1093/jxb/ery214

Liu S, Zheng L, Jia J, Guo J, Zheng M, Zhao J, Shao J, Liu X, An L, Yu F, Qi Y (2019) Chloroplast translation elongation factor EF-Tu/SVR11 is involved in var2-mediated leaf variegation and leaf development in arabidopsis [original research]. Front Plant Sci 10. https://doi.org/10.3389/fpls.2019.00295

Lokdarshi A, von Arnim AG (2022) Review: emerging roles of the signaling network of the protein kinase GCN2 in the plant stress response. Plant Sci 320:111280. https://doi.org/10.1016/j.plantsci.2022.111280

Lokdarshi A, Guan J, Urquidi Camacho RA, Cho SK, Morgan PW, Leonard M, Shimono M, Day B, von Arnim AG (2020a) Light activates the translational regulatory kinase GCN2 via reactive oxygen species emanating from the chloroplast. Plant Cell 32(4):1161–1178. https://doi.org/10.1105/tpc.19.00751

Lokdarshi A, Morgan PW, Franks M, Emert Z, Emanuel C, von Arnim AG (2020b) Light-dependent activation of the GCN2 kinase under cold and salt stress is mediated by the photosynthetic status of the chloroplast. Front Plant Sci 11:431. https://doi.org/10.3389/fpls.2020.00431

Lokdarshi A, von Arnim AG, Akuoko TK (2022) Modulation of GCN2 activity under excess light stress by osmoprotectants and amino acids. Plant Signal Behav 17(1):2115747. https://doi.org/10.1080/15592324.2022.2115747

Maier RM, Neckermann K, Igloi GL, Kössel H (1995) Complete sequence of the maize chloroplast genome: gene content, hotspots of divergence and fine tuning of genetic information by transcript editing. J Mol Biol 251(5):614–628. https://doi.org/10.1006/jmbi.1995.0460

Maier UG, Zauner S, Woehle C, Bolte K, Hempel F, Allen JF, Martin WF (2013) Massively convergent evolution for ribosomal protein gene content in plastid and mitochondrial genomes. Genome Biol Evol 5(12):2318–2329. https://doi.org/10.1093/gbe/evt181

Majeran W, Friso G, Asakura Y, Qu X, Huang M, Ponnala L, Watkins KP, Barkan A, van Wijk KJ (2011) Nucleoid-enriched proteomes in developing plastids and chloroplasts from Maize leaves: a new conceptual framework for nucleoid functions. Plant Physiol 158(1):156–189. https://doi.org/10.1104/pp.111.188474

Manavski N, Schmid LM, Meurer J (2018) RNA-stabilization factors in chloroplasts of vascular plants. Essays Biochem 62(1):51–64. https://doi.org/10.1042/ebc20170061

Manuell AL, Quispe J, Mayfield SP (2007) Structure of the chloroplast ribosome: novel domains for translation regulation. PLoS Biol 5(8):e209. https://doi.org/10.1371/journal.pbio.0050209

Margulies MM (1966) Effect of chloramphenicol on formation of chloroplast structure and protein during greening of etiolated leaves of Phaseolus vulgaris. Plant Physiol 41(6):992–1003. https://doi.org/10.1104/pp.41.6.992

Marín-Navarro J, Manuell AL, Wu J, Mayfield SP (2007) Chloroplast translation regulation. Photosynth Res 94(2-3):359–374. https://doi.org/10.1007/s11120-007-9183-z

Martin W, Rujan T, Richly E, Hansen A, Cornelsen S, Lins T, Leister D, Stoebe B, Hasegawa M, Penny D (2002) Evolutionary analysis of Arabidopsis, cyanobacterial, and chloroplast genomes reveals plastid phylogeny and thousands of cyanobacterial genes in the nucleus. Proc Natl Acad Sci USA 99(19):12246–12251. https://doi.org/10.1073/pnas.182432999

McDermott JJ, Watkins KP, Williams-Carrier R, Barkan A (2019) Ribonucleoprotein capture by in vivo expression of a designer pentatricopeptide repeat protein in Arabidopsis. Plant Cell 31(8):1723–1733. https://doi.org/10.1105/tpc.19.00177

Merchante C, Stepanova AN, Alonso JM (2017) Translation regulation in plants: an interesting past, an exciting present and a promising future. Plant J 90(4):628–653. https://doi.org/10.1111/tpj.13520

Meurer J, Grevelding C, Westhoff P, Reiss B (1998) The PAC protein affects the maturation of specific chloroplast mRNAs in Arabidopsis thaliana. Mol Gen Genet 258(4):342–351. https://doi.org/10.1007/s004380050740

Meurer J, Lezhneva L, Amann K, Gödel M, Bezhani S, Sherameti I, Oelmüller R (2002) A peptide chain release factor 2 affects the stability of UGA-containing transcripts in Arabidopsis chloroplasts. Plant Cell 14(12):3255–3269. https://doi.org/10.1105/tpc.006809

Meurer J, Schmid LM, Stoppel R, Leister D, Brachmann A, Manavski N (2017) PALE CRESS binds to plastid RNAs and facilitates the biogenesis of the 50S ribosomal subunit. Plant J 92(3):400–413. https://doi.org/10.1111/tpj.13662

Miura E, Kato Y, Matsushima R, Albrecht V, Laalami S, Sakamoto W (2007) The balance between protein synthesis and degradation in chloroplasts determines leaf variegation in Arabidopsis yellow variegated mutants. Plant Cell 19(4):1313–1328. https://doi.org/10.1105/tpc.106.049270

Motohashi R, Yamazaki T, Myouga F, Ito T, Ito K, Satou M, Kobayashi M, Nagata N, Yoshida S, Nagashima A, Tanaka K, Takahashi S, Shinozaki K (2007) Chloroplast ribosome release factor 1 (AtcpRF1) is essential for chloroplast development. Plant Mol Biol 64(5):481–497. https://doi.org/10.1007/s11103-007-9166-7

Mudd EA, Sullivan S, Gisby MF, Mironov A, Kwon CS, Chung WI, Day A (2008) A 125 kDa RNase E/G-like protein is present in plastids and is essential for chloroplast development and autotrophic growth in Arabidopsis. J Exp Bot 59(10):2597–2610. https://doi.org/10.1093/jxb/ern126

Mulo P, Pursiheimo S, Hou C-X, Tyystjärvi T, Aro E-M (2003) Multiple effects of antibiotics on chloroplast and nuclear gene expression. Funct Plant Biol 30. https://doi.org/10.1071/FP03149

Nakagawa S, Niimura Y, Gojobori T (2017) Comparative genomic analysis of translation initiation mechanisms for genes lacking the Shine-Dalgarno sequence in prokaryotes. Nucleic Acids Res 45(7):3922–3931. https://doi.org/10.1093/nar/gkx124

Nawaz G, Kang H (2017) Chloroplast- or mitochondria-targeted DEAD-box RNA helicases play essential roles in organellar RNA metabolism and abiotic stress responses [review]. Front Plant Sci 8. https://doi.org/10.3389/fpls.2017.00871

Nesbit AD, Whippo C, Hangarter RP, Kehoe DM (2015) Translation initiation factor 3 families: what are their roles in regulating cyanobacterial and chloroplast gene expression? Photosynth Res 126(1):147–159. https://doi.org/10.1007/s11120-015-0074-4

Nishimura K, Ashida H, Ogawa T, Yokota A (2010) A DEAD box protein is required for formation of a hidden break in Arabidopsis chloroplast 23S rRNA. Plant J 63(5):766–777. https://doi.org/10.1111/j.1365-313X.2010.04276.x

Nishiyama Y, Allakhverdiev SI, Yamamoto H, Hayashi H, Murata N (2004) Singlet oxygen inhibits the repair of photosystem II by suppressing the translation elongation of the D1 protein in Synechocystis sp. PCC 6803. Biochemistry 43(35):11321–11330. https://doi.org/10.1021/bi036178q

Ostheimer GJ, Williams-Carrier R, Belcher S, Osborne E, Gierke J, Barkan A (2003) Group II intron splicing factors derived by diversification of an ancient RNA-binding domain. EMBO J 22(15):3919–3929. https://doi.org/10.1093/emboj/cdg372

Owttrim GW (2006) RNA helicases and abiotic stress. Nucleic Acids Res 34(11):3220–3230. https://doi.org/10.1093/nar/gkl408

Owttrim GW (2013) RNA helicases: diverse roles in prokaryotic response to abiotic stress. RNA Biol 10(1):96–110. https://doi.org/10.4161/rna.22638

Paieri F, Tadini L, Manavski N, Kleine T, Ferrari R, Morandini P, Pesaresi P, Meurer J, Leister D (2018) The DEAD-box RNA helicase RH50 Is a 23S-4.5S rRNA maturation factor that functionally overlaps with the plastid signaling factor GUN1. Plant Physiol 176(1):634–648. https://doi.org/10.1104/pp.17.01545

Park YJ, Cho HK, Jung HJ, Ahn CS, Kang H, Pai HS (2011) PRBP plays a role in plastid ribosomal RNA maturation and chloroplast biogenesis in Nicotiana benthamiana. Planta 233(6):1073–1085. https://doi.org/10.1007/s00425-011-1362-7

Pesaresi P (2011) Studying translation in Arabidopsis chloroplasts. Methods Mol Biol 774:209–224. https://doi.org/10.1007/978-1-61779-234-2_14

Pesaresi P, Varotto C, Meurer J, Jahns P, Salamini F, Leister D (2001) Knock-out of the plastid ribosomal protein L11 in Arabidopsis: effects on mRNA translation and photosynthesis. Plant J 27(3):179–189. https://doi.org/10.1046/j.1365-313x.2001.01076.x

Petracek ME, Dickey LF, Huber SC, Thompson WF (1997) Light-regulated changes in abundance and polyribosome association of ferredoxin mRNA are dependent on photosynthesis. Plant Cell 9(12):2291–2300. https://doi.org/10.1105/tpc.9.12.2291

Pfalz J, Bayraktar OA, Prikryl J, Barkan A (2009) Site-specific binding of a PPR protein defines and stabilizes 5′ and 3' mRNA termini in chloroplasts. EMBO J 28(14):2042–2052. https://doi.org/10.1038/emboj.2009.121

Pfeiffer A, Janocha D, Dong Y, Medzihradszky A, Schöne S, Daum G, Suzaki T, Forner J, Langenecker T, Rempel E, Schmid M, Wirtz M, Hell R, Lohmann JU (2016) Integration of light and metabolic signals for stem cell activation at the shoot apical meristem. elife 5. https://doi.org/10.7554/eLife.17023

Pinker F, Bonnard G, Gobert A, Gutmann B, Hammani K, Sauter C, Gegenheimer PA, Giegé P (2013) PPR proteins shed a new light on RNase P biology. RNA Biol 10(9):1457–1468. https://doi.org/10.4161/rna.25273

Prikryl J, Rojas M, Schuster G, Barkan A (2011) Mechanism of RNA stabilization and translational activation by a pentatricopeptide repeat protein. Proc Natl Acad Sci 108(1):415–420. https://doi.org/10.1073/pnas.1012076108

Prud'homme-Généreux A, Beran RK, Iost I, Ramey CS, Mackie GA, Simons RW (2004) Physical and functional interactions among RNase E, polynucleotide phosphorylase and the cold-shock protein, CsdA: evidence for a 'cold shock degradosome'. Mol Microbiol 54(5):1409–1421. https://doi.org/10.1111/j.1365-2958.2004.04360.x

Pulido P, Zagari N, Manavski N, Gawronski P, Matthes A, Scharff LB, Meurer J, Leister D (2018) CHLOROPLAST RIBOSOME ASSOCIATED supports translation under stress and interacts with the ribosomal 30S subunit. Plant Physiol 177(4):1539–1554. https://doi.org/10.1104/pp.18.00602

Reiter RS, Coomber SA, Bourett TM, Bartley GE, Scolnik PA (1994) Control of leaf and chloroplast development by the Arabidopsis gene pale cress. Plant Cell 6(9):1253–1264. https://doi.org/10.1105/tpc.6.9.1253

Reuveni S, Ehrenberg M, Paulsson J (2017) Ribosomes are optimized for autocatalytic production. Nature 547(7663):293–297. https://doi.org/10.1038/nature22998

Robaglia C, Thomas M, Meyer C (2012) Sensing nutrient and energy status by SnRK1 and TOR kinases. Curr Opin Plant Biol 15(3):301–307. https://doi.org/10.1016/j.pbi.2012.01.012
Robles P, Quesada V (2019) Transcriptional and post-transcriptional regulation of organellar gene expression (OGE) and its roles in plant salt tolerance. Int J Mol Sci 20(5):1056. https://www.mdpi.com/1422-0067/20/5/1056
Rodriguez M, Parola R, Andreola S, Pereyra C, Martínez-Noël G (2019) TOR and SnRK1 signaling pathways in plant response to abiotic stresses: do they always act according to the "yin-yang" model? Plant Sci 288:110220. https://doi.org/10.1016/j.plantsci.2019.110220
Ruppel N, Hangarter R (2007) Mutations in a plastid-localized elongation factor G alter early stages of plastid development in Arabidopsis thaliana. BMC Plant Biol 7:37. https://doi.org/10.1186/1471-2229-7-37
Sadhu L, Kumar K, Kumar S, Dass A, Pathak R, Bhardwaj A, Pandey P, Van Cuu N, Rawat BS, Reddy VS (2023) Chloroplasts evolved an additional layer of translational regulation based on non-AUG start codons for proteins with different turnover rates. Sci Rep 13(1):896. https://doi.org/10.1038/s41598-022-27347-9
Sauert M, Temmel H, Moll I (2015) Heterogeneity of the translational machinery: variations on a common theme. Biochimie 114:39–47. https://doi.org/10.1016/j.biochi.2014.12.011
Scharff LB, Childs L, Walther D, Bock R (2011) Local absence of secondary structure permits translation of mRNAs that lack ribosome-binding sites. PLoS Genet 7(6):e1002155. https://doi.org/10.1371/journal.pgen.1002155
Scharff LB, Ehrnthaler M, Janowski M, Childs LH, Hasse C, Gremmels J, Ruf S, Zoschke R, Bock R (2017) Shine-Dalgarno sequences play an essential role in the translation of plastid mRNAs in tobacco. Plant Cell 29(12):3085–3101. https://doi.org/10.1105/tpc.17.00524
Schmid L-M, Manavski N, Chi W, Meurer J (2023) Chloroplast ribosome biogenesis factors. Plant Cell Physiol 65(4):516–536. https://doi.org/10.1093/pcp/pcad082
Schmitz-Linneweber C, Williams-Carrier RE, Williams-Voelker PM, Kroeger TS, Vichas A, Barkan A (2006) A pentatricopeptide repeat protein facilitates the trans-splicing of the maize chloroplast rps12 pre-mRNA. Plant Cell 18(10):2650–2663. https://doi.org/10.1105/tpc.106.046110
Schröter Y, Steiner S, Matthäi K, Pfannschmidt T (2010) Analysis of oligomeric protein complexes in the chloroplast sub-proteome of nucleic acid-binding proteins from mustard reveals potential redox regulators of plastid gene expression. Proteomics 10(11):2191–2204. https://doi.org/10.1002/pmic.200900678
Schult K, Meierhoff K, Paradies S, Töller T, Wolff P, Westhoff P (2007) The nuclear-encoded factor HCF173 is involved in the initiation of translation of the psbA mRNA in Arabidopsis thaliana. Plant Cell 19(4):1329–1346. https://doi.org/10.1105/tpc.106.042895
Schwarz Z, Kössel H (1979) Sequencing of the 3′-terminal region of a 16S rRNA gene from Zea mays chloroplast reveals homology with E. coli 16S rRNA. Nature 279(5713):520–522. https://doi.org/10.1038/279520a0
Schwarz Z, Kössel H (1980) The primary structure of 16S rDNA from Zea mays chloroplast is homologous to E. coli 16S rRNA. Nature 283(5749):739–742. https://doi.org/10.1038/283739a0
Semrad K, Green R, Schroeder R (2004) RNA chaperone activity of large ribosomal subunit proteins from Escherichia coli. RNA 10(12):1855–1860. https://doi.org/10.1261/rna.7121704
Shajani Z, Sykes MT, Williamson JR (2011) Assembly of bacterial ribosomes. Annu Rev Biochem 80:501–526. https://doi.org/10.1146/annurev-biochem-062608-160432
Shapira M, Lers A, Heifetz PB, Irihimovitz V, Osmond CB, Gillham NW, Boynton JE (1997) Differential regulation of chloroplast gene expression in Chlamydomonas reinhardtii during photoacclimation: light stress transiently suppresses synthesis of the Rubisco LSU protein while enhancing synthesis of the PS II D1 protein. Plant Mol Biol 33(6):1001–1011. https://doi.org/10.1023/a:1005814800641
Sharma MR, Wilson DN, Datta PP, Barat C, Schluenzen F, Fucini P, Agrawal RK (2007) Cryo-EM study of the spinach chloroplast ribosome reveals the structural and functional roles of plastid-specific ribosomal proteins. Proc Natl Acad Sci USA 104(49):19315–19320. https://doi.org/10.1073/pnas.0709856104

Sharma MR, Dönhöfer A, Barat C, Marquez V, Datta PP, Fucini P, Wilson DN, Agrawal RK (2010) PSRP1 is not a ribosomal protein, but a ribosome-binding factor that is recycled by the ribosome-recycling factor (RRF) and elongation factor G (EF-G). J Biol Chem 285(6):4006–4014. https://doi.org/10.1074/jbc.M109.062299

Sharwood RE, Halpert M, Luro S, Schuster G, Stern DB (2011) Chloroplast RNase J compensates for inefficient transcription termination by removal of antisense RNA. RNA 17(12):2165–2176. https://doi.org/10.1261/rna.028043.111

Shi Z, Fujii K, Kovary KM, Genuth NR, Röst HL, Teruel MN, Barna M (2017) Heterogeneous ribosomes preferentially translate distinct subpools of mRNAs genome-wide. Mol Cell 67(1):71–83.e77. https://doi.org/10.1016/j.molcel.2017.05.021

Shine J, Dalgarno L (1974) The 3′-terminal sequence of Escherichia coli 16S ribosomal RNA: complementarity to nonsense triplets and ribosome binding sites. Proc Natl Acad Sci USA 71(4):1342–1346. https://doi.org/10.1073/pnas.71.4.1342

Shteiman-Kotler A, Schuster G (2000) RNA-binding characteristics of the chloroplast S1-like ribosomal protein CS1. Nucleic Acids Res 28(17):3310–3315. https://doi.org/10.1093/nar/28.17.3310

Siddell SG, Ellis RJ (1975) Protein synthesis in chloroplasts. Characteristics and products of protein synthesis in vitro in etioplasts and developing chloroplasts from pea leaves. Biochem J 146(3):675–685. https://doi.org/10.1042/bj1460675

Sijben-Müller G, Hallick RB, Alt J, Westhoff P, Herrmann RG (1986) Spinach plastid genes coding for initiation factor IF-1, ribosomal protein S11 and RNA polymerase alpha-subunit. Nucleic Acids Res 14(2):1029–1044. https://doi.org/10.1093/nar/14.2.1029

Singh BN, Mishra RN, Agarwal PK, Goswami M, Nair S, Sopory SK, Reddy MK (2004) A pea chloroplast translation elongation factor that is regulated by abiotic factors. Biochem Biophys Res Commun 320(2):523–530. https://doi.org/10.1016/j.bbrc.2004.05.192

Song Y, Alyafei MS, Masmoudi K, Jaleel A, Ren M (2021) Contributions of TOR signaling on photosynthesis. Int J Mol Sci 22(16). https://doi.org/10.3390/ijms22168959

Sosso D, Canut M, Gendrot G, Dedieu A, Chambrier P, Barkan A, Consonni G, Rogowsky PM (2012) PPR8522 encodes a chloroplast-targeted pentatricopeptide repeat protein necessary for maize embryogenesis and vegetative development. J Exp Bot 63(16):5843–5857. https://doi.org/10.1093/jxb/ers232

Sreedharan SP, Beck CM, Spremulli LL (1985) Euglena gracilis chloroplast elongation factor Tu. Purification and initial characterization. J Biol Chem 260(5):3126–3131

Stein DB, Palmer JD, Thompson WF (1986) Structural evolution and flip-flop recombination of chloroplast DNA in the fern genus Osmunda. Curr Genet 10(11):835–841. https://doi.org/10.1007/BF00418530

Stoppel R, Meurer J (2012) The cutting crew – ribonucleases are key players in the control of plastid gene expression. J Exp Bot 63(4):1663–1673. https://doi.org/10.1093/jxb/err401

Stoppel R, Lezhneva L, Schwenkert S, Torabi S, Felder S, Meierhoff K, Westhoff P, Meurer J (2011) Recruitment of a ribosomal release factor for light- and stress-dependent regulation of petB transcript stability in Arabidopsis chloroplasts. Plant Cell 23(7):2680–2695. https://doi.org/10.1105/tpc.111.085324

Stoppel R, Manavski N, Schein A, Schuster G, Teubner M, Schmitz-Linneweber C, Meurer J (2012) RHON1 is a novel ribonucleic acid-binding protein that supports RNase E function in the Arabidopsis chloroplast. Nucleic Acids Res 40(17):8593–8606. https://doi.org/10.1093/nar/gks613

Sudhamsu J, Lee GI, Klessig DF, Crane BR (2008) The structure of YqeH. An AtNOS1/AtNOA1 ortholog that couples GTP hydrolysis to molecular recognition. J Biol Chem 283(47):32968–32976. https://doi.org/10.1074/jbc.M804837200

Sugita M (2022) An overview of pentatricopeptide repeat (PPR) proteins in the moss physcomitrium patens and their role in organellar gene expression. Plants (Basel) 11(17). https://doi.org/10.3390/plants11172279

Sugiura M (1992) The chloroplast genome. Plant Mol Biol 19(1):149–168. https://doi.org/10.1007/bf00015612
Sugiura M (2014) Plastid mRNA translation. Methods Mol Biol 1132:73–91. https://doi.org/10.1007/978-1-62703-995-6_4
Sukhoverkov KV, Breese KJ, Debowski AW, Murcha MW, Stubbs KA, Mylne JS (2022) Inhibition of chloroplast translation as a new target for herbicides. RSC Chem Biol 3(1):37–43. https://doi.org/10.1039/d1cb00192b
Sun Y, Zerges W (2015) Translational regulation in chloroplasts for development and homeostasis. Biochim Biophys Acta (BBA) – Bioenerg 1847(9):809–820. https://doi.org/10.1016/j.bbabio.2015.05.008
Sun X, Singla-Rastogi M, Wang J, Zhao C, Wang X, Li P (2024) The uS10c-BPG2 module mediates ribosomal RNA processing in chloroplast nucleoids. Nucleic Acids Res. https://doi.org/10.1093/nar/gkae339
Swift K, Chotewutmontri P, Belcher S, Williams-Carrier R, Barkan A (2020) Functional analysis of PSRP1, the chloroplast homolog of a cyanobacterial ribosome hibernation factor. Plants (Basel) 9(2). https://doi.org/10.3390/plants9020209
Tadini L, Ferrari R, Lehniger MK, Mizzotti C, Moratti F, Resentini F, Colombo M, Costa A, Masiero S, Pesaresi P (2018) Trans-splicing of plastid rps12 transcripts, mediated by AtPPR4, is essential for embryo patterning in Arabidopsis thaliana. Planta 248(1):257–265. https://doi.org/10.1007/s00425-018-2896-8
Tan J, Tan Z, Wu F, Sheng P, Heng Y, Wang X, Ren Y, Wang J, Guo X, Zhang X, Cheng Z, Jiang L, Liu X, Wang H, Wan J (2014) A novel chloroplast-localized pentatricopeptide repeat protein involved in splicing affects chloroplast development and abiotic stress response in rice. Mol Plant 7(8):1329–1349. https://doi.org/10.1093/mp/ssu054
Tiller N, Bock R (2014) The translational apparatus of plastids and its role in plant development. Mol Plant 7(7):1105–1120. https://doi.org/10.1093/mp/ssu022
Tiller N, Weingartner M, Thiele W, Maximova E, Schöttler MA, Bock R (2012) The plastid-specific ribosomal proteins of Arabidopsis thaliana can be divided into non-essential proteins and genuine ribosomal proteins. Plant J 69(2):302–316. https://doi.org/10.1111/j.1365-313X.2011.04791.x
Timmis JN, Ayliffe MA, Huang CY, Martin W (2004) Endosymbiotic gene transfer: organelle genomes forge eukaryotic chromosomes. Nat Rev Genet 5(2):123–135. https://doi.org/10.1038/nrg1271
Toriu M, Horie M, Kumaki Y, Yoneyama T, Kore-eda S, Mitsuyama S, Yoshida K, Hisabori T, Nishiyama Y (2023) Chloroplast translation factor EF-Tu of Arabidopsis thaliana can be inactivated via oxidation of a specific cysteine residue. Biochem J 480(5):307–318. https://doi.org/10.1042/bcj20220609
Tripathy BC, Oelmüller R (2012) Reactive oxygen species generation and signaling in plants. Plant Signal Behav 7(12):1621–1633. https://doi.org/10.4161/psb.22455
Uniacke J, Zerges W (2009) Chloroplast protein targeting involves localized translation in Chlamydomonas. Proc Natl Acad Sci USA 106(5):1439–1444. https://doi.org/10.1073/pnas.0811268106
Urquidi Camacho RA, Lokdarshi A, von Arnim AG (2020) Translational gene regulation in plants: a green new deal. Wiley Interdiscip Rev RNA 11(6):e1597. https://doi.org/10.1002/wrna.1597
Vashisht AA, Tuteja N (2006) Stress responsive DEAD-box helicases: a new pathway to engineer plant stress tolerance. J Photochem Photobiol B 84(2):150–160. https://doi.org/10.1016/j.jphotobiol.2006.02.010
Verstraeten N, Fauvart M, Versées W, Michiels J (2011) The universally conserved prokaryotic GTPases. Microbiol Mol Biol Rev 75(3):507–542., second and third pages of table of contents. https://doi.org/10.1128/mmbr.00009-11
Walter M, Piepenburg K, Schöttler MA, Petersen K, Kahlau S, Tiller N, Drechsel O, Weingartner M, Kudla J, Bock R (2010) Knockout of the plastid RNase E leads to defective RNA processing and chloroplast ribosome deficiency. Plant J 64(5):851–863. https://doi.org/10.1111/j.1365-313X.2010.04377.x

Wang P, Zhao Y, Li Z, Hsu C-C, Liu X, Fu L, Hou Y-J, Du Y, Xie S, Zhang C, Gao J, Cao M, Huang X, Zhu Y, Tang K, Wang X, Tao WA, Xiong Y, Zhu J-K (2018) Reciprocal regulation of the TOR kinase and ABA receptor balances plant growth and stress response. Mol Cell 69(1):100–112.e106. https://doi.org/10.1016/j.molcel.2017.12.002
Wang X, An Y, Xu P, Xiao J (2021) Functioning of PPR proteins in organelle RNA metabolism and chloroplast biogenesis. Front Plant Sci 12:627501. https://doi.org/10.3389/fpls.2021.627501
Watkins KP, Williams-Carrier R, Chotewutmontri P, Friso G, Teubner M, Belcher S, Ruwe H, Schmitz-Linneweber C, van Wijk KJ, Barkan A (2020) Exploring the proteome associated with the mRNA encoding the D1 reaction center protein of Photosystem II in plant chloroplasts. Plant J 102(2):369–382. https://doi.org/10.1111/tpj.14629
Wen B, Liu Y, Wang P, Wu T, Zhang S, Shan X, Lu J (2012) Toxic effects of chlortetracycline on maize growth, reactive oxygen species generation and the antioxidant response. J Environ Sci (China) 24(6):1099–1105. https://doi.org/10.1016/s1001-0742(11)60901-7
Whitfeld PR, Leaver CJ, Bottomley W, Atchison B (1978) Low-molecular-weight (4.5S) ribonucleic acid in higher-plant chloroplast ribosomes. Biochem J 175(3):1103–1112. https://doi.org/10.1042/bj1751103
Williams-Carrier R, Brewster C, Belcher SE, Rojas M, Chotewutmontri P, Ljungdahl S, Barkan A (2019) The Arabidopsis pentatricopeptide repeat protein LPE1 and its maize ortholog are required for translation of the chloroplast psbJ RNA. Plant J 99(1):56–66. https://doi.org/10.1111/tpj.14308
Wobbe L, Blifernez O, Schwarz C, Mussgnug JH, Nickelsen J, Kruse O (2009) Cysteine modification of a specific repressor protein controls the translational status of nucleus-encoded LHCII mRNAs in *Chlamydomonas*. Proc Natl Acad Sci 106(32):13290–13295. https://doi.org/10.1073/pnas.0900670106
Wu W, Liu S, Ruwe H, Zhang D, Melonek J, Zhu Y, Hu X, Gusewski S, Yin P, Small ID, Howell KA, Huang J (2016) SOT1, a pentatricopeptide repeat protein with a small MutS-related domain, is required for correct processing of plastid 23S–4.5S rRNA precursors in Arabidopsis thaliana. Plant J 85(5):607–621. https://doi.org/10.1111/tpj.13126
Wu H-YL, Ai Q, Teixeira RT, Nguyen PHT, Song G, Montes C, Elmore JM, Walley JW, Hsu PY (2023) Improved super-resolution ribosome profiling reveals prevalent translation of upstream ORFs and small ORFs in Arabidopsis. Plant Cell 36(3):510–539. https://doi.org/10.1093/plcell/koad290
Wurzinger B, Nukarinen E, Nägele T, Weckwerth W, Teige M (2018) The SnRK1 kinase as central mediator of energy signaling between different organelles. Plant Physiol 176(2):1085–1094. https://doi.org/10.1104/pp.17.01404
Xie X, Zhou Q, Lin D, Guo J, Bao Y (2011) Toxic effect of tetracycline exposure on growth, antioxidative and genetic indices of wheat (Triticum aestivum L.). Environ Sci Pollut Res Int 18(4):566–575. https://doi.org/10.1007/s11356-010-0398-8
Xiong Y, McCormack M, Li L, Hall Q, Xiang C, Sheen J (2013) Glucose-TOR signalling reprograms the transcriptome and activates meristems. Nature 496(7444):181–186. https://doi.org/10.1038/nature12030
Xu T, Lee K, Gu L, Kim JI, Kang H (2013) Functional characterization of a plastid-specific ribosomal protein PSRP2 in Arabidopsis thaliana under abiotic stress conditions. Plant Physiol Biochem 73:405–411. https://doi.org/10.1016/j.plaphy.2013.10.027
Yamaguchi K, Subramanian AR (2000) The plastid ribosomal proteins. Identification of all the proteins in the 50 S subunit of an organelle ribosome (chloroplast). J Biol Chem 275(37):28466–28482. https://doi.org/10.1074/jbc.M005012200
Yamaguchi K, Subramanian AR (2003) Proteomic identification of all plastid-specific ribosomal proteins in higher plant chloroplast 30S ribosomal subunit. Eur J Biochem 270(2):190–205. https://doi.org/10.1046/j.1432-1033.2003.03359.x
Yamaguchi K, von Knoblauch K, Subramanian AR (2000) The plastid ribosomal proteins. Identification of all the proteins in the 30 S subunit of an organelle ribosome (chloroplast). J Biol Chem 275(37):28455–28465. https://doi.org/10.1074/jbc.M004350200

Yamamoto T, Burke J, Autz G, Jagendorf AT (1981) Bound ribosomes of pea chloroplast thylakoid membranes: location and release in vitro by high salt, puromycin, and RNase. Plant Physiol 67(5):940–949. https://doi.org/10.1104/pp.67.5.940
Yamane K, Yano K, Kawahara T (2006) Pattern and rate of indel evolution inferred from whole chloroplast intergenic regions in sugarcane, maize and rice. DNA Res 13(5):197–204. https://doi.org/10.1093/dnares/dsl012
Zhang X, Han Y, Han X, Zhang S, Xiong L, Chen T (2023a) Peptide chain release factor DIG8 regulates plant growth by affecting ROS-mediated sugar transportation in Arabidopsis thaliana. Front Plant Sci 14:1172275. https://doi.org/10.3389/fpls.2023.1172275
Zhang Y, Tian L, Lu C (2023b) Chloroplast gene expression: Recent advances and perspectives. Plant Commun 4(5):100611. https://doi.org/10.1016/j.xplc.2023.100611
Zheng M, Liu X, Liang S, Fu S, Qi Y, Zhao J, Shao J, An L, Yu F (2016) Chloroplast translation initiation factors regulate leaf variegation and development. Plant Physiol 172(2):1117–1130. https://doi.org/10.1104/pp.15.02040
Zhou W, Lu Q, Li Q, Wang L, Ding S, Zhang A, Wen X, Zhang L, Lu C (2017) PPR-SMR protein SOT1 has RNA endonuclease activity. Proc Natl Acad Sci 114(8):E1554–E1563. https://doi.org/10.1073/pnas.1612460114
Zhu A, Guo W, Gupta S, Fan W, Mower JP (2016) Evolutionary dynamics of the plastid inverted repeat: the effects of expansion, contraction, and loss on substitution rates. New Phytol 209(4):1747–1756. https://doi.org/10.1111/nph.13743
Ziehe D, Dünschede B, Schünemann D (2017) From bacteria to chloroplasts: evolution of the chloroplast SRP system. Biol Chem 398(5-6):653–661. https://doi.org/10.1515/hsz-2016-0292
Zoschke R, Barkan A (2015) Genome-wide analysis of thylakoid-bound ribosomes in maize reveals principles of cotranslational targeting to the thylakoid membrane. Proc Natl Acad Sci USA 112(13):E1678–E1687. https://doi.org/10.1073/pnas.1424655112
Zoschke R, Bock R (2018) Chloroplast translation: structural and functional organization, operational control, and regulation. Plant Cell 30(4):745–770. https://doi.org/10.1105/tpc.18.00016
Zoschke R, Kroeger T, Belcher S, Schöttler MA, Barkan A, Schmitz-Linneweber C (2012) The pentatricopeptide repeat-SMR protein ATP4 promotes translation of the chloroplast atpB/E mRNA. Plant J 72(4):547–558. https://doi.org/10.1111/j.1365-313X.2012.05081.x
Zybailov B, Rutschow H, Friso G, Rudella A, Emanuelsson O, Sun Q, van Wijk KJ (2008) Sorting signals, N-terminal modifications and abundance of the chloroplast proteome. PLoS One 3(4):e1994. https://doi.org/10.1371/journal.pone.0001994

Chapter 5
Chloroplast Stress Signals: Control of Retrograde Signaling, Chloroplast Turn-Over, and Cell Fate Decisions

Jesse D. Woodson, Matthew D. Lemke, David W. Tano, and Snigdha Rai

Abstract Chloroplasts (photosynthetic plastids) are semiautonomous organelles that contain their own small genomes. The proteomes of chloroplasts, however, are a mixture of plastid and nuclear-encoded proteins. Chloroplasts perform photosynthesis, which is prone to damaging the organelles, leading to the production of reactive oxygen species (ROS) that damage the cell under environmental stresses. Thus, for the cell to maintain proper chloroplast function, efficient photosynthesis, and avoid ROS damage, it relies on complex crosstalk between the chloroplast, the nucleus, other organelles within the cell, and the cytoplasm in between. This communication involves retrograde signals from chloroplasts to control nuclear gene expression, programmed cell death (PCD), and chloroplast degradation. Here we review these signals with an emphasis on the roles of the ROS singlet oxygen ($^1O_2$) and plastid gene expression. We cover (1) recent work on understanding how multiple $^1O_2$ signaling pathways can be initiated within stressed chloroplasts, (2) how individualized post-translational regulatory systems allow chloroplasts to control their proteomes and degradation, and (3) how chloroplast signals ultimately control cell fate decisions, such as PCD, senescence, and vacuole-mediated degradation of chloroplasts (chloroplast quality control). Together, this chapter discusses how chloroplasts can act as environmental sensors for the cell and allow plants to acclimate to stress and thrive in dynamic environments.

Keywords Cellular degradation · Chloroplast · Organelle gene expression · Photosynthesis · Reactive oxygen species · Senescence · Signaling · Singlet oxygen

---

Matthew D. Lemke, David W. Tano and Snigdha Rai contributed equally with all other contributors.

J. D. Woodson (✉) · M. D. Lemke · D. W. Tano · S. Rai
The School of Plant Sciences, University of Arizona, Tucson, AZ, USA
e-mail: jessewoodson@arizona.edu

T. M. Burch-Smith (ed.), *Chloroplast Gene Expression: Regulation, Stress Signaling and Biotechnology*, Nucleic Acids and Molecular Biology 37,
https://doi.org/10.1007/978-3-031-70098-9_5

5.1 Introduction

Chloroplasts are specialized plastid organelles that generate chemical energy and assimilate carbon through photosynthesis. They are also responsible for the photo-assimilation of nitrogen and the synthesis of lipids, amino acids, phytohormones, nucleotides, vitamins, and secondary metabolites. Chloroplasts are semi-autonomous and maintain their own genome which is a left over from their prokaryotic ancestors. According to the endosymbiotic theory, chloroplasts were originally free-living cyanobacteria that became integrated into eukaryotic cells and entered into a symbiotic relationship ~1.2 billion years ago (Cavalier-Smith 2000). Later, most of the prokaryotic cyanobacterial genes were lost or migrated to the host cell nucleus, shrinking the chloroplast genome to approximately 100 genes. As a result, >95% of the chloroplast proteome (~3000 proteins) is encoded in the nucleus, translated into "pre-proteins" in the cytosol, and imported into chloroplasts (Abdallah et al. 2000). However, the chloroplast genome has still retained *16S*, *23S*, and *5S* rRNA genes, a full set of tRNA genes, and genes encoding subunits of a prokaryotic-type RNA polymerase, the large subunit of RuBisCO, and core components of photosystem I (PSI), photosystem II (PSII), ATP synthase, and the cytochrome b_6f complex (Green 2011). The retainment of these genes in the chloroplast is presumed to be beneficial, as it allows for rapid on-site gene expression for photosynthesis, which is dynamic and prone to generating reactive oxygen species (ROS) (Allen 2015).

Having spatially separated genomes, however, creates an immense challenge for the cell to properly maintain the proteomes of chloroplasts. This is achieved through bi-directional communication between the chloroplast and the nucleus. Most chloroplast functions are controlled by anterograde (nucleus-to-plastid) signals, including plastid division, DNA replication, and protein import. At the same time, the chloroplasts can signal back to the nucleus to regulate gene expression using retrograde signals. More than 40 retrograde signals have been proposed and they can affect photosynthesis, stress responses, development, and flowering (Chan et al. 2016). These generally fall into two major categories. Biogenic retrograde signals are used by developing chloroplasts to co-ordinate the expression of photosynthesis-related genes in the nucleus during photomorphogenesis (reviewed in (Chan et al. 2016)). On the other hand, operational signals are used by mature chloroplasts to send signals to the cell during biotic and abiotic stresses (Pogson et al. 2008). These signals are poorly understood, but can involve ROS (Dogra and Kim 2019), the redox state of the chloroplast (Dietz 2003), and metabolites (de Souza et al. 2017). These signals regulate hundreds of nuclear genes and cell fate decisions, including programed cell death (PCD), chloroplast degradation, and senescence (Chan et al. 2016; D'Alessandro et al. 2020; Woodson 2022). Here, we will review these signals, with a special emphasis on those controlled by the ROS singlet oxygen ($^1O_2$), which has emerged as a specific chloroplast signal to control these outcomes.

Although essential for plants (and life on the planet), photosynthesis is a liability for chloroplasts, cells, and any "green tissue" (e.g., leaves, shoots). The high-energy chemistry of light capture for photosynthesis is very sensitive to stress and can rapidly lead to the production of ROS within chloroplasts. Stresses such as excess light (EL) (Triantaphylides et al. 2008), drought (Estavillo et al. 2011), extreme temperatures (Sun and Guo 2016), salinity (Suo et al. 2017), and pathogen attack (Lu and Yao 2018) can all perturb photosynthesis and lead to ROS production in the chloroplast. This ROS can then oxidize nearby macromolecules, such as DNA, proteins, and lipids. Chloroplasts have safety mechanisms in place to deal with excess ROS, including antioxidants (e.g., carotenoids and tocopherols) and ROS-scavenging enzymes (e.g., peroxiredoxins and ascorbate peroxidases), but these systems can be overwhelmed (Asada 2006).

Chloroplasts can produce ROS in a variety of ways. At PSII, this involves the over-excitation of chlorophyll molecules. These photosensitive molecules can absorb light energy and pass it to nearby ground-state molecular oxygen ($^3O_2$) to create singlet oxygen ($^1O_2$) (Triantaphylides and Havaux 2009). If the photosynthetic electron transport chain is over-reduced due to the stresses mentioned above, excited chlorophylls cannot donate electrons to photochemistry. This provides more time for these excited chlorophylls to pass the energy to $^3O_2$ (Asada 2006). Electrons are not transferred during this process, only excitation energy, reversing the spin direction of one of the unpaired electrons of $^3O_2$ (Sharma et al. 2012). Other tetrapyrrole molecules (i.e., chlorophylls and hemes and their biosynthesis and breakdown intermediates), which are mostly produced within chloroplasts (Tanaka et al. 2011), can also lead to $^1O_2$ production in the light. There are no enzymes to metabolize $^1O_2$, and plants rely on antioxidants such as carotenoids and tocopherols to absorb the excess excitation energy through physical or thermal means (Triantaphylides and Havaux 2009).

If PSI cannot move electrons to the electron acceptor ferredoxin due to over-reduction of the ferredoxin pool, then $^3O_2$ can act as an electron acceptor and form superoxide (O_2^-) (Lima-Melo et al. 2021). Superoxide dismutase (SOD) then catalyzes the disproportionation of O_2^- to produce hydrogen peroxide (H_2O_2). From here, enzymes such as peroxiredoxin and ascorbate peroxidase can convert H_2O_2 into water. However, if these enzymes cannot efficiently detoxify O_2^- and H_2O_2, then hydroxyl radicals (OH^-) can form (Sharma et al. 2012). Plant enzymes that metabolize OH^- have not been identified. Instead, plants employ flavonoid antioxidants and proline to quench this ROS (Das and Roychoudhury 2014).

5.2 Chloroplast Singlet Oxygen Signaling Pathways

For years it was assumed ROS were strictly toxic and "unwanted" metabolites. However, more recent research has revealed that these molecules can activate signaling cascades, and in plants, they can send retrograde signals from chloroplasts to reprogram nuclear gene expression, initiate PCD, and induce chloroplast degradation (Dogra and Kim 2019; Woodson 2022). Although we have a strong understanding of the production of ROS, the signaling mechanisms are poorly understood.

As mentioned above, environmental stress can lead to photosynthetically-produced ROS. However, the caveat of studying these stresses is they may produce multiple types of ROS in multiple cellular compartments (Choudhury et al. 2017; Rosenwasser et al. 2013). To circumvent this problem, researchers have studied mutants that conditionally produce specific types of ROS to understand their signaling (Ramel et al. 2013; Woodson et al. 2015; op den Camp et al. 2003). By controlling when and where ROS is made, researchers have tools to study how chloroplasts signal.

While H_2O_2 produced in chloroplasts clearly has signaling abilities (for reviews, see (Breeze and Mullineaux 2022; Van Aken and Van Breusegem 2015)), this section of the chapter will focus on the role of chloroplast $^1O_2$ for a few reasons. First, under some stresses, $^1O_2$ is expected to be the dominant cellular ROS (Triantaphylides et al. 2008) and is primarily produced within chloroplasts, unlike H_2O_2 which can be produced in multiple organelles. Second, $^1O_2$ is most strongly associated with cellular degradation in photosynthetic tissue (D'Alessandro et al. 2020). Third, the extremely short half-life of $^1O_2$ (~0.5–1.0 μs) (Ogilby 2010) means that the bulk of this ROS is likely confined within the chloroplast in which it was produced, necessitating secondary messengers to initiate cellular responses.

5.2.1 The *flu* Mutant

Much of the initial work involving chloroplast $^1O_2$ retrograde signaling used the Arabidopsis *fluorescent in blue light* (*flu*) mutant (Meskauskiene et al. 2001). Under constant light conditions, growth of the *flu* mutant is comparable to that of wild-type (wt) plants. However, in the dark, *flu* mutants accumulate the tetrapyrrole chlorophyll precursor protochlorophyllide (Pchlide) in thylakoid membranes (Przybyla et al. 2008) (Fig. 5.1). Like chlorophyll, Pchlide is a photosensitive molecule that can accept light energy and pass it to $^3O_2$ to produce $^1O_2$. This leads to the regulation of hundreds of nuclear genes followed by cell death and growth arrest (op den Camp et al. 2003). Microarray analyses showed that this response was distinct from chloroplast H_2O_2 stress caused by methyl viologen (paraquat) treatment. Later studies have referred to these genes as *SINGLET OXYGEN RESPONSE GENES* (*SORGS*) (Dogra et al. 2017). Therefore, the retrograde signaling occurring in *flu* mutants is specific, and due to $^1O_2$ production in the chloroplast.

Although chloroplast $^1O_2$ was believed to be detrimental to plant cells, the above analysis suggested that the cellular response to $^1O_2$ may be due to signaling rather than toxicity. To test this possibility, a forward genetic screen was performed to identify suppressors of $^1O_2$ signaling and cell death in the *flu* mutant. One mutation, *executor1* (*ex1*), was shown to abrogate cell death and growth arrest and to reverse a significant amount of the gene expression exhibited by the *flu* mutant (Lee et al. 2007; Wagner et al. 2004). Importantly, *flu ex1* mutants still accumulated similar levels of Pchlide and $^1O_2$ compared to *flu*, suggesting that the cell death and growth arrest are caused by a genetic signal controlled by the EX1 protein (Fig. 5.2). Later

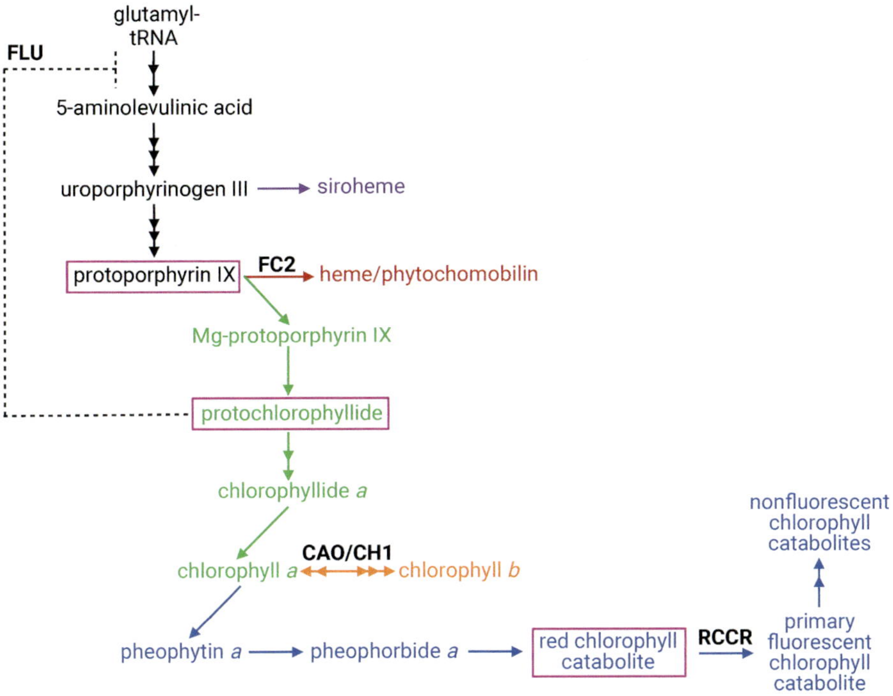

Fig. 5.1 The chloroplast tetrapyrrole biosynthesis pathway in plants. A simplified model of the tetrapyrrole biosynthesis pathway in chloroplasts, with relevant enzymes and intermediates depicted. Displayed are shared intermediates of the tetrapyrrole pathway (black) and branch points to synthesize siroheme (purple), heme/phytochromobilin (red), and chlorophyll (green). Once chlorophyll is synthesized, it can be converted reversibly from chlorophyll *a* to *b* via the chlorophyll cycle (orange). Finally, chlorophyll can be catabolized using the chlorophyll degradation pathway (blue). Marked on the figure are important enzymes affected by mutations mentioned within the main text that result in the production of $^1O_2$. Figure adapted from (Tanaka et al. 2011). FLU fluorescent in blue light, FC2 plastid ferrochelatase 2, CAO/CH1 chlorophyll *a* oxygenase/chlorina 1, RCCR red chlorophyll catabolite reductase. Created with BioRender.com

studies showed that this cell death included many hallmarks of PCD, e.g., membrane blebbing and DNA nicking (Ambastha et al. 2020). Together, this was some of the first evidence that chloroplast ROS genetically controls PCD.

EX1 resides in the thylakoid grana margins, the site of chlorophyll synthesis and PSII repair (Dogra et al. 2017; Wang et al. 2016). Furthermore, when the *flu* mutant produces $^1O_2$, the EX1 protein is oxidized at tryptophan 643 (Trp643). If this amino acid is substituted with lysine or alanine, EX1 is not oxidized and does not transduce the retrograde signal (Dogra et al. 2019). EX1 must also be subsequently cleaved by the chloroplast metalloprotease FtsH2 for signal transduction to occur. $^1O_2$-induced gene expression and PCD are reversed when an *ftsh2* mutation is introduced into the *flu* background (Dogra et al. 2017; Wang et al. 2016). In total, this work suggests EX1 acts as a sensor for $^1O_2$ and initiates retrograde signaling to control PCD and nuclear gene expression.

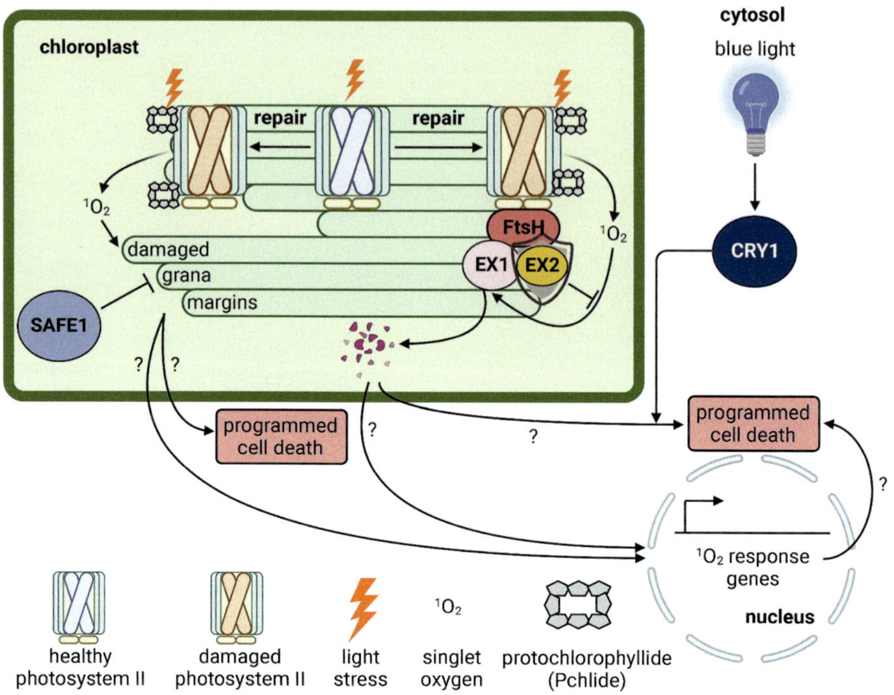

Fig. 5.2 $^1O_2$ signaling pathways activated in the *Arabidopsis thaliana flu* mutant. $^1O_2$-signaling pathways identified in the *Arabidopsis thaliana flu* mutant. In the dark, *flu* mutants accumulate the photosensitive chloroplast precursor protochlorophyllide (Pchlide). In the light, this produces singlet oxygen ($^1O_2$) in the chloroplast thylakoid membranes and oxidizes the EX1 protein, which is localized to the grana margins. Once oxidized, EX1 is degraded by the metalloprotease FtsH, which leads to a signal to control retrograde signaling and cell death. The EX1 homolog EX2 can also be oxidized by $^1O_2$ and subsequently degraded by FtsH. However, this does not lead to signaling. Instead, this may act as a buffer to reduce EX1 oxidation and signaling. In a separate pathway, the SAFE1 protein prevents $^1O_2$ damage to the grana margins of the thylakoid membranes, mitigating $^1O_2$ signals. In both cases, the source of $^1O_2$ is also hypothesized to be from damaged photosystem II (PSII) or tetrapyrrole intermediates, which are repaired and synthesized in the grana margins, respectively. Blue light activates a chloroplast $^1O_2$ signal to control cell death through a Cryptochrome (CRY1)-dependent mechanism. This pathway may be at least partially independent from EX1 signaling as it has a minimal impact on $^1O_2$ signals to control nuclear gene expression. Created with BioRender.com

An EX1 homolog, EX2, was originally thought to be redundant and play a secondary role to EX1 in the *flu* background (Lee et al. 2007). However, during a far-red block of chloroplast development, EX2 had a dominant role in $^1O_2$ signaling in young seedlings (Page et al. 2017). Recent work has suggested a more complex scenario. Like EX1, EX2 is oxidized at a conserved tryptophan residue (Trp530) and is also subsequently degraded by FtsH2 (Dogra et al. 2022). Excitingly, the oxidation and degradation of EX2 acts antagonistically to that of EX1 signaling, preventing cell death and transcriptomic changes by preventing oxidation and degradation of EX1. Thus, EX2 acts as a buffer to prevent EX1 from prematurely activating $^1O_2$ signaling and PCD (Fig. 5.2).

Additional genetic suppressors of PCD in the *flu* mutant have been identified, including the mutations *singlet oxygen-linked death activator 8* (*soldat8*) and *soldat10*. *soldat8* was mapped to *SIGMA FACTOR 6* (*SIG6*) (Coll et al. 2009), which encodes a protein responsible for targeting the Plastid-Encoded RNA Polymerase (PEP) to DNA. PEP is the major plastid polymerase that broadly regulates photosynthetic gene expression, relying on sigma factors for promoter specificity (Lysenko 2007). However, of the transcripts analyzed, the *soldat8* mutation only reduced plastid transcripts for *psbA* (encoding the D1 protein of the PSII reaction center) and *rrn16* (encoding *16S* rRNA). *soldat10* was mapped to *MITOCHONDRIAL TRANSCRIPTION TERMINATION FACTOR 10* (*mTERF10*) (Meskauskiene et al. 2009). Despite the name, mTERF proteins are targeted to mitochondria and chloroplasts in plants, where the gene families encoding them are greatly expanded, with 35 in *Arabidopsis thaliana* and 31 in *Zea mays* compared to ~4 in animals (Wobbe 2020). In the plant organelles, mTERF proteins regulate the expression of a gene (or a small number of genes) via multiple mechanisms, which include RNA splicing, transcription termination, and transcription initiation (discussed in Chap. 2). The *soldat10* mutation led to a decrease in plastid rRNA transcripts (*16S* and *23S*), which was likely responsible for the lowered plastid protein levels observed. In both *soldat8* and *soldat10*, these defects in plastid expression were correlated to increased stress acclimation. *soldat8* mutants were tolerant to increased light intensities (Coll et al. 2009), and *soldat10* mutants exhibited a constitutive stress response (Meskauskiene et al. 2009). These studies hypothesized that by altering plastid homeostasis through changes in plastid gene expression, these mutations were able to antagonize $^1O_2$-induced PCD. Alternatively, a plastid-encoded factor that is reduced in *soldat8* and *soldat10* mutants may be involved in $^1O_2$ signaling.

Interestingly, blue light was necessary to trigger PCD in *flu* protoplasts (Danon et al. 2005). This was supported by the requirement of CRYPTOCHROME 1 (CRY1) for this signaling: the *cry1* mutation suppressed the *flu* cell death phenotype (Danon et al. 2005) (Fig. 5.2). Cryptochromes are photoreceptor proteins responsible for sensing blue light in the plant's environment, helping to entrain the circadian clock and maintain appropriate developmental phenotypes (hypocotyl elongation, flowering time, etc.) (Yu et al. 2010). More recently, we confirmed the role of CRY1 in controlling PCD in *flu* mutant cotyledons (Tano et al. 2022). To understand the impact of CRY1 on the $^1O_2$ retrograde signal, a microarray analysis comparing the expression between *flu* and *flu cry1* mutants was performed (Danon et al. 2005). Of the 588 genes induced ≥three-fold in the *flu* mutant compared to wild type, only 34 were reversed in expression after the introduction of the *cry1* mutation. Because the *cry1* mutation had a minimal effect on the transcriptome, these results provide evidence that the retrograde signals for PCD and control of nuclear gene expression can be at least partially uncoupled and may represent separate $^1O_2$ signaling pathways (see below).

To find other $^1O_2$ signaling pathways, a forward genetic screen was used to identify mutations that restored PCD in the *flu ex1* double mutant. One mutation was mapped to *SAFEGUARD1* (*SAFE1*). This gene encodes a protein that protects chloroplast grana margins from $^1O_2$, and without it, thylakoids are damaged, and plastoglobules become enlarged in the *flu* background (Wang et al. 2020) (Fig. 5.2). As the *safe1* mutation also increased $^1O_2$ sensitivity and signaling in the *flu* background, it was hypothesized that SAFE1 represented a pathway independent of EX1 and EX2. How these pathways may function separately, or if they converge downstream to control similar nuclear genes, is still not clear.

5.2.2 The ch1 Mutant

A second Arabidopsis mutant that can conditionally produce $^1O_2$ is *chlorina1* (*ch1*) (Ramel et al. 2013). The *ch1* mutant cannot convert chlorophyll *a* into chlorophyll *b* due to the absence of the CHLOROPHYLL *A* OXYGENASE enzyme (CAO) (Espineda et al. 1999) (Fig. 5.1). Because of this, the protective PSII antenna complex cannot form as chlorophyll *b* holds this structure together. Therefore, the reaction center is left unprotected, allowing excited chlorophyll molecules to easily pass excess energy to $^3O_2$ to produce $^1O_2$ (Ramel et al. 2013). As such, the *ch1* produces excess $^1O_2$ when exposed to EL, leading to lipid peroxidation, inhibition of PSII, PCD, and retrograde signaling to control nuclear gene expression (Ramel et al. 2013).

As in *flu* mutants, these physiological changes are due to genetic signals in response to $^1O_2$, but the mechanisms appear to be complex. At least part of this signal is due to OXIDATIVE INDUCIBLE KINASE 1 (OXI1), a nuclear-localized Ser/Thr kinase that was previously shown to respond to H_2O_2 during basal defense against pathogens (Rentel et al. 2004). The *ch1 oxi1* double mutant tolerates EL conditions, with less lipid peroxidation and PSII inhibition (Shumbe et al. 2016). The role of OXI1 is not specific to *ch1*, as single *oxi1* mutants are also more tolerant to EL than wt plants. The phytohormone jasmonic acid (JA) also plays a role in EL-stressed *ch1* plants. JA levels are increased in *ch1* plants after EL stress and blocking this accumulation with the *dde2* mutation attenuates lipid peroxidation and PCD (Ramel et al. 2013).

EL stress also induces an acclimation signal in *ch1* plants. If *ch1* plants are exposed to mild light stress that does not induce PCD, they become more tolerant to subsequent EL stress and produce less JA. This acclimation signal appears to be controlled by volatile oxidation products of the carotenoid β-carotene, which is associated with PSII and in direct contact with $^1O_2$ produced under EL conditions. One of these volatile derivatives, β-cyclocitral (β-cc), was shown to produce a very similar transcript profile in wt compared to the $^1O_2$-stressed *flu* mutant (>80% and >90% overlap for up and down-regulated genes, respectively) (Ramel et al. 2012). Furthermore, if plants are first treated with β-cc, they avoid EL-induced photooxidative damage (Ramel et al. 2012). Ultimately, β-cc activates the SCARECROW Like14 (SCL14)-dependent xenobiotic detoxification pathway, removing EL-induced reactive carbonyl molecules before they can cause severe damage (D'Alessandro et al. 2018). How β-cc signals is not clear, but it partly involves the transcription factor METHYLENE BLUE SENSITIVITY 1 (MBS1): *mbs1* mutants are unable to acclimate to EL stress (Shumbe et al. 2017).

In vivo, β-cc can be oxidized to the water-soluble metabolite β-cyclocitric acid (β-cca), which can induce drought tolerance (D'Alessandro et al. 2019). Although the mechanism by with β-cca achieves this is unknown, it is independent of stomatal regulation. Amazingly, β-cca can be exogenously applied through irrigation and can induce drought tolerance in a wide range of species, including pepper (*capsicum*), pansy flower (*Viola tricolor*), and tomato (*Solanum lycopersicum*). Overall, these studies show how chloroplast ROS has the potential to prime plants for different types of abiotic stress.

5.2.3 The acd2 Mutant

Another Arabidopsis $^1O_2$-producing mutant is *accelerated cell death 2* (*acd2*), which spontaneously produces $^1O_2$ and lesions under regular growing conditions (Mach et al. 2001). The *acd2* mutant lacks a functional red chlorophyll catabolite reductase (RCCR) involved in chlorophyll catabolism (Tanaka et al. 2011) (Fig. 5.1). As such, *acd2* mutants accumulate chlorophyll breakdown products such as red chlorophyll catabolite (RCC) (Pruzinská et al. 2007). Like other tetrapyrroles mentioned herein, these molecules can absorb light energy and generate $^1O_2$. Interestingly, *acd2* mutants accumulate at least some of this $^1O_2$ in mitochondria, and mitochondria-targeted ACD2 can complement the *acd2* lesion phenotype (Pattanayak et al. 2012). This work suggests a model where RCC may travel from chloroplasts to mitochondria to produce $^1O_2$ that also has signaling abilities.

5.2.4 The fc2 Mutant

A fourth Arabidopsis $^1O_2$ over-producing mutant is *plastid ferrochelatase 2* (*fc2*), which conditionally accumulates the tetrapyrrole intermediate protoporphyrin-IX (Proto). Proto is located at the branching point of tetrapyrrole synthesis and can be used to produce chlorophyll or heme (Fig. 5.1). In *fc2* mutants, Proto accumulation occurs at dawn under diurnal light cycling conditions, the time when the tetrapyrrole pathway is most activated (Woodson et al. 2015; Papenbrock et al. 2001). The lack of FC2 creates a bottleneck in the tetrapyrrole pathway, and Proto (the substrate of FC2) rapidly accumulates within 10 minutes. Like Pchlide, Proto is extremely photosensitive and this leads to a concurrent production of $^1O_2$. Within 30 min after dawn, the increase in $^1O_2$ leads to a retrograde signal to change the expression of hundreds of nuclear genes (Woodson et al. 2015). After three hours, chloroplasts are degraded. After eight hours, cell death occurs, and photosynthetic tissues begin to bleach and die. Under constant light conditions, *fc2* seedlings only accumulate small levels of Proto and $^1O_2$, which does not activate cell death, but does lead to selective chloroplast degradation (discussed below) (Fisher et al. 2022). Thus, the *fc2* mutant is an ideal system to conditionally investigate chloroplast $^1O_2$ signals that control chloroplast degradation and cell death (Fig. 5.3).

To identify chloroplast $^1O_2$ signaling factors, we performed a forward genetic screen to identify second-site suppressor mutations that blocked cell death in *fc2* mutants. We identified 24 such *plastid ferrochelatase 2 suppressor* (*fts*) mutations (Woodson et al. 2015). Several of these mutations affect plastid gene expression, suggesting that the expression of a plastid gene (or set of genes) is important for $^1O_2$ signaling and that a product of the plastid genome may be a signaling component (Fig. 5.3). We mapped one mutation, *fts32*, to *mTERF9* (Alamdari et al. 2020), which encodes a chloroplast-localized mTERF protein that associates with the 30S ribosome and is important for chloroplast ribosome assembly (Méteignier et al. 2021). We mapped *fts3* and *fts38* to *PENTATRICOPEPTIDE REPEAT CONTAINING PROTEIN 30* (*PPR30*) (Alamdari et al. 2020). PPR proteins comprise a large group of proteins in plants (between 400–1000 members) (see Chap. 1). Targeted to the chloroplast and mitochondria, PPR proteins bind an RNA transcript (or small group of transcripts) and affect their expression by influencing editing, stability, processing, and translation (Barkan and Small 2014). As such, PPR proteins affect photosynthesis, cellular respiration, and plant development.

We demonstrated that the *mterf9* and *ppr30* mutations reduced the expression of PEP-targeted genes within the chloroplast and led to delayed chloroplast development. However, this block in $^1O_2$ signaling appeared to be specific as other mutations known to affect plastid gene expression did not lead to *fts* phenotypes and block PCD in the *fc2* mutant. Together, these results indicate that the RNA targets of PPR30 (and possibly mTERF9) may encode a $^1O_2$ signaling factor (Fig. 5.3). It is not known what gene(s) PPR30 may target, but null *ppr30* alleles are embryo lethal suggesting that they are needed to express an essential gene.

We also mapped *fts39* to *CYTIDINE TRIPHOSPHATE SYNTHASE 2* (*CTPS2*) (Alamdari et al. 2021), which encodes one of five enzymes (CTPS1–5) responsible for the de novo synthesis of cytoplasmic CTP and dCTP for RNA and DNA synthesis, respectively (Fig. 5.3). CTP and dCTP are also cofactors for CDP- and dCDP-choline synthesis, respectively, which are intermediates of phospholipid biosynthesis (Hong et al. 2018). Like *ppr30* and *mterf9* mutations, we observed that *ctps2* led to a broad reduction in plastid gene expression and chloroplast development, but this was accompanied by a decrease in plastid DNA content. When we fed deoxycytidine (a salvageable precursor for dCTP, but not CTP (Witte and Herde 2020)) to *fc2 ctps2* seedlings, plastid DNA levels, plastid transcript levels, and PCD were restored (Alamdari et al. 2021). Together, this suggested that *fc2 ctps2* mutants were deficient in dCTP for plastid DNA synthesis, which was responsible their lowered plastid gene expression and block in the $^1O_2$ signal. Surprisingly, the *ctps2* mutation did not affect mitochondrial DNA content or RNA transcripts, suggesting its effect on chloroplasts was specific and that chloroplast signaling can potentially be controlled by metabolite allocation in the cytoplasm.

Interestingly, CTPS2 belongs to a monophyletic clade containing CTPS1, CTPS4, and CTPS5. When we over-produced any of these CTPS enzymes in the *fc2*

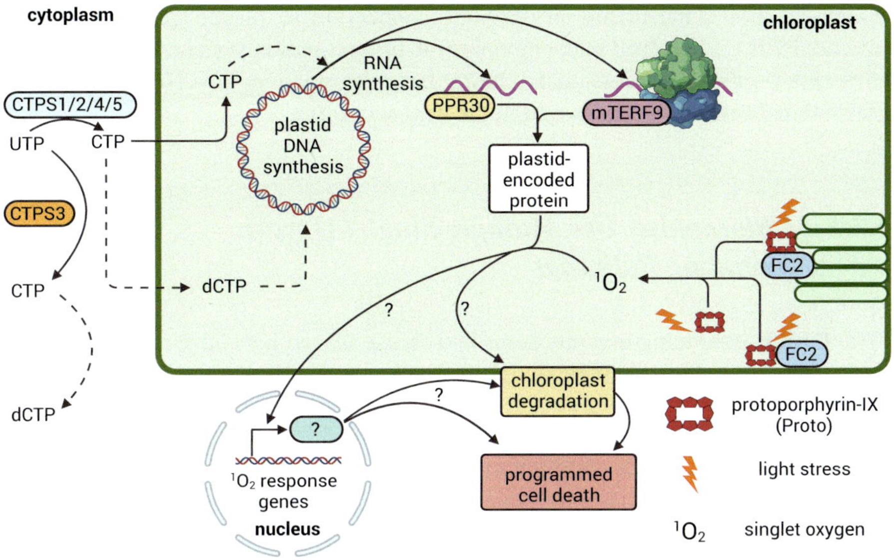

Fig. 5.3 $^1O_2$ signaling pathways activated in the *plastid ferrochelatase 2* mutant. Shown is a model depicting singlet oxygen ($^1O_2$) signaling in the *plastid ferrochelatase 2* (*fc2*) mutant to control nuclear gene expression, chloroplast degradation, and programmed cell death (PCD). Under diurnal light cycling conditions, *fc2* mutants accumulate protoporphyrin-IX (Proto), the photosensitive substrate of the FC2 enzyme. In the light, Proto leads to the generation of $^1O_2$, which can induce signals to control nuclear gene expression, chloroplast degradation, and PCD. $^1O_2$ signaling is dependent on chloroplast-localized Pentatricopeptide Repeat-Containing Protein 30 (PPR30) and "mitochondrial" Transcription Termination Factor 9 (mTERF9). These proteins are predicted to regulate plastid gene expression: PPR30 may bind an RNA target and affect its stability, processing, editing, or translation and mTERF9 likely affects translation through associations with the 30S ribosome subunit. Thus, a gene product(s) regulated by PPR30 and/or mTERF9 is hypothesized to be part of the $^1O_2$ signaling pathway. Plastid gene expression is also dependent on sufficient plastid DNA content, which can be limited by dCTP levels. De novo dCTP is synthesized from UTP, in part, by cytoplasmic cytidine triphosphate synthase (CTP) enzymes. CTPS1/2/4/5 is a conserved clade of these enzymes with a role in providing this metabolite to chloroplasts. Created with BioRender.com

ctps2 double mutant, they restored the *fc2* PCD phenotype, $^1O_2$ stress marker gene expression, and plastid DNA and transcript levels (Alamdari et al. 2021). Conserved CTPS3, however, did not restore these phenotypes despite being a bona fide CTPS enzyme *in vitro* (Daumann et al. 2018). Together, these results suggest CTPS1/2/4/5 and CTPS3 have different roles within the cell, with the former having the capability to produce CTP (and thus dCTP) for plastid DNA synthesis and, thus, chloroplast signaling. Due to its effect on plastid gene expression, it is expected that the *ctps2* mutation blocks the same signal as *ppr30*/*mterf9* mutations (Fig. 5.3).

We demonstrated that another mutation, *fts29,* blocked chloroplast degradation without affecting Proto or $^1O_2$ levels, indicating that a signal was affected. We mapped *fts29* to *PLANT U-BOX 4* (*PUB4*) (Woodson et al. 2015), which encodes a U-Box containing E3 ubiquitin (Ub)-protein ligase. Unlike the other FTS proteins discussed above, PUB4 has no obvious connection to plastid gene expression. Instead, PUB4 is involved in the cytoplasmic ubiquitination system, suggesting that chloroplast $^1O_2$ signaling and chloroplast degradation involves such post-translational regulation (discussed in the next section).

5.2.5 Chloroplasts Use Multiple Singlet Oxygen Signaling Pathways

With the $^1O_2$-producing mutants described above, researchers have multiple genetic tools to analyze chloroplast ROS signaling. However, it is unclear if these mutants initiate the same $^1O_2$ pathways or if separate pathways exist. To test this, we compared publicly available transcriptomic data for *flu*, *fc2*, and *ch1* mutants and found varying degrees of overlap between these mutants, despite the data originating from distinct growing conditions and tissue types (seedlings vs. adult leaves) (Tano et al. 2022). Each mutant shared a core transcriptomic response, suggesting a specific set of genes is induced in response to chloroplast $^1O_2$ stress. For upregulated genes, gene-ontology terms of transcription and general stress were enriched. For down-regulated genes, gene-ontology terms of photosynthesis and tetrapyrrole synthesis were enriched. Together, this suggested that chloroplast $^1O_2$ generally signals for a whole cell stress response while also reducing chloroplast function to mitigate further $^1O_2$ production.

If these mutants elicit the same $^1O_2$ signaling pathways, then a genetic suppressor in one background should prevent $^1O_2$ signaling in a second background (Tano et al. 2022). To this end, we generated several mutant combinations and measured their response to $^1O_2$ stress. We observed that the *pub4* mutation, which was first found to suppress PCD in *fc2*, also blocked $^1O_2$-induced PCD in *ch1* and *acd2* mutants. Moreover, the *oxi1* mutation suppressed PCD in the *fc2* mutant, but only in the adult life stage. This suggested that *fc2* and *ch1* (and possibly *acd2*) share a genetic mechanism for $^1O_2$ signaling. The *flu* suppressor mutations (*ex1* and *cry1*) did not suppress PCD or retrograde signaling in the *fc2* background, while *pub4* and *oxi1* did not suppress PCD in the *flu* background. Interestingly, we observed that an *ex1 ex2* combination blocked $^1O_2$ signaling in *fc2*. This appeared to be indirect as *ex1 ex2* also reduced $^1O_2$ levels, likely by a decrease in tetrapyrrole synthesis.

Overall, we concluded that there exist at least two $^1O_2$ signaling pathways: one shared by *fc2* and *ch1* (and possibly *acd2*) and one exclusive to *flu*. This work agrees with previous reports showing that *ex1* is not involved in $^1O_2$ signaling in *ch1* (Ramel et al. 2013) or *acd2* (Pattanayak et al. 2012). Together, this may indicate that the site of $^1O_2$ production or accumulation determines the type of signal generated. In *flu* mutants, $^1O_2$ likely originates in the grana margins, the site of EX1/2 localization. In *ch1* mutants, $^1O_2$ originates from PSII in the grana core. It is not known where Proto or $^1O_2$ is produced in *fc2* mutants. However, in pea (*Pisum sativum*) and beet (*Beta vulgaris*), Proto accumulates in the chloroplast stroma and envelope, respectively

(Mohapatra and Tripathy 2002, 2007). In any case, having multiple $^1O_2$ pathways may allow plants to fine-tune their physiology to match the environment. Future work will likely explore how these separate stresses are initiated and propagated.

5.3 A Third Class of Chloroplast Retrograde Signal: Post-translational Regulation

The classic retrograde signals described above are useful when most of the chloroplasts within a cell experience similar levels of stress and collectively signal to the nucleus to induce a stress response or regulate development. However, what if only a few chloroplasts in a cell are stressed? Such a situation may occur if a single chloroplast accumulates excess ROS due to differences in age, damage, position within the cell/tissue (e.g., full sunlight vs. shade), or in the abundance of antioxidant reserves. In such cases, how can the nucleus or cell distinguish between healthy and stressed chloroplasts? The retrograde signals described above would not suffice as there is no way for the nucleus to 'know' which one of the cell's chloroplasts has released these signals. This could not be ignored as a damaged chloroplast would be a ROS generator and waste resources. This gives rise to a third type of retrograde signaling that involves post-translational mechanisms to regulate chloroplast proteomes or induce chloroplast degradation. Such chloroplast quality control systems would ensure a healthy population of chloroplasts within cells (about 120 chloroplasts in Arabidopsis mesophyll cells (Pyke and Leech 1994)). Chloroplasts contain many essential nutrients and house >80% of the leaf's nitrogen content. Thus, chloroplast turnover also allows for important nutrient redistribution (Makino and Osmond 1991).

5.3.1 *The Eukaryotic Ubiquitination System*

Although chloroplasts are prokaryotic in origin, they have evolved to use the eukaryotic ubiquitination system to help control chloroplast quality control, the regulation of their proteomes, and plastid transitions. Ubiquitination occurs in eukaryotic cells (yeast, animals, and plants) and is one of the most abundant types of post-translational modifications, involved in cellular processes such as signaling, circadian rhythms, self-incompatibility, photomorphogenesis, and defense against biotic/abiotic stresses (Callis 2014). This system uses ubiquitin (Ub), a conserved 8.5 kDa (76-amino-acid) protein that attaches to target proteins through its carboxy-terminal glycine residue (Sullivan et al. 2003). Ub is covalently attached to a substrate protein's lysine residues via the step-wise activity of E1 (Ub-activating), E2 (Ub-conjugating) and E3 (Ub-protein ligase) enzymes (Fig. 5.4).

E1 (or Ub-activating enzyme) couples ATP hydrolysis to activate Ub (Fig. 5.4), which is then transferred to a cysteine residue on an E2 Ub-conjugating enzyme via transesterification. The E2-Ub complex then transfers the Ub onto a lysine residue of a target (substrate) protein, catalyzed by E3 Ub-protein ligase (Smalle and Vierstra 2004). The Arabidopsis genome encodes two E1 proteins (AtUBA1 and AtUBA2), thirty-seven E2 proteins, and over 1300 E3 proteins. Ub itself contains seven lysine residues (K6, K11, K27, K29, K31, K48, and K63), all of which can be ubiquitinated. Depending on the E2/E3 complex and the Ub lysine being used, a target protein can be mono- or poly-ubiquitinated, and along with the length and topology of the Ub chain, this determines the fate of the protein. The most common fate is degradation via the 26S proteasome, but ubiquitination can also affect protein localization, function, and interactions (Callis 2014).

The E3 ligases are decisive in selecting target proteins and hence provide substrate specificity for ubiquitination (Callis 2014). Based on their structure and function, E3 ligases can be classified into four types: HECT (Homology to E6-AP C Terminus), U-box, RING (Really Interesting New Gene), and multi-subunit Cullin-RING box 1-Ligases (CRLs). CRL types are comprised of four subfamilies: APC (anaphase-promoting complex), BTB (Bric-a-brac–Tramtrack–Broad complex), DDB (DNA Damage-Binding domain-containing), and SCF (S phase kinase-associated protein 1–Cullin 1–F-box). The ability of E3's from each group to interact with specific protein targets is due, in part, to the mix-and-match assembly of multiple peptide modules within E3 proteins. These modules may contain various protein-protein interaction motifs, as in the case of the BTB E3 family (Vierstra

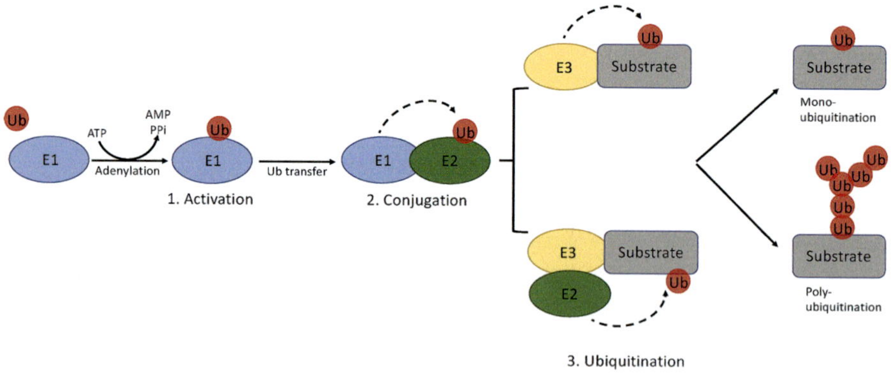

Fig. 5.4 The enzyme cascade for protein ubiquitination. Ubiquitination is achieved by a cascade of three sequential steps (1) Activation: Ubiquitin (Ub) is transferred to an E1 (Ub-activating) protein in an ATP-dependent manner. (2) Conjugation: Activated Ub is transferred to an E2 (Ub-conjugating) protein. (3) Ubiquitination: The Ub is covalently attached to the target protein by an E3 (Ub-protein ligase) protein, leading to mono-ubiquitination or poly-ubiquitination. E3s interact with both E2 ~ Ub and the substrate to which Ub is to be transferred, thus providing specificity to the system. Some E3 ligases transfer Ub directly to the substrate (HECT type E3s) while others mediate the transfer of Ub from E2 to the substrate (RING type E3s). Created with BioRender.com

2009). Some E3 ligases are discussed below to elucidate the scope and scale of these proteins in affecting post-translational chloroplast signaling.

5.3.2 Singlet Oxygen-Induced Chloroplast Degradation

As discussed above, we identified the E3 Ub-protein ligase PUB4 as being involved in $^1O_2$ signaling in *fc2* mutants (Woodson et al. 2015). The PUB (plant U-box) proteins are a small subfamily of E3 ligases that have a highly-conserved U-box domain, which interacts with E2 proteins, and is involved in ubiquitination. The U-box is a modified RING-finger domain, where the conserved ring binding sites (cysteine and histidine residues) are replaced by hydrogen bonds (Aravind and Koonin 2000). The *pub4* allele found in our genetic screen (referred to as *pub4–6*) leads to a single amino acid substitution (Gly255 to Arg) within the U-box, likely affecting U-box function and the ability to interact with the E2 enzyme. PUB4 also has five ARMADILLO (ARM) repeat domains at the C-terminus (Wang et al. 2013a), which are predicted to facilitate protein-protein interactions (Samuel et al. 2006).

The identification of PUB4 prompted us to investigate if protein ubiquitination may play a role during $^1O_2$ signaling. Indeed, when we grew *fc2* mutants in $^1O_2$-producing light cycling conditions, their chloroplast envelopes became ubiquitinated, as demonstrated by anti-Ub immunoblotting and immunogold labeling (Woodson et al. 2015). This ubiquitination was reduced by the *pub4–6* mutation and chloroplasts were not degraded in *fc2 pub4–6* mutants. However, these chloroplasts appeared abnormal, with irregular membranes and shapes, suggesting they were damaged, but not being degraded. A similar pattern was observed in the variegated Arabidopsis *gun1 ftsH5* mutant that contains bleached sectors in its leaves (Jeran et al. 2021). Like *fc2*, this mutant also had damaged chloroplasts associated with Ub. Introgression of a *pub4* mutation into the *gun1 ftsh5* background suppressed the ubiquitination of damaged chloroplasts and variegated phenotype. As in *fc2 pub4*, these chloroplasts still appeared stressed and abnormal, indicating that PUB4-dependent degradation of damaged chloroplasts is active in *gun1 ftsh5* cotyledons and is the basis of their variegated phenotype. Together, these results suggest the presence of a chloroplast quality control pathway involving Ub, where damaged chloroplasts can be selectively removed from photosynthetic cells for stress acclimation and longevity. In line with this, we observed that *pub4–6* mutants senesce early, suggesting that chloroplasts (or cells) in this mutant are compromised in the long-term (Woodson et al. 2015; Tano et al. 2022).

Evidence for such Ub-mediated selective organelle degradation can be inferred from animal systems. Some neurological diseases, such as Parkinson's, can be caused by a cell's inability to selectively degrade damaged mitochondria, a process called mitophagy. This involves two key proteins: a Ser/Thr kinase (PINK) that is normally imported into mitochondria and a cytosolic RING-type E3 ligase (Parkin) (Youle and Narendra 2011). When mitochondria become damaged and depolarized, the import of PINK1 into mitochondria is blocked, leading to its accumulation on the outer mitochondrial membrane. There, PINK1 then recruits Parkin to damaged mitochondria by phosphorylation on Ser65 (Kondapalli et al. 2012). Phosphorylated Parkin ubiquitinates several outer mitochondrial proteins, which then "mark" damaged mitochondria for engulfment by autophagosomes and degradation in lysosomes (Lazarou et al. 2015). Homologs of these proteins have not been identified in plants, but the possibility that plant cells use a similar post-translational mechanism to identify and degrade damaged chloroplasts is intriguing in light of the observation that $^1O_2$-damaged chloroplasts are ubiquitinated prior to turn-over.

Such in vivo ubiquitination targets of PUB4 have not been identified (although PUB4 can ubiquitinate non-activated BIK1 in vitro (Yu et al. 2022), yet PUB4 has been shown to play diverse roles in plants. It is reported to function in asymmetric cell division and proliferation in the root and shoot meristems (Kinoshita et al. 2015), cytokinin signaling (Wang et al. 2017), chitin signaling (Desaki et al. 2019), pattern-triggered immunity defense responses (Wang et al. 2022), and degeneration of tapetal cells during pollen development (Wang et al. 2013a). This suggests that PUB4 might be targeting and ubiquitinating multiple proteins in a cell depending on the cell type and/or the environmental conditions. Alternatively, PUB4-dependent chloroplast degradation may have a broader impact than originally assumed.

5.3.3 *The CHIP E3 Ligase*

Another U-box-containing E3 ligase, the Carboxyl terminus of Hsp70-interacting protein (CHIP), has an important role in plant stress responses and is involved in the turnover of chloroplast proteins that accumulate in the cytoplasm (Zhang et al. 2021). CHIP contains a U-box domain at its C-terminus and tetratricopeptide repeats (TPR) for chaperone binding at its N-terminus (Zhang et al. 2005, 2015). The TPR domain allows CHIP to use chaperone interactors (Hsp70/Hsp90) to recognize misfolded proteins that accumulate in the cytoplasm. These misfolded proteins are then ubiquitinated by CHIP and degraded via the 26S proteasome and include light-harvesting complex proteins (LHCs) and the RuBisCO small subunit (RbcS) (Zhang et al. 2021). In addition, CHIP can regulate the accumulation of chloroplast proteases (FtsH and ClpP), thereby indirectly controlling the chloroplast proteome (Shen et al. 2007a, b). Although a clear connection exists between photosynthesis and CHIP, it is unknown how much of CHIP function is controlled by chloroplast signals.

5.3.4 Chloroplast Protein Import and the Chloroplast-Associated Protein Degradation System

Ubiquitination has also been demonstrated to control chloroplast proteomes by modulating how cytoplasmic proteins are imported into chloroplasts, which provides another post-translational mechanism by which chloroplasts can respond to photo-oxidative stress. Nuclear-encoded proteins are first synthesized as precursors in the cytoplasm and contain N-terminal targeting sequences called transit peptides (cTP). Import of most precursor proteins into the chloroplast is facilitated by the TOC (translocon on the outer chloroplast membrane) and TIC (translocon on the inner chloroplast membrane) translocation complexes within the outer and inner chloroplast envelope membranes, respectively (Fig. 5.5). As such, these complexes mediate the import of polypeptides to the chloroplast envelope membranes and thylakoid membranes, and stroma (reviewed in (Paila et al. 2015; Jarvis and Lopez-Juez 2013)). The TOC complex consists of the Toc159 and Toc33 subunits, which identify and bind to the transit peptides (cTP) of the precursor proteins, and the Toc75 subunit, which forms the protein-conducting channel (Jarvis and Lopez-Juez 2013). The components of the TIC complex are still under debate, but Tic20, Tic21, and/or Tic110 are likely to be the major pore-forming subunits in the inner envelope membrane (Paila et al. 2015; Ganesan et al. 2018). Stromal chaperones such as Hsp93, cpHsp70, and Hsp90C act as motors, directing the preprotein through the translocon complex. Soon after the chloroplast precursor proteins are guided into the chloroplast through the TOC/TIC channel, the transit peptide is cleaved by the stromal processing peptidase (SPP). Next, the protein is folded, assembled into functional complexes, or guided to the inner envelope or thylakoid membranes.

Chloroplast-associated protein degradation (CHLORAD) is a recently described regulatory system involving ubiquitination and 26S proteasomal degradation to selectively remove and degrade chloroplast outer envelope membrane (OEM) proteins, including TOC components. CHLORAD components were first identified in a forward genetic screen in Arabidopsis for suppressors of the pale-yellow *toc33* loss of function mutant, *ppi1*, that displays delayed chloroplast development (Ling et al. 2012). *SP1* (*SUPPRESSOR OF PPI1 LOCUS 1*) encodes a chloroplast outer envelope membrane (OEM)-localized RING-type E3 Ub-protein ligase that directs the ubiquitination of Toc159, Toc33, and Toc75, which are subsequently removed by the 26S proteasome. SP1 also has a critical role in plant responses to abiotic stress (Ling and Jarvis 2015). The *sp1* mutant is hypersensitive to salt, osmotic, and oxidative stresses caused by H_2O_2 accumulation (due to methyl viologen treatments). In contrast, plants overexpressing *SP1* were more tolerant to these stresses than wild type plants. Under these conditions, SP1 ubiquitinates and depletes the Toc apparatus, to limit the import of proteins forming the photosynthetic apparatus. This attenuates photosynthetic activity and reduces the potential for ROS production and photo-oxidative damage.

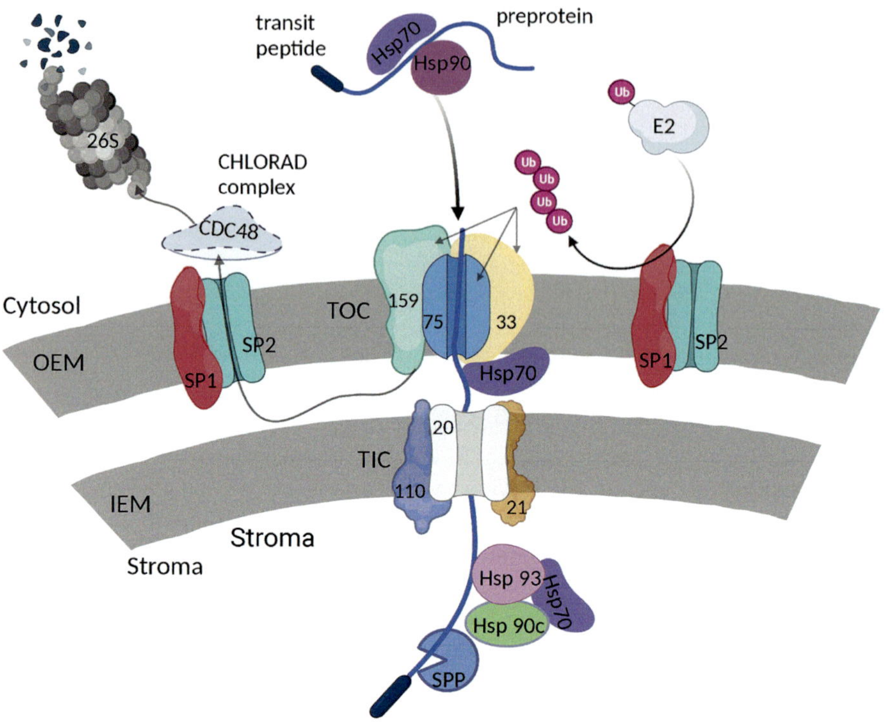

Fig. 5.5 Model for the control of protein import into chloroplasts. A schematic illustration of preprotein transport across the chloroplast envelope membranes; and chloroplast-associated protein degradation (CHLORAD)-mediated proteolytic regulation of the outer membrane translocases. The outer envelope membrane (OEM) has the Translocon of the Outer membrane of the Chloroplast (TOC) complex formed by Toc159, Toc34 and Toc75, and the inner envelope membrane (IEM) has the Translocon of the Inner membrane of the Chloroplast (TIC) complex mainly formed by Tic110, Tic20 and Tic21. Individual components are identified by their predicted molecular weights. Cytosolic proteins Hsp70 and Hsp90 form a 'guidance complex' that direct preproteins to the TOC apparatus and maintain the import competent state of the preprotein by preventing their aggregation. Chaperones Hsp93, Hsp90c and Hsp70 act as motors to "pull" the preprotein across the TIC complex. Upon crossing the TIC complex, the transit peptide of the preprotein is removed by the stromal processing peptidase (SPP). The folding, assembly, and transport of the preprotein inside the chloroplast is facilitated by group of chaperons and chaperonins. The E3 ubiquitin (Ub)-protein ligase SP1, along with SP2 (a β-barrel channel protein) are present in the OEM. SP1 ubiquitinates its substrates (i.e., TOC complex proteins), which are subsequently retrotranslocated from the OEM with help of SP2 and CDC48 (a cytosolic ATPase) enabling degradation of the substrates by the 26S proteasome in the cytosol. Adapted from (Ling et al. 2019; Andrès et al. 2010). Created with BioRender.com

Once ubiquitinated, TOC proteins are removed from chloroplast envelopes by two other proteins, SP2 (an Omp85-type β-barrel channel protein, also identified as a *SUPPRESSOR OF PPI1 LOCUS*) and CDC48 (a cytosolic AAA+ chaperone ATPase that serves as a molecular "motor" to remove ubiquitinated proteins from the chloroplast OEM) (Ling et al. 2019). SP2 and CDC48 co-operate with SP1 to extract the ubiquitinated components to the cytosol enabling its degradation by the 26S proteasome. Thus, SP1, SP2, and Cdc48 together constitute the CHLORAD system for the post-translational regulation of chloroplast outer envelope protein accumulation.

The CHLORAD proteolytic pathway is also reported to facilitate the transition of chloroplasts to chromoplasts (specialized plastids for pigment synthesis and storage) during fruit ripening in tomato (*Solanum lycopersicum*) (Ling et al. 2021). Knockdown of the tomato *SP1* orthologs *slSP1* and *slSP1-Like2* (*slSPL2*) delayed leaf senescence and fruit ripening, while overexpression of *slSP1* accelerated these processes. This was attributed to the roles of SP1 and SPL2 in altering the import of photosynthesis proteins into chloroplasts by degrading TOC components. Thus, the CHLORAD system has an additional role in providing stress responses and can be involved in plastid transitions (chloroplast to chromoplast) during fruit ripening.

5.3.5 The Role of SUMOylation During Chloroplast Stress

Another form of post-translational modification that regulates protein turnover is SUMOylation, which involves the attachment of a conserved 11 kDa Ub-related modifier (SUMO) protein to lysine residues of target proteins. SUMOylation affects protein stability and plays a critical role in plant development and stress responses (Accossato et al. 2020). Abscisic acid signaling and the development of roots, seeds, embryos, and meristems are some of the biological processes associated with SUMOylation in plants (Miura et al. 2009; Ishida et al. 2009; Kwak et al. 2019; Srivastava et al. 2020). In Arabidopsis, over 1000 proteins, including transcription factors, coactivators/repressors, and chromatin modifiers, can be SUMOylated (Rytz et al. 2018). Recently, SUMOylation has been shown to regulate the stability of TOC complex proteins, thus regulating the chloroplast proteome via protein import (Watson et al. 2021). In this study using Arabidopsis *ppi1* mutants, it was demonstrated that SUMO modifications control the levels of the TOC complex (TOC33) in outer envelope membranes, raising the possibility that SUMOylation may help to control the CHLORAD pathway in ubiquitinating and destroying the TOC complex. Another member of TOC, TOC159, has a SUMO3-interacting motif (SIM), and SUMOylation of TOC159 is reported to partially stabilize it under stress conditions triggered by low gibberellic acid levels (a growth-regulating phytohormone) (Accossato et al. 2020). Together, these results demonstrate how the control of protein import into chloroplasts is complex, but its regulation may allow cells to modulate stress responses and developmental transitions.

Ub-mediated proteolysis and SUMOylation has been extensively studied in animals and humans, but their roles in plants are relatively less understood, particularly how they control organelle function and quality control. This potential role for post-translational modifications, however, is particularly exciting because chloroplasts are the major site for cellular ROS production and cells need ways to quickly modulate their functions. However, there is still much we do not understand. In the case of PUB4, the mechanism by which this protein acts on chloroplasts (direct vs. indirect) is still unknown. SP1 is known to target the outer envelope proteins, including Toc33 and Toc159, but it is not known how these events are initiated or regulated. New pathways, like the SUMO system, are only beginning to be explored and their full impact on chloroplast biology is probably not fully understood. It is therefore required to use new tools and "omics" approaches to (i) identify additional factors involved in chloroplast quality control and (ii) to further investigate these E3 ligases and their interaction partners. Such insight will provide better clarity on the mechanism of their action under normal and stress conditions, both of which are central to plant growth and development.

5.4 Chloroplasts as Regulators of Cell Fate Decisions

5.4.1 Chloroplast-Controlled Programmed Cell Death

As described above, stressed chloroplasts can send signals to regulate cell physiology. How such signals control cell fate decisions and the downstream events that promote PCD, senescence, and chloroplast quality control is of great interest (Fig. 5.6). PCD is a fundamental cell fate decision that the cell must make when harsh cellular conditions go past a point of no return, and as such, must trigger an irreversible process that leads to its death to benefit overall plant fitness. While many extracellular and intracellular factors regulate PCD, the chloroplast plays a key role in directing PCD signals in response to biotic and abiotic environmental stresses (Ambastha et al. 2015). Chloroplast-induced PCD signals can be initiated by changes in ROS homeostasis that can lead to the oxidation of metabolites, lipids, proteins, and nucleotides (Tripathy and Oelmuller 2012). Generally, PCD can help plants resist pathogen infection (Jorgensen et al. 2017), prevent the spread of cytotoxic elements (such as ROS and heavy metals) (Huang et al. 2019; Sychta et al. 2021), mitigate water loss (D'Alessandro et al. 2020), and destroy dysfunctional cells (Burke et al. 2020). As PCD is an irreversible cell fate decision, its activation must be tightly controlled to ensure it is only activated under the appropriate cellular conditions. In cases of intermediate cellular damage and stress levels, such damage may be offset by the activation of quality control mechanisms (degradation of dysfunctional macromolecules and organelles), cellular repair systems, and cell cycle arrest mechanisms.

While plants have many pathways for PCD, some involve chloroplast ROS as an initial signal (Woodson 2022).

Besides $^1O_2$, H_2O_2 is also known to induce PCD (Van Aken and Van Breusegem 2015; Ye et al. 2021). For instance, during pathogen attack, chloroplast-generated H_2O_2 travels to the nucleus via specialized structures called stromules to trigger PCD (Caplan et al. 2015). However, $^1O_2$ strongly correlates to stress-induced PCD in photosynthetic tissues (Triantaphylides et al. 2008). Also, as discussed above, $^1O_2$

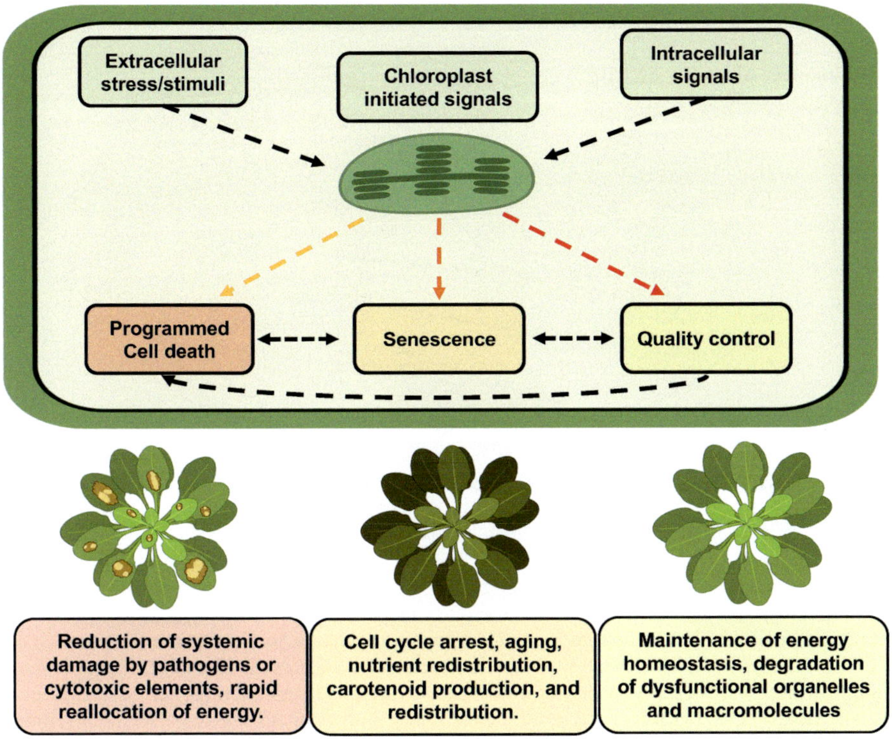

Fig. 5.6 Chloroplasts play critical roles in cell fate decisions. During stress, chloroplasts act as communication hubs to direct cell fate decisions, including programmed cell death (PCD), senescence, and quality control. PCD may be triggered to mitigate water loss, prevent the spread of pathogens, or destroy dysfunctional cells. Senescence mechanisms allow for nutrient redistribution. Quality control mechanisms remove dysfunctional chloroplast or chloroplast components to support plant fitness, but may also induce PCD. Because these outcomes are complex and poorly understood in plants, significant overlap likely exists between these outcomes. Created with BioRender.com

has a short half-life and therefore it is not likely to leave the chloroplast in which it was produced. Therefore, $^1O_2$-induced PCD must involve a signaling pathway and/or secondary messengers.

Recent work has demonstrated that the endoplasmic reticulum (ER) is involved in the chloroplast $^1O_2$ signal for PCD. In the *ch1* mutant (or EL-stressed wild type plants), PCD is activated through the nuclear-localized AGC2 kinase OXI1, a positive regulator of PCD (described above) (Shumbe et al. 2016) (Fig. 5.7). However, expression of *OXI1* itself is negatively regulated by the ER-localized proteins DEFENDER AGAINST CELL DEATH 1 and 2 (DAD1 and DAD2) (Beaugelin et al. 2019) and the expression of *DAD1* is induced in response to EL and $^1O_2$ (Beaugelin et al. 2019).

A major form of ER stress involves the unfolded protein response (UPR), which can be induced by the cytotoxic accumulation of misfolded proteins (Graner 2013).

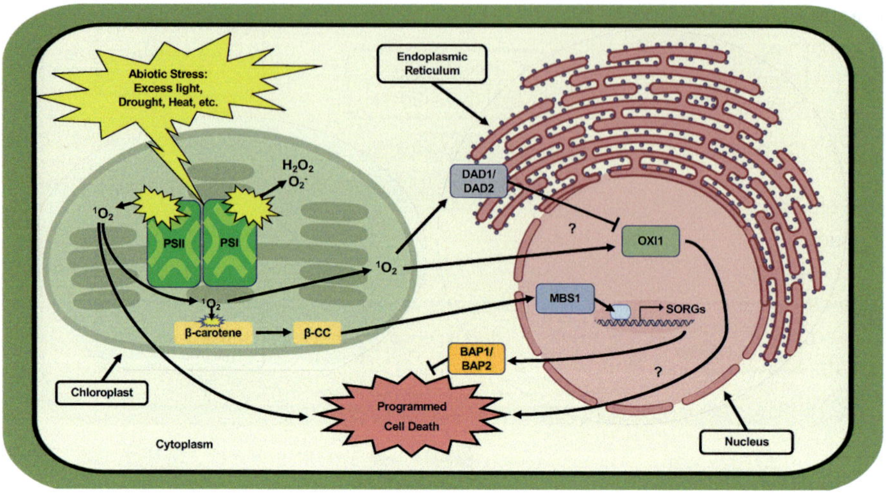

Fig. 5.7 Control of programmed cell death by chloroplast $^1O_2$ signals. A model depicting pathways by which chloroplast singlet oxygen ($^1O_2$) can control programmed cell death (PCD) is shown. Abiotic stress can lead to the production of reactive oxygen species (ROS) in chloroplasts; $^1O_2$ at photosystem II (PSII), and superoxide (O_2^-) and hydrogen peroxide (H_2O_2) at PSI. $^1O_2$ can act as a signaling molecule through multiple pathways to regulate *Singlet Oxygen Response Genes* (*SORGs*) and genes involved with PCD. β-Carotene can be oxidized by $^1O_2$, forming β-cyclocitral (β-cc), which can act as a secondary messenger to trigger stress acclimation. METHYLENE BLUE SENSITIVE 1 (MBS1) acts downstream of β-cc and positively regulates SORGs to control acclimation and detoxify ROS. Two proteins encoded by *SORGs*, BON ASSOCIATED PROTEIN 1 and 2 (BAP1 and BAP2), can negatively modulate the expression of PCD genes. $^1O_2$ can also lead to the induction of *DEFENDER AGAINST APOPTOTIC CELL DEATH 1* and *2* (*DAD1* and *DAD2*), which encode ER-localized proteins that negatively regulate OXIDATIVE SIGNAL INDUCIBLE1 (OXI1), a positive regulator of PCD. Created with BioRender.com

Multiple forms of stress can activate the UPR, which results in the induction of genes encoding chaperones that prevent such protein misfolding. One such group of chaperones, HEAT SHOCK PROTEIN 70 (HSP70) proteins, can regulate protein transport and facilitate the degradation of misfolded proteins to prevent their toxic aggregation (Sung et al. 2001). The ER's UPR has also been shown to play a role in $^1O_2$ signaling. In this case, a null mutation of *BINDING PROTEIN 3* (*BIP3*), a gene encoding an HSP70 family chaperone protein, leads to decreased tolerance of *ch1* to EL and $^1O_2$ stress. Conversely, treating *ch1* mutants with a chemical chaperone, tauroursodeoxycholic acid, led to increased tolerance to EL and $^1O_2$ (Beaugelin et al. 2020). Therefore, $^1O_2$-mediated PCD may be regulated by a UPR or UPR-like response in the ER, further highlighting the role of chloroplasts as central signaling hubs that control PCD and retrograde signaling through complex subcellular crosstalk.

$^1O_2$-induced PCD can also arise from the miscoordination of nuclear and plastid gene expression as shown in the *lesion simulating disease 1* (*lsd1*) mutant that accumulates leaf lesions (Dietrich et al. 1997). LSD1 is a pathogen response-related protein containing a zinc-finger motif. In *lsd1* mutants, salicylic acid (SA) levels increase and induce the expression of *SIGMA FACTOR BINDING PROTEIN 1* (*SIB1*), which encodes a transcription factor that is dually localized to the chloroplast and nucleus (Lv et al. 2019). This leads to the up-regulation of *photosynthesis-associated nuclear genes* (*PhANGs*) (in the nucleus) and down-regulation of *photosynthesis-associated plastid genes* (*PhAPGs*) (in the plastid). This misregulation leads to the disruption of the stoichiometry of photosynthetic complexes encoded by both genomes, producing chloroplast $^1O_2$ in the light. This $^1O_2$ leads to a PCD signal that is EX1 dependent: *lsd1 ex1* mutants suffer fewer lesions than *lsd1* mutants (Lv et al. 2019). Collectively, this suggests a possible mechanism by which cells could intentionally use their chloroplasts to initiate PCD in response to pathogen attack.

In some cases, chloroplast $^1O_2$ leads to the induction of negative regulators of PCD (Fig. 5.7). In addition to PCD suppressors DAD1/2 mentioned above, $^1O_2$ induces expression of *BON ASSOCIATION PROTEIN 1* and 2 (*BAP1* and *BAP2*) (Woodson et al. 2015; op den Camp et al. 2003), which encode phospholipid-binding proteins that are negative regulators of PCD and retrograde signaling (Yang et al. 2007) (Fig. 5.7). Through ß-cc and MBS1,$^1O_2$ can also induce genes involved in stress acclimation (discussed above) and mitigate $^1O_2$ signaling and PCD (Shumbe et al. 2017). The simultaneous induction of positive and negative regulators of PCD may be a safety mechanism to ensure that PCD is not prematurely activated.

Interestingly, $^1O_2$ may be able to trigger PCD if it accumulates outside of chloroplasts. Studies using the photo-sensitizer Rose Bengal demonstrated that cytoplasmic $^1O_2$ can also affect nuclear gene expression by oxidizing RNA and affecting translation (Koh et al. 2021). The connection to PCD is not clear, but underscores the possibility that severely damaged chloroplasts may leak ROS into the cytoplasm to induce retrograde signaling, which may control the expression of PCD genes such as *ENHANCED DISEASE SUSCEPTIBILITY* 1 (*EDS1*) and OXI1, which encode positive regulators of PCD (Shumbe et al. 2016; Ochsenbein et al. 2006).

A second study used the vacuole-accumulating $^1O_2$-producing photosensitizer Acridine Orange to produce $^1O_2$ in the vacuole. This treatment correlated with vacuolar membrane breakdown and cell death. Membrane breakdown of lytic compartments is a plausible mechanism for PCD, as it can release hydrolases and pro-apoptotic proteases, leading to broad cellular degradation and eventual cell death (Su et al. 2020). Whether $^1O_2$ in the vacuole leads to cell death via a signal or is purely due to mechanical failure of this compartment remains unclear. However, overexpression of *SERPIN1*, which encodes an inhibitor of pro-apoptotic proteases, suppresses this form of $^1O_2$-induced vacuole-mediated cell death, suggesting this may represent a type of PCD (Koh et al. 2016). While the central vacuole may not generate $^1O_2$ under normal conditions, wholesale chloroplast degradation that can occur in *fc2* (Woodson et al. 2015) and *flu* mutants (Kim et al. 2012), may load this compartment with tetrapyrroles leading to a burst of $^1O_2$. These observations highlight $^1O_2$ as a signaling molecule with a broader range of action than originally assumed (Fig. 5.7).

5.4.2 Chloroplasts and Senescence

In plants, PCD is an immediate "self-destruction" system to prevent systemic damage or to quickly dispose of dysfunctional or infected cells (van Doorn and Woltering 2004). In contrast, senescence is the gradual decline of cell function. In this case, eventual cellular degradation and cell death is a checkpoint that plant cells can use to reallocate and mobilize nutrients from dysfunctional cells to help maintain the maximal function of healthy cells and tissue (Munoz-Espin and Serrano 2014). Senescence in aging cells often correlates with cellular degradation and the accumulation of ROS, with O_2^- and H_2O_2 playing predominant roles in this process (Jajic et al. 2015; Khanna-Chopra 2012). $^1O_2$ production also increases in aging senescent photosynthetic tissue in plants (Jajic et al. 2015; Springer et al. 2015), likely due to the impairment of cellular function in regulating photosynthesis and quenching mechanisms in the chloroplast during senescence. In senescing plant cells, chloroplasts transition to gerontoplasts (senescing plastids), characterized by chlorophyll degradation, thylakoid breakdown, and plastoglobule swelling (Dominguez and Cejudo 2021). Little is known about the role of gerontoplasts in leaf senescence, but they are likely fated for final turnover so that these nutrient-rich compartments may be recycled. As such, the dismantling of chloroplasts and their transition to gerontoplasts would allow nutrient redistribution to younger leaves and sink tissues. While senescence often involves developmental signals, there is some evidence that chloroplast signals are involved. In *fc2* mutants, we have observed that chloroplast $^1O_2$ leads to the accumulation of transcripts involved in senescence and chloroplast disassembly and plastids that resemble gerontoplasts, with swollen sizes and enlarged plastoglobules (Fisher et al. 2022). The SP1 protein also promotes leaf senescence during prolonged dark treatments, suggesting that control of chloroplast function is central to cell senescence pathways (Ling et al. 2012).

5.4.3 Chloroplast Degradation and Quality Control

Chloroplasts can also signal their own degradation and turnover. These degradation events often correlate and overlap with cellular stress, PCD, and senescence, as evidenced by chloroplast degradation being triggered by UVB (Izumi et al. 2017), EL (Nakamura et al. 2018), $^1O_2$ (Woodson et al. 2015), salt and drought stress (Wang and Blumwald 2014), and nutrient deprivation (Ishida et al. 2008). Such chloroplast quality control pathways can be used to fulfill two purposes: (1) to maintain a healthy population of chloroplasts performing efficient photosynthesis, and (2) to recycle the nutrient-rich organelles when repair is too costly. There appears to exist multiple pathways by which chloroplasts may be degraded, and, in most cases, chloroplast nutrients may be remobilized via the central vacuole (Zhuang and Jiang 2019) (Fig. 5.8). Degradation can be selective (only a few chloroplasts within a healthy cell) or wholesale (all chloroplasts followed by cell death or senescence).

Chloroplast degradation pathways can be classified as autophagy-dependent or -independent, and they can degrade chloroplasts piecemeal or whole. Autophagy (or self-eating) is a fundamental and highly conserved process in eukaryotes that can proceed by a wide range of mechanisms. Canonical autophagy involves the transport of cytosolic cargo to the central vacuole (plants/yeast) or lysosomes (animals) for turnover and remobilization (Kim et al. 2012). This transport mechanism commonly involves core autophagy (ATG) proteins that direct the tagging of cargo with ATG8 (a Ub-like protein), form autophagosomes (ATG8-associated, double-membrane structures) and phagophores (partially formed autophagosomes), and direct the cargo to the vacuole/lysosome (Geng and Klionsky 2008). The conjugation of cargo proteins with ATG8 involves Ub-protein ligase-like machinery, allowing for both selective and nonselective transport of target cargo (Mizushima 2020).

5.4.4 Autophagy-Dependent Chloroplast Transport and Degradation

Most of the research into autophagy-related processes in plants has been informed heavily by what is known about the function of orthologous proteins in animals and yeast. Autophagy has, however, been shown in plants to be important for growth, stress responses, and energy metabolism (Su et al. 2020). ATG-dependent systems may proceed either by macroautophagy (the canonical form of autophagy involving the complete envelopment of cytosolic cargo by an autophagosome) or by ATG-dependent microautophagy (partial envelopment/association of cytosolic cargo by a phagophore) (Schuck 2020). In macroautophagy, the autophagosome fuses with the vacuole/lysosome membrane to deliver its cargo to the targeted cellular compartment. ATG-dependent microautophagy involves the "pushing" of cytosolic cargo into a vacuolar/lysosomal invagination followed by a final fusing of the phagophore with the vacuole/lysosome membrane (Schuck 2020). ATG-independent microautophagy mechanisms that do not rely on autophagosome formation also exist. In these cases, cytosolic components are transported into the central vacuole/lysosome via invagination events involving a final pinching-off and vesicular fission event as the delivery mechanism of transport into the vacuole/lysosome (Schuck 2020). This form of microautophagy strongly resembles endosomal sorting complexes required for transport (ESCRT)-mediated membrane remodeling and clathrin-mediated endocytosis (Schuck 2020). Few examples of ATG-independent microautophagy exist in plants, but Arabidopsis uses a similar process to transport anthocyanin pigments into the central vacuole (Chanoca et al. 2015).

In some cases, ATG-dependent macroautophagy can be employed to dismantle and degrade chloroplasts. Such processes depend on autophagosomes (or phagophores) and allow for the transport of chloroplast components and whole chloroplasts

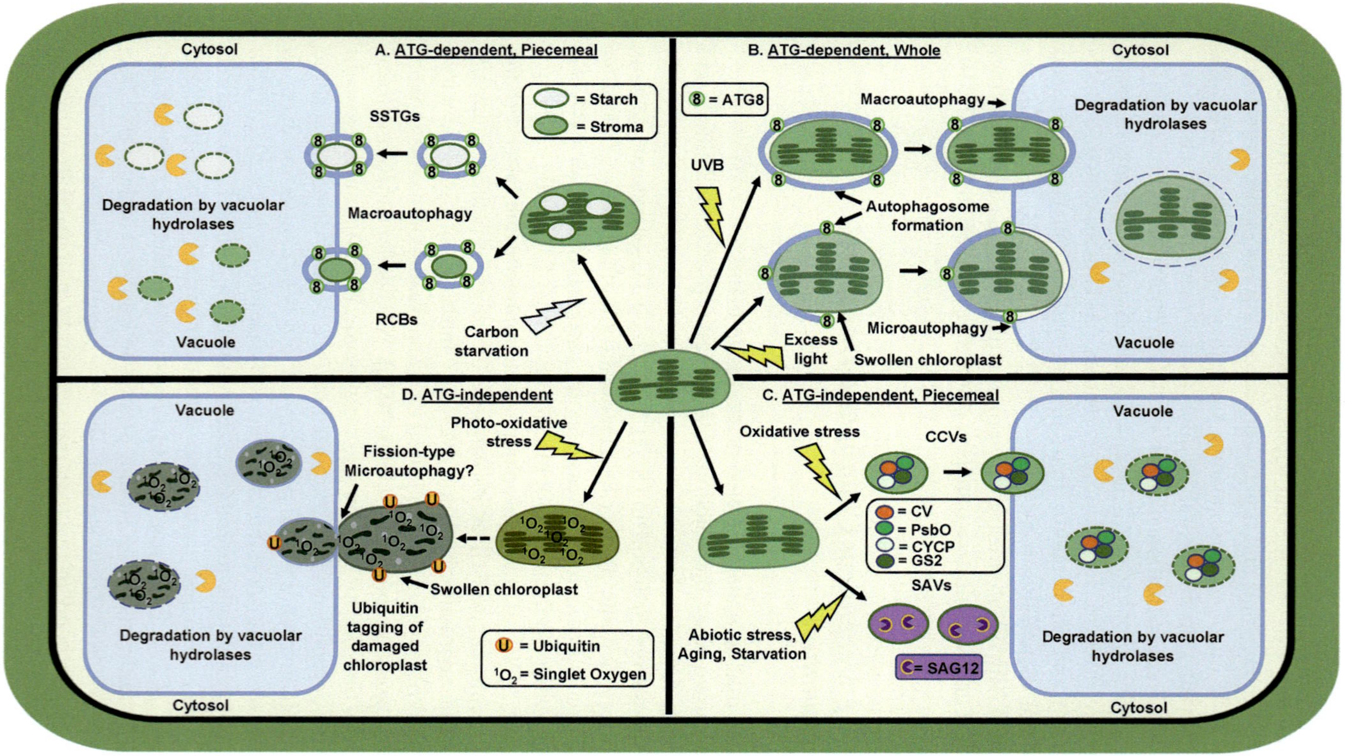

Fig. 5.8 Chloroplast degradation pathways. Models for multiple chloroplast degradation pathways in plant cells are presented. These proceed by ATG-dependent and -independent mechanisms. (**a**) Piecemeal ATG-dependent macroautophagy may be employed to transport small starch granule-like structures (SSTG)s and Rubisco-containing bodies (RCB)s from the chloroplast. (**b**) ATG-dependent macroautophagy and microautophagy can be activated to transport whole UVB or EL-damaged chloroplasts, respectively, to the central vacuole via mechanisms dependent on autophagosome formation. (**c**) Piecemeal ATG-independent chloroplast degradation can be induced by oxidative stress, abiotic stress, aging, and starvation and can lead to the formation of CV-containing vesicles (CCVs) and senescence-associated vacuoles (SAV)s from chloroplasts. (**d**) ATG-independent chloroplast degradation can be induced by $^1O_2$ and can lead to the ubiquitin (Ub)-tagging of damaged chloroplasts that begin degradation in the cytosol and are then transported into the vacuole by "blebbing" events. This pathway resembles fission-type microautophagy. Created with BioRender.com

and to the central vacuole via piecemeal and wholesale mechanisms, respectively. In piecemeal pathways (Fig. 5.8a), Rubisco-containing bodies (RCBs) have been shown to bud from the chloroplast and associate with autophagosomes in response to carbon deprivation caused by prolonged dark starvation. These RCBs were observed to accumulate in the central vacuole in mesophyll cells in wild type plants treated with concanamycin A (a vacuolar ATPase inhibitor that prevents vacuolar degradation) (Ishida et al. 2008). This phenomenon was lost in *atg5* and *atg7* mutants lacking the core autophagy machinery required for autophagosome formation, showing that RCBs are transported to the vacuole via an ATG-dependent mechanism. Similarly, starch is remobilized via starch granule-like structures (SSGLs) and transported to the central vacuole. This process is also blocked in *atg* mutants (Wang et al. 2013b).

Whole chloroplasts can also be transported to the central vacuole via an ATG-dependent process called chlorophagy (Fig. 5.8b). This can be observed under nutrient deprivation (Wada et al. 2009) and in response to chloroplast stress (Izumi et al. 2017). Such chlorophagy can be induced by wrapping individual leaves in aluminum foil for three-five days to initiate carbon starvation. At this point, chloroplasts are moved to the central vacuole, presumably to recycle nutrients (Wada et al. 2009). This process involves autophagosomes, as chloroplast movement is blocked in mutants lacking *ATG4*. Furthermore, in a separate study using a GFP-ATG8 marker, autophagosomes were observed to associate with dark-starved chloroplasts prior to senescence (Lemke et al. 2021).

A similar process occurs after UVB stress. Chloroplasts are surrounded by GFP-ATG8-labeled autophagosomes before being moved to the central vacuole. Again, this process is blocked in *atg* mutants lacking the ability to form autophagosomes (Izumi et al. 2017). EL stress can also induce the vacuolar transport of chloroplasts via ATG-dependent microautophagy, where phagophores associate with swelling chloroplasts and facilitate their transport to the central vacuole via vacuolar invagination and presumed fusion of the phagophore with the vacuolar membrane (Nakamura et al. 2018). This process may specifically target swollen, membrane-damaged chloroplasts, as ATG-dependent micro-chlorophagy can be suppressed by increased osmotic pressure in the plant cell (via mannitol treatment) or by the overproduction of VIPP1, a thylakoid membrane maintenance protein (Nakamura et al. 2018).

5.4.5 ATG-Independent Chloroplast Transport and Degradation

ATG-independent systems for chloroplast degradation also exist. For piecemeal degradation, lytic compartments from chloroplast-derived senescence-associated vacuoles (SAVs) can form (Fig. 5.8c). Here, SAG12 (a senescence-associated cysteine protease) was observed to accumulate in cytosolic vesicles in an

ATG-independent process (Martinez et al. 2008; Otegui et al. 2005). These vacuoles are abundant in senescing leaves, characterized by strong proteolytic activity, and can contain chloroplast stromal proteins (Rubisco complex proteins and glutamine synthetase), suggesting a role in the degradation of chloroplasts during senescence. These SAVs can be stained with LysoTracker Red, a marker for acidic lytic compartments, confirming high levels of hydrolase activity in these compartments. Unlike RCBs, SAVs do not appear to fuse with the central vacuole.

Another pathway for ATG-independent, piecemeal degradation of chloroplasts involves the CHLOROPAST VESICULATION (CV) protein, which is induced by drought, salinity, and H_2O_2 (Wang and Blumwald 2014). CV localizes to the thylakoid membranes and interacts with components of the photosynthetic machinery (CYCP and PsbO of PSI and PSII, respectively). CV-containing vesicles (CCVs) that derive from chloroplasts can be transported to the central vacuole in an ATG-independent manner. CCVs still form in *atg5* mutants and fail to co-localize with GFP-ATG8-containing structures (Fig. 5.8c). LysoTracker Red treatment fails to stain these CCVs, suggesting that these are not lytic compartments and, thus, are independent of SAVs.

In rice (*Oryza sativa*), silencing of *CV* delays chloroplast turnover and stress-induced senescence and increases nitrogen photo-assimilation rates (Sade et al. 2018). This is at least partly due to the ability of CV to target the degradation of GLUTAMINE SYNTHASE 2 (GS2), a key enzyme in nitrogen assimilation in the chloroplast stroma. Elevated CO_2 can also induce *CV* in rice, leading to the degradation of chloroplasts, mitochondria, and peroxisomes (Umnajkitikorn et al. 2020). This may be a direct effect in the case of peroxisomes, as CV can transport PEX11–1, a protein involved in peroxisome proliferation, to the central vacuole. Such a "stay-green phenotype" (a delay in chlorosis and maintenance of photosynthetically active tissue during senescence) during stress is often associated with increased grain yield (Kamal et al. 2019). The mechanisms behind this are unknown, but this work suggests control of chloroplast degradation pathways may offer tools to control such phenotypes.

When *fc2* mutants are grown under permissive constant light conditions, we observed that individual chloroplasts were degraded, likely due to a slight increase in Proto and $^1O_2$ (Fisher et al. 2022). These chloroplasts were swollen with compressed grana and enlarged plastoglobules. Unlike cycling light conditions, this did not lead to PCD. Instead, this process was selective, and healthy chloroplasts were observed next to damaged and degrading chloroplasts in the same cells. These degrading chloroplasts were also observed to sometimes "bleb" and protrude into the central vacuole (Woodson et al. 2015; Fisher et al. 2022; Lemke et al. 2021) (Fig. 5.8d). These blebbing events resembled fission-type microautophagy and endosomal trafficking, where cargo is transported into the vacuole/lysosome or endosome, respectively, via invagination of vacuolar, lysosomal, or endosomal membranes that encapsulate cargo prior to the internal vesiculation of such cargo (Schuck 2020). We hypothesize that this process leads to the degradation of whole chloroplasts as *fc2* mutants have fewer total chloroplasts per cell area (Woodson et al. 2015). However, it is difficult to confirm that whole (rather than partial) chloroplasts enter the central vacuole due to their extensive degradation. Nonetheless, we observed blebs of varying sizes (ranging from 14% to 202% of the associated chloroplasts' size) in 3D transmission electron microscopy (TEM) analyses, and nearly a third of degrading chloroplasts interacted with bleb structures in the *fc2* mutant (Fisher et al. 2022). This process involves PUB4, as the *pub4–6* allele also blocked selective chloroplast degradation and blebbing suggesting that chloroplast turnover may be upstream of PCD (Fisher et al. 2022).

As chloroplast degradation can proceed via autophagy, we tested if the core autophagy machinery was necessary for chloroplast turnover in the *fc2* mutant. To this end, we generated *fc2 atg* double mutants that cannot form autophagosomes (Lemke et al. 2021). Interestingly, these mutants still exhibited chloroplast degradation and blebbing. As such, the core autophagy machinery is dispensable to the $^1O_2$-induced vacuolar transport of degrading chloroplasts observed in *fc2*. The *atg* mutations also failed to block PCD and retrograde signals in stressed *fc2* seedlings. However, we observed chloroplasts with altered ultrastructures in the *fc2 atg*5 and *fc2 atg7* double mutants, suggesting that autophagy may have complex interactions with chloroplast development and ROS (Lemke et al. 2021).

The above results suggested that $^1O_2$-induced selective chloroplast degradation is independent of chlorophagy. This hypothesis is further supported by the observation that these processes occur at different rates. $^1O_2$-induced selective chloroplast degradation can be initiated within three hours (Woodson et al. 2015), while EL- and UVB-induced chloroplast take one and three days, respectively (Izumi et al. 2017; Nakamura et al. 2018). $^1O_2$-induced selective chloroplast degradation also begins in the cytoplasm and chloroplasts within the central vacuole are difficult to recognize due to their advanced state of degradation (Fisher et al. 2022). During chlorophagy, however, chloroplasts within the central vacuole retain thylakoid membranes and appear relatively intact (Izumi et al. 2017). Chlorophagy is also strongly correlated to H_2O_2 (Izumi et al. 2017), while H_2O_2 production appears negligible in *fc2* mutants (Alamdari et al. 2020). Finally, the PUB4 protein, which is necessary for $^1O_2$-induced selective chloroplast degradation, is dispensable for EL-induced chlorophagy (Kikuchi et al. 2020). Interestingly, *pub4 atg* double mutants were particularly sensitive to nitrogen and carbon starvation, suggesting that chlorophagy and PUB4-mediated chloroplast degradation are parallel pathways involved in chloroplast turnover (Kikuchi et al. 2020).

If canonical autophagy pathways do not mediate $^1O_2$-induced selective chloroplast degradation, then what machinery is involved? A clue was inferred from our transcript analysis of *fc2* mutants showing that genes predicted to be involved in fission-type microautophagy were induced (Schuck 2020; Lemke et al. 2021). Very little is known about this type of ATG-independent autophagy, particularly in plants. Thus, a deeper understanding of the mechanisms driving the transport of $^1O_2$-damaged chloroplasts to the vacuole may not only shed light on chloroplast quality control pathways, but on a poorly understood class of cellular degradation.

The existence of many chloroplast degradation pathways suggests that they are generally beneficial to plant fitness in response to a dynamic environment. With such a limited knowledge base of chloroplast degradation, plant autophagy and microautophagy, and their roles in stress responses, there remain many open questions: What are the specific upstream signals and the signaling mechanisms involved in the different classes of chloroplast degradation? How much of the signaling originates from the chloroplasts themselves? How does chloroplast degradation benefit plant fitness, and can these pathways be manipulated to increase crop yield, possibly by controlling "stay green" traits? In total, these questions highlight the necessity for continued research into chloroplast degradation, which will provide a better understanding of the role of chloroplast stress signaling. This may allow for the manipulation of these complex systems to produce crop plants that are more tolerant to stress and more resistant to pathogens, all of which can limit crop quality and yield.

Acknowledgments The authors acknowledge the Division of Chemical Sciences, Geosciences, and Biosciences, Office of Basic Energy Sciences of the U.S. Department of Energy grant [DE-SC0019573]. This work is also supported by the NIH BMCB Training grant [T32 GM136536] awarded to MDL and the University of Arizona University Fellows Award awarded to DWT. The authors have no conflict of interest to declare.

References

Abdallah F, Salamini F, Leister D (2000) A prediction of the size and evolutionary origin of the proteome of chloroplasts of Arabidopsis. Trends Plant Sci 5:141–142

Accossato S, Kessler F, Shanmugabalaji V (2020) SUMOylation contributes to proteostasis of the chloroplast protein import receptor TOC159 during early development. elife 9:e60968

Alamdari K, Fisher KE, Sinson AB, Chory J, Woodson JD (2020) Roles for the chloroplast-localized PPR Protein 30 and the "Mitochondrial" Transcription Termination Factor 9 in chloroplast quality control. Plant J 103:735–751

Alamdari K, Fisher KE, Tano DW, Rai S, Palos KR, Nelson ADL, Woodson JD (2021) Chloroplast quality control pathways are dependent on plastid DNA synthesis and nucleotides provided by cytidine triphosphate synthase two. New Phytol 231:1431–1448

Allen JF (2015) Why chloroplasts and mitochondria retain their own genomes and genetic systems: colocation for redox regulation of gene expression. Proc Natl Acad Sci 112:10231–10238

Ambastha V, Tripathy BC, Tiwari BS (2015) Programmed cell death in plants: a chloroplastic connection. Plant Signal Behav 10:e989752

Ambastha V, Chauhan G, Tiwari BS, Tripathy BC (2020) Execution of programmed cell death by singlet oxygen generated inside the chloroplasts of *Arabidopsis thaliana*. Protoplasma 257:841–851

Andrès C, Agne B, Kessler F (2010) The TOC complex: preprotein gateway to the chloroplast. Biochim Biophys Acta (BBA) – Mol Cell Res 1803:715–723

Aravind L, Koonin EV (2000) The U box is a modified RING finger – a common domain in ubiquitination. Curr Biol 10:R132–R134

Asada K (2006) Production and scavenging of reactive oxygen species in chloroplasts and their functions. Plant Physiol 141:391–396

Barkan A, Small I (2014) Pentatricopeptide repeat proteins in plants. Annu Rev Plant Biol 65:415–442

Beaugelin I, Chevalier A, D'Alessandro S, Ksas B, Novák O, Strnad M, Forzani C, Hirt H, Havaux M, Monnet F (2019) OXI1 and DAD regulate light-induced cell death antagonistically through jasmonate and salicylate levels. Plant Physiol 180:1691–1708

Beaugelin I, Chevalier A, D'Alessandro S, Ksas B, Havaux M (2020) Endoplasmic reticulum-mediated unfolded protein response is an integral part of singlet oxygen signalling in plants. Plant J 102:1266–1280

Breeze E, Mullineaux PM (2022) The passage of H_2O_2 from chloroplasts to their associated nucleus during retrograde signalling: reflections on the role of the nuclear envelope. Plants (Basel) 11:552

Burke R, Schwarze J, Sherwood OL, Jnaid Y, McCabe PF, Kacprzyk J (2020) Stressed to death: the role of transcription factors in plant programmed cell death induced by abiotic and biotic stimuli. Front Plant Sci 11:1235

Callis J (2014) The ubiquitination machinery of the ubiquitin system. Arabidopsis Book/American Society of Plant Biologists 12:e0174

op den Camp RG, Przybyla D, Ochsenbein C, Laloi C, Kim C, Danon A, Wagner D, Hideg E, Gobel C, Feussner I, Nater M, Apel K (2003) Rapid induction of distinct stress responses after the release of singlet oxygen in Arabidopsis. Plant Cell 15:2320–2332

Caplan JL, Kumar AS, Park E, Padmanabhan MS, Hoban K, Modla S, Czymmek K, Dinesh-Kumar SP (2015) Chloroplast stromules function during innate immunity. Dev Cell 34:45–57

Cavalier-Smith T (2000) Membrane heredity and early chloroplast evolution. Trends Plant Sci 5:174–182

Chan KX, Phua SY, Crisp P, McQuinn R, Pogson BJ (2016) Learning the languages of the chloroplast: retrograde signaling and beyond. Annu Rev Plant Biol 67:25–53

Chanoca A, Kovinich N, Burkel B, Stecha S, Bohorquez-Restrepo A, Ueda T, Eliceiri KW, Grotewold E, Otegui MS (2015) Anthocyanin vacuolar inclusions form by a microautophagy mechanism. Plant Cell 27:2545–2599

Choudhury FK, Rivero RM, Blumwald E, Mittler R (2017) Reactive oxygen species, abiotic stress and stress combination. Plant J 90:856–867

Coll NS, Danon A, Meurer J, Cho WK, Apel K (2009) Characterization of *soldat8*, a suppressor of singlet oxygen-induced cell death in Arabidopsis seedlings. Plant Cell Physiol 50:707–718

D'Alessandro S, Mizokami Y, Légeret B, Havaux M (2019) The apocarotenoid β-cyclocitric acid elicits drought tolerance in plants. iScience 19:461–473

D'Alessandro S, Beaugelin I, Havaux M (2020) Tanned or sunburned: how excessive light triggers plant cell death. Mol Plant 13:1545–1555

D'Alessandro S, Ksas B, Havaux M (2018) Decoding beta-cyclocitral-mediated retrograde signaling reveals the role of a detoxification response in plant tolerance to photooxidative stress. Plant Cell 30:2495–2511

Danon A, Miersch O, Felix G, Camp RG, Apel K (2005) Concurrent activation of cell death-regulating signaling pathways by singlet oxygen in *Arabidopsis thaliana*. Plant J 41:68–80

Das K, Roychoudhury A (2014) Reactive oxygen species (ROS) and response of antioxidants as ROS-scavengers during environmental stress in plants. Front Environ Sci 2:1

Daumann M, Hickl D, Zimmer D, DeTar RA, Kunz HH, Mohlmann T (2018) Characterization of filament-forming CTP synthases from *Arabidopsis thaliana*. Plant J 96:316–328

Desaki Y, Takahashi S, Sato K, Maeda K, Matsui S, Yoshimi I, Miura T, Jumonji JI, Takeda J, Yashima K, Kohari M, Suenaga T, Terada H, Narisawa T, Shimizu T, Yumoto E, Miyamoto K, Narusaka M, Narusaka Y, Kaku H, Shibuya N (2019) PUB4, a CERK1-interacting ubiquitin ligase, positively regulates MAMP-triggered immunity in Arabidopsis. Plant Cell Physiol 60:2573–2583

Dietrich RA, Richberg MH, Schmidt R, Dean C, Dangl JL (1997) A novel zinc finger protein is encoded by the Arabidopsis *LSD1* gene and functions as a negative regulator of plant cell death. Cell 88:685–694

Dietz KJ (2003) Redox control, redox signaling, and redox homeostasis in plant cells. Int Rev Cytol 228:141–193

Dogra V, Kim C (2019) Singlet oxygen metabolism: from genesis to signaling. Front Plant Sci 10:1640

Dogra V, Duan J, Lee KP, Lv S, Liu R, Kim C (2017) FtsH2-dependent proteolysis of EXECUTER1 is essential in mediating singlet oxygen-triggered retrograde signaling in *Arabidopsis thaliana*. Front Plant Sci 8:1145

Dogra V, Li M, Singh S, Li M, Kim C (2019) Oxidative post-translational modification of EXECUTER1 is required for singlet oxygen sensing in plastids. Nat Commun 10:1–12

Dogra V, Singh RM, Li M, Li M, Singh S, Kim C (2022) EXECUTER2 modulates the EXECUTER1 signalosome through its singlet oxygen-dependent oxidation. Mol Plant 15:438–453

Dominguez F, Cejudo FJ (2021) Chloroplast dismantling in leaf senescence. J Exp Bot 72:5905–5918

van Doorn WG, Woltering EJ (2004) Senescence and programmed cell death: substance or semantics? J Exp Bot 55:2147–2153

Espineda CE, Linford AS, Devine D, Brusslan JA (1999) *AtCAO* gene, encoding chlorophyll *a* oxygenase, is required for chlorophyll *b* synthesis in *Arabidopsis thaliana*. Proc Natl Acad Sci USA 96:10507–10511

Estavillo GM, Crisp PA, Pornsiriwong W, Wirtz M, Collinge D, Carrie C, Giraud E, Whelan J, David P, Javot H, Brearley C, Hell R, Marin E, Pogson BJ (2011) Evidence for a SAL1-PAP chloroplast retrograde pathway that functions in drought and high light signaling in Arabidopsis. Plant Cell 23:3992–4012

Fisher KE, Krishnamoorthy P, Joens MS, Chory J, Fitzpatrick JAJ, Woodson JD (2022) Singlet oxygen leads to structural changes to chloroplasts during their degradation in the *Arabidopsis thaliana plastid ferrochelatase two* mutant. Plant Cell Physiol 63:248–264

Ganesan I, Shi LX, Labs M, Theg SM (2018) Evaluating the functional pore size of chloroplast TOC and TIC protein translocons: import of folded proteins. Plant Cell 30:2161–2173

Geng J, Klionsky DJ (2008) The Atg8 and Atg12 ubiquitin-like conjugation systems in macroautophagy. 'Protein modifications: beyond the usual suspects' review series. EMBO Rep 9:859–864

Graner MW (2013) The unfolded protein response in glioblastomas: passing the stress test. CNS Oncol 2:469–472

Green BR (2011) Chloroplast genomes of photosynthetic eukaryotes. Plant J 66:34–44

Hong Y, Yuan S, Sun L, Wang X, Hong Y (2018) Cytidinediphosphate-diacylglycerol synthase 5 is required for phospholipid homeostasis and is negatively involved in hyperosmotic stress tolerance. Plant J 94:1038–1050

Huang H, Ullah F, Zhou DX, Yi M, Zhao Y (2019) Mechanisms of ROS regulation of plant development and stress responses. Front Plant Sci 10:800

Ishida H, Yoshimoto K, Izumi M, Reisen D, Yano Y, Makino A, Ohsumi Y, Hanson MR, Mae T (2008) Mobilization of Rubisco and stroma-localized fluorescent proteins of chloroplasts to the vacuole by an ATG gene-dependent autophagic process. Plant Physiol 148:142–155

Ishida T, Fujiwara S, Miura K, Stacey N, Yoshimura M, Schneider K, Adachi S, Minamisawa K, Umeda M, Sugimoto K (2009) SUMO E3 ligase HIGH PLOIDY2 regulates endocycle onset and meristem maintenance in Arabidopsis. Plant Cell 21:2284–2297

Izumi M, Ishida H, Nakamura S, Hidema J (2017) Entire photodamaged chloroplasts are transported to the central vacuole by autophagy. Plant Cell 29:377–394

Jajic I, Sarna T, Strzalka K (2015) Senescence, stress, and reactive oxygen species. Plants (Basel) 4:393–411

Jarvis P, Lopez-Juez E (2013) Biogenesis and homeostasis of chloroplasts and other plastids. Nat Rev Mol Cell Biol 14:787–802

Jeran N, Rotasperti L, Frabetti G, Calabritto A, Pesaresi P, Tadini L (2021) The PUB4 E3 ubiquitin ligase is responsible for the variegated phenotype observed upon alteration of chloroplast protein homeostasis in Arabidopsis cotyledons. Genes (Basel) 12:1387

Jorgensen I, Rayamajhi M, Miao EA (2017) Programmed cell death as a defence against infection. Nat Rev Immunol 17:151–164

Kamal NM, Alnor Gorafi YS, Abdelrahman M, Abdellatef E, Tsujimoto H (2019) Stay-green trait: a prospective approach for yield potential, and drought and heat stress adaptation in globally important cereals. Int J Mol Sci 20:5837

Khanna-Chopra R (2012) Leaf senescence and abiotic stresses share reactive oxygen species-mediated chloroplast degradation. Protoplasma 249:469–481

Kikuchi Y, Nakamura S, Woodson JD, Ishida H, Ling Q, Hidema J, Jarvis RP, Hagihara S, Izumi M (2020) Chloroplast autophagy and ubiquitination combine to manage oxidative damage and starvation responses. Plant Physiol 183:1531–1544

Kim C, Meskauskiene R, Zhang S, Lee KP, Lakshmanan AM, Blajecka K, Herrfurth C, Feussner I, Apel K (2012) Chloroplasts of Arabidopsis are the source and a primary target of a plant-specific programmed cell death signaling pathway. Plant Cell 24:3026–3039

Kinoshita A, ten Hove CA, Tabata R, Yamada M, Shimizu N, Ishida T, Yamaguchi K, Shigenobu S, Takebayashi Y, Iuchi S, Kobayashi M, Kurata T, Wada T, Seo M, Hasebe M, Blilou I, Fukuda H, Scheres B, Heidstra R, Kamiya Y, Sawa S (2015) A plant U-box protein, PUB4, regulates asymmetric cell division and cell proliferation in the root meristem. Development 142:444–453

Koh E, Carmieli R, Mor A, Fluhr R (2016) Singlet oxygen-induced membrane disruption and serpin-protease balance in vacuolar-driven cell death. Plant Physiol 171:1616–1625

Koh E, Cohen D, Brandis A, Fluhr R (2021) Attenuation of cytosolic translation by RNA oxidation is involved in singlet oxygen-mediated transcriptomic responses. Plant Cell Environ 44:3597–3615

Kondapalli C, Kazlauskaite A, Zhang N, Woodroof HI, Campbell DG, Gourlay R, Burchell L, Walden H, Macartney TJ, Deak M, Knebel A, Alessi DR, Muqit MM (2012) PINK1 is activated by mitochondrial membrane potential depolarization and stimulates Parkin E3 ligase activity by phosphorylating Serine 65. Open Biol 2:120080

Kwak JS, Kim SI, Park SW, Song JT, Seo HS (2019) E3 SUMO ligase AtSIZ1 regulates the cruciferin content of Arabidopsis seeds. Biochem Biophys Res Commun 519:761–766

Lazarou M, Sliter DA, Kane LA, Sarraf SA, Wang C, Burman JL, Sideris DP, Fogel AI, Youle RJ (2015) The ubiquitin kinase PINK1 recruits autophagy receptors to induce mitophagy. Nature 524:309–314

Lee KP, Kim C, Landgraf F, Apel K (2007) EXECUTER1- and EXECUTER2-dependent transfer of stress-related signals from the plastid to the nucleus of *Arabidopsis thaliana*. Proc Natl Acad Sci USA 104:10270–10275

Lemke MD, Fisher EM, Kozlowska MA, Tano DW, Woodson JD (2021) The core autophagy machinery is not required for chloroplast singlet oxygen-mediated cell death in the Arabidopsis *plastid ferrochelatase two* mutant. BMC Plant Biol 21:342

Lima-Melo Y, Kılıç M, Aro E-M, Gollan PJ (2021) Photosystem I inhibition, protection and signalling: knowns and unknowns. Front Plant Sci 12:791124–791124

Ling Q, Jarvis P (2015) Regulation of chloroplast protein import by the ubiquitin E3 ligase SP1 is important for stress tolerance in plants. Curr Biol 25:2527–2534

Ling Q, Huang W, Baldwin A, Jarvis P (2012) Chloroplast biogenesis is regulated by direct action of the ubiquitin-proteasome system. Science 338:655–659

Ling Q, Broad W, Trösch R, Töpel M, Demiral ST, Lymperopoulos P, Baldwin A, Jarvis RP (2019) Ubiquitin-dependent chloroplast-associated protein degradation in plants. Science 363:eaav4467

Ling Q, Sadali NM, Soufi Z, Zhou Y, Huang B, Zeng Y, Rodriguez-Concepcion M, Jarvis RP (2021) The chloroplast-associated protein degradation pathway controls chromoplast development and fruit ripening in tomato. Nat Plants 7:655–666

Lu Y, Yao J (2018) Chloroplasts at the crossroad of photosynthesis, pathogen infection and plant defense. Int J Mol Sci 19:3900

Lv R, Li Z, Li M, Dogra V, Lv S, Liu R, Lee KP, Kim C (2019) Uncoupled expression of nuclear and plastid photosynthesis-associated genes contributes to cell death in a lesion mimic mutant. Plant Cell 31:210–230

Lysenko EA (2007) Plant sigma factors and their role in plastid transcription. Plant Cell Rep 26:845–859

Mach JM, Castillo AR, Hoogstraten R, Greenberg JT (2001) The Arabidopsis-accelerated cell death gene *ACD2* encodes red chlorophyll catabolite reductase and suppresses the spread of disease symptoms. Proc Natl Acad Sci USA 98:771–776

Makino A, Osmond B (1991) Effects of nitrogen nutrition on nitrogen partitioning between chloroplasts and mitochondria in pea and wheat. Plant Physiol 96:355–362

Martinez DE, Costa ML, Gomez FM, Otegui MS, Guiamet JJ (2008) 'Senescence-associated vacuoles' are involved in the degradation of chloroplast proteins in tobacco leaves. Plant J 56:196–206

Meskauskiene R, Nater M, Goslings D, Kessler F, op den Camp R., Apel K. (2001) FLU: a negative regulator of chlorophyll biosynthesis in *Arabidopsis thaliana*. Proc Natl Acad Sci USA 98:12826–12831

Meskauskiene R, Würsch M, Laloi C, Vidi P-A, Coll NS, Kessler F, Baruah A, Kim C, Apel K (2009) mutation in the Arabidopsis mTERF-related plastid protein SOLDAT10 activates retrograde signaling and suppresses $^1O_2$-induced cell death. Plant J 60:399–410

Méteignier L, Ghandour R, Zimmerman A, Kuhn L, Meurer J, Zoschke R, Hammani K (2021) Arabidopsis mTERF9 protein promotes chloroplast ribosomal assembly and translation by establishing ribonucleoprotein interactions in vivo. Nucleic Acids Res 49:1114–1132

Miura K, Lee J, Jin JB, Yoo CY, Miura T, Hasegawa PM (2009) Sumoylation of ABI5 by the Arabidopsis SUMO E3 ligase SIZ1 negatively regulates abscisic acid signaling. Proc Natl Acad Sci USA 106:5418–5423

Mizushima N (2020) The ATG conjugation systems in autophagy. Curr Opin Cell Biol 63:1–10

Mohapatra A, Tripathy BC (2002) Detection of protoporphyrin IX in envelope membranes of pea chloroplasts. Biochem Biophys Res Commun 299:751–754

Mohapatra A, Tripathy BC (2007) Differential distribution of chlorophyll biosynthetic intermediates in stroma, envelope and thylakoid membranes in Beta vulgaris. Photosynth Res 94:401–410

Munoz-Espin D, Serrano M (2014) Cellular senescence: from physiology to pathology. Nat Rev Mol Cell Biol 15:482–496

Nakamura S, Hidema J, Sakamoto W, Ishida H, Izumi M (2018) Selective elimination of membrane-damaged chloroplasts via microautophagy. Plant Physiol 177:1007–1026

Ochsenbein C, Przybyla D, Danon A, Landgraf F, Göbel C, Imboden A, Feussner I, Apel K (2006) The role of EDS1 (enhanced disease susceptibility) during singlet oxygen-mediated stress responses of Arabidopsis. Plant J 47:445–456

Ogilby PR (2010) Singlet oxygen: there is indeed something new under the sun. Chem Soc Rev 39:3181–3209

Otegui MS, Noh YS, Martinez DE, Vila Petroff MG, Staehelin LA, Amasino RM, Guiamet JJ (2005) Senescence-associated vacuoles with intense proteolytic activity develop in leaves of Arabidopsis and soybean. Plant J 41:831–844

Page MT, McCormac AC, Smith AG, Terry MJ (2017) Singlet oxygen initiates a plastid signal controlling photosynthetic gene expression. New Phytol 213:1168–1180

Paila YD, Richardson LGL, Schnell DJ (2015) New insights into the mechanism of chloroplast protein import and its integration with protein quality control, organelle biogenesis and development. J Mol Biol 427:1038–1060

Papenbrock J, Mishra S, Mock HP, Kruse E, Schmidt EK, Petersmann A, Braun HP, Grimm B (2001) Impaired expression of the plastidic ferrochelatase by antisense RNA synthesis leads to a necrotic phenotype of transformed tobacco plants. Plant J 28:41–50

Pattanayak GK, Venkataramani S, Hortensteiner S, Kunz L, Christ B, Moulin M, Smith AG, Okamoto Y, Tamiaki H, Sugishima M, Greenberg JT (2012) Accelerated cell death 2 suppresses mitochondrial oxidative bursts and modulates cell death in Arabidopsis. Plant J 69:589–600

Pogson BJ, Woo NS, Forster B, Small ID (2008) Plastid signalling to the nucleus and beyond. Trends Plant Sci 13:602–609

Pruzinská A, Anders I, Aubry S, Schenk N, Tapernoux-Lüthi E, Müller T, Kräutler B, Hörtensteiner S (2007) In vivo participation of red chlorophyll catabolite reductase in chlorophyll breakdown. Plant Cell 19:369–387

Przybyla D, Gobel C, Imboden A, Hamberg M, Feussner I, Apel K (2008) Enzymatic, but not non-enzymatic, $^1O_2$-mediated peroxidation of polyunsaturated fatty acids forms part of the EXECUTER1-dependent stress response program in the *flu* mutant of *Arabidopsis thaliana*. Plant J 54:236–248

Pyke KA, Leech RM (1994) A genetic analysis of chloroplast division and expansion in *Arabidopsis thaliana*. Plant Physiol 104:201–207

Ramel F, Birtic S, Ginies C, Soubigou-Taconnat L, Triantaphylides C, Havaux M (2012) Carotenoid oxidation products are stress signals that mediate gene responses to singlet oxygen in plants. Proc Natl Acad Sci USA 109:5535–5540

Ramel F, Ksas B, Akkari E, Mialoundama AS, Monnet F, Krieger-Liszkay A, Ravanat JL, Mueller MJ, Bouvier F, Havaux M (2013) Light-induced acclimation of the Arabidopsis *chlorina1* mutant to singlet oxygen. Plant Cell 25:1445–1462

Rentel MC, Lecourieux D, Ouaked F, Usher SL, Petersen L, Okamoto H, Knight H, Peck SC, Grierson CS, Hirt H, Knight MR (2004) OXI1 kinase is necessary for oxidative burst-mediated signalling in Arabidopsis. Nature 427:858–861

Rosenwasser S, Fluhr R, Joshi JR, Leviatan N, Sela N, Hetzroni A, Friedman H (2013) ROSMETER: a bioinformatic tool for the identification of transcriptomic imprints related to reactive oxygen species type and origin provides new insights into stress responses. Plant Physiol 163:1071–1083

Rytz TC, Miller MJ, McLoughlin F, Augustine RC, Marshall RS, Juan YT, Charng YY, Scalf M, Smith LM, Vierstra RD (2018) SUMOylome profiling reveals a diverse array of nuclear targets modified by the SUMO ligase SIZ1 during heat stress. Plant Cell 30:1077–1099

Sade N, Umnajkitikorn K, Rubio Wilhelmi MDM, Wright M, Wang S, Blumwald E (2018) Delaying chloroplast turnover increases water-deficit stress tolerance through the enhancement of nitrogen assimilation in rice. J Exp Bot 69:867–878

Samuel MA, Salt JN, Shiu SH, Goring DR (2006) Multifunctional arm repeat domains in plants. Int Rev Cytol 253:1–26

Schuck S (2020) Microautophagy – distinct molecular mechanisms handle cargoes of many sizes. J Cell Sci 133:jcs246322

Sharma P, Jha AB, Dubey RS, Pessarakli M (2012) Reactive oxygen species, oxidative damage, and antioxidative defense mechanism in plants under stressful conditions. J Bot 2012:1–26

Shen G, Adam Z, Zhang H (2007a) The E3 ligase AtCHIP ubiquitylates FtsH1, a component of the chloroplast FtsH protease, and affects protein degradation in chloroplasts. Plant J 52:309–321

Shen G, Yan J, Pasapula V, Luo J, He C, Clarke AK, Zhang H (2007b) The chloroplast protease subunit ClpP4 is a substrate of the E3 ligase AtCHIP and plays an important role in chloroplast function. Plant J 49:228–237

Shumbe L, Chevalier A, Legeret B, Taconnat L, Monnet F, Havaux M (2016) Singlet oxygen-induced cell death in Arabidopsis under high-light stress is controlled by OXI1 kinase. Plant Physiol 170:1757–1771

Shumbe L, D'Alessandro S, Shao N, Chevalier A, Ksas B, Bock R, Havaux M (2017) METHYLENE BLUE SENSITIVITY 1 (MBS1) is required for acclimation of Arabidopsis to singlet oxygen and acts downstream of beta-cyclocitral. Plant Cell Environ 40:216–226

Smalle J, Vierstra RD (2004) The ubiquitin 26S proteasome proteolytic pathway. Annu Rev Plant Biol 55:555–590

de Souza A, Wang JZ, Dehesh K (2017) Retrograde signals: integrators of interorganellar communication and orchestrators of plant development. Annu Rev Plant Biol 68:85–108

Springer A, Acker G, Bartsch S, Bauerschmitt H, Reinbothe S, Reinbothe C (2015) Differences in gene expression between natural and artificially induced leaf senescence in barley. J Plant Physiol 176:180–191

Srivastava M, Srivastava AK, Orosa-Puente B, Campanaro A, Zhang C, Sadanandom A (2020) SUMO conjugation to BZR1 enables brassinosteroid signaling to integrate environmental cues to shape plant growth. Curr Biol 30:1410–1423.e1413

Su T, Li X, Yang M, Shao Q, Zhao Y, Ma C, Wang P (2020) Autophagy: an intracellular degradation pathway regulating plant survival and stress response. Front Plant Sci 11:164

Sullivan JA, Shirasu K, Deng XW (2003) The diverse roles of ubiquitin and the 26S proteasome in the life of plants. Nat Rev Genet 4:948–958

Sun AZ, Guo FQ (2016) Chloroplast retrograde regulation of heat stress responses in plants. Front Plant Sci 7:398

Sung D-Y, Kaplan F, Guy CL (2001) Plant Hsp70 molecular chaperones: protein structure, gene family, expression and function. Physiol Plant 113:443–451

Suo J, Zhao Q, David L, Chen S, Dai S (2017) Salinity response in chloroplasts: insights from gene characterization. Int J Mol Sci 18:1011

Sychta K, Slomka A, Kuta E (2021) Insights into plant programmed cell death induced by heavy metals-discovering a terra incognita. Cells 10:65

Tanaka R, Kobayashi K, Masuda T (2011) Tetrapyrrole metabolism in *Arabidopsis thaliana*. Arabidopsis Book/American Society of Plant Biologists 9:e0145

Tano DW, Kozlowska MA, Easter RA, Woodson JD (2022) Multiple pathways mediate chloroplast singlet oxygen stress signaling. Plant Mol Biol. https://doi.org/10.1007/s11103-022-01319-z

Triantaphylides C, Havaux M (2009) Singlet oxygen in plants: production, detoxification and signaling. Trends Plant Sci 14:219–228

Triantaphylides C, Krischke M, Hoeberichts FA, Ksas B, Gresser G, Havaux M, Van Breusegem F, Mueller MJ (2008) Singlet oxygen is the major reactive oxygen species involved in photooxidative damage to plants. Plant Physiol 148:960–968

Tripathy BC, Oelmuller R (2012) Reactive oxygen species generation and signaling in plants. Plant Signal Behav 7:1621–1633

Umnajkitikorn K, Sade N, Rubio Wilhelmi MDM, Gilbert ME, Blumwald E (2020) Silencing of OsCV (chloroplast vesiculation) maintained photorespiration and N assimilation in rice plants grown under elevated CO(2). Plant Cell Environ 43:920–933

Van Aken O, Van Breusegem F (2015) Licensed to kill: mitochondria, chloroplasts, and cell death. Trends Plant Sci 20:754–766

Vierstra RD (2009) The ubiquitin-26S proteasome system at the nexus of plant biology. Nat Rev Mole Cell Biol 10:385–397

Wada S, Ishida H, Izumi M, Yoshimoto K, Ohsumi Y, Mae T, Makino A (2009) Autophagy plays a role in chloroplast degradation during senescence in individually darkened leaves. Plant Physiol 149:885–893

Wagner D, Przybyla D, Op den Camp R, Kim C, Landgraf F, Lee KP, Wursch M, Laloi C, Nater M, Hideg E, Apel K (2004) The genetic basis of singlet oxygen-induced stress responses of *Arabidopsis thaliana*. Science 306:1183–1185

Wang S, Blumwald E (2014) Stress-induced chloroplast degradation in Arabidopsis is regulated via a process independent of autophagy and senescence-associated vacuoles. Plant Cell 26:4875–4888

Wang H, Lu Y, Jiang T, Berg H, Li C, Xia Y (2013a) The Arabidopsis U-box/ARM repeat E3 ligase AtPUB4 influences growth and degeneration of tapetal cells, and its mutation leads to conditional male sterility. Plant J 74:511–523

Wang Y, Yu B, Zhao J, Guo J, Li Y, Han S, Huang L, Du Y, Hong Y, Tang D, Liu Y (2013b) Autophagy contributes to leaf starch degradation. Plant Cell 25:1383–1399

Wang L, Kim C, Xu X, Piskurewicz U, Dogra V, Singh S, Mahler H, Apel K (2016) Singlet oxygen- and EXECUTER1-mediated signaling is initiated in grana margins and depends on the protease FtsH2. Proc Natl Acad Sci USA 113:E3792–E3800

Wang Y, Wu Y, Yu B, Yin Z, Xia Y (2017) EXTRA-LARGE G PROTEINs interact with E3 ligases PUB4 and PUB2 and function in cytokinin and developmental processes. Plant Physiol 173:1235–1246

Wang L, Leister D, Guan L, Zheng Y, Schneider K, Lehmann M, Apel K, Kleine T (2020) The Arabidopsis SAFEGUARD1 suppresses singlet oxygen-induced stress responses by protecting grana margins. Proc Natl Acad Sci PNAS 117:6918–6927
Wang Y, Wu Y, Zhong H, Chen S, Wong KB, Xia Y (2022) Arabidopsis PUB2 and PUB4 connect signaling components of pattern-triggered immunity. New Phytol 233:2249–2265
Watson SJ, Li N, Ye Y, Wu F, Ling Q, Jarvis RP (2021) Crosstalk between the chloroplast protein import and SUMO systems revealed through genetic and molecular investigation in Arabidopsis. elife 10:e60960
Witte CP, Herde M (2020) Nucleotide metabolism in plants. Plant Physiol 182:63–78
Wobbe L (2020) The molecular function of plant mTERFs as key regulators of organellar gene expression. Plant Cell Physiol 61:2004–2017
Woodson JD (2022) Control of chloroplast degradation and cell death in response to stress. Trends Biochem Sci 47:851–864
Woodson JD, Joens MS, Sinson AB, Gilkerson J, Salome PA, Weigel D, Fitzpatrick JA, Chory J (2015) Ubiquitin facilitates a quality-control pathway that removes damaged chloroplasts. Science 350:450–454
Yang H, Yang S, Li Y, Hua J (2007) The Arabidopsis *BAP1* and *BAP2* genes are general inhibitors of programmed cell death. Plant Physiol 145:135–146
Ye C, Zheng S, Jiang D, Lu J, Huang Z, Liu Z, Zhou H, Zhuang C, Li J (2021) Initiation and execution of programmed cell death and regulation of reactive oxygen species in plants. Int J Mol Sci 22:12942
Youle RJ, Narendra DP (2011) Mechanisms of mitophagy. Nat Rev Mol Cell Biol 12:9–14
Yu X, Liu H, Klejnot J, Lin C (2010) The cryptochrome blue light receptors. Arabidopsis Book 8:e0135
Yu G, Derkacheva M, Rufian JS, Brillada C, Kowarschik K, Jiang S, Derbyshire P, Ma M, DeFalco TA, Morcillo RJL, Stransfeld L, Wei Y, Zhou JM, Menke FLH, Trujillo M, Zipfel C, Macho AP (2022) The Arabidopsis E3 ubiquitin ligase PUB4 regulates BIK1 and is targeted by a bacterial type-III effector. EMBO J 41:e107257. https://doi.org/10.15252/embj.2020107257
Zhang M, Windheim M, Roe SM, Peggie M, Cohen P, Prodromou C, Pearl LH (2005) Chaperoned ubiquitylation – crystal structures of the CHIP U box E3 ubiquitin ligase and a CHIP-Ubc13-Uev1a complex. Mol Cell 20:525–538
Zhang H, Amick J, Chakravarti R, Santarriaga S, Schlanger S, McGlone C, Dare M, Nix JC, Scaglione KM, Stuehr DJ, Misra S, Page RC (2015) A bipartite interaction between Hsp70 and CHIP regulates ubiquitination of chaperoned client proteins. Structure 23:472–482
Zhang Y, Xia G, Zhu Q (2021) Conserved and unique roles of chaperone-dependent E3 ubiquitin ligase CHIP in plants. Front Plant Sci 12:699756
Zhuang X, Jiang L (2019) Chloroplast degradation: multiple routes into the vacuole. Front Plant Sci 10:359

Chapter 6
The Genetic Engineering Toolbox for Transformation of Higher Plant Plastids

Alessandro Occhialini

Abstract Plastids are membrane-bound organelles that assume a wide range of essential functions in the plant kingdom, ranging from photosynthesis to coordinating with other compartments in important biosynthesis of metabolites. During the evolution of plant cells, this class of organelles originated from an event of endosymbiosis between a cyanobacterium and an early eukaryotic cell. After this first incorporation, this prokaryotic progenitor progressively lost the peculiar characteristic of an autonomous genetic system, evolving into an organelle with essential functions in what we know today as plant cells. However, modern plastids still preserve several prokaryotic-like features, including the retention of a genome (the plastome), and other aspects related to the transcription and translation machineries that enable plastids to express genes arranged in operons. These characteristics, along with the central role of plastids in many endogenous metabolic pathways, make this group of organelles valuable targets for plant genetic engineering applications. In this chapter we describe the fundamental aspects of plastid genetic engineering, from the design of transformation vectors for delivery of transgenes, to methods for production of stable transgenic (transplastomic) plants in tissue culture. This chapter will also provide a description of the current progress in designing novel genetic tools and future challenges in plastid genetic engineering.

Keywords Plastids · Genetic engineering · Plastome · Transplastomic plants · Homologous recombination · Site-specific integration

A. Occhialini (✉)
Department of Plant Sciences, University of Tennessee, Knoxville, TN, USA

Center for Agricultural Synthetic Biology (CASB), University of Tennessee, Knoxville, TN, USA
e-mail: aocchial@utk.edu

T. M. Burch-Smith (ed.), *Chloroplast Gene Expression: Regulation, Stress Signaling and Biotechnology*, Nucleic Acids and Molecular Biology 37,
https://doi.org/10.1007/978-3-031-70098-9_6

6.1 Introduction

Plastids belong to a large group of endosymbiotic organelles containing their own genome (plastome) that assume many specialized functions in plant cells. This group of organelles originates from the same organelle precursor (proplastid) contained in meristematic cells that differentiate into compartments with distinct functions in different plant cells and organs (Gould et al. 2008; McFadden 2014).

Compared to their cyanobacterial ancestor, the genome contained in plastids has undergone a progressive size reduction due to a process of plastid-to-nucleus gene flow with loss of function of an independent living system. Therefore, actual plastids are defined as "semi-autonomous" organelles with the majority of plastid-specific proteins being encoded by the nuclear genome and imported in these organelles by sorting mechanisms mediated by the presence of chloroplast-targeting peptides (cTPs) at the amino termini of proteins imported in plastids (Soll and Schleiff 2004; Bruce 2000).

Proplastid differentiation into organelles with specialized functions is regulated by complex nucleus-to-plastid signaling mechanisms modulated by environmental factors in which exposure to light irradiation plays a central role. In fact, exposure to light triggers proplastid differentiation into chloroplasts, the photosynthetic organelles of green tissue, and chromoplasts, a type of plastids with biosynthetic and storage functions related to important metabolites in fruits. On the contrary, proplastids not exposed to light are subjected to other developmental pathways that lead to differentiation into leucoplasts and other derivative plastids in storage tissues involved in biosynthesis of essential molecules such as starch (amyloplasts), proteins (proteinoplasts), and lipids (elaioplasts) (Jarvis and López-Juez 2013; Choi et al. 2021; Sierra et al. 2023).

The central role of plastids in plant metabolism has attracted the scientific community working in plant genetic engineering for many years. From the first attempt to transform plastids isolated from plant tissues and the production of the first transgenic (transplastomic) plant with transgenes stably integrated in the plastome (Daniell and McFadden 1987; Svab et al. 1990; Svab and Maliga 1993), plastids have been considered valuable targets for large numbers of potential biotech applications for over 30 years (Jin and Daniell 2015). Since the concept of plastid genetic engineering was developed, plastids have had the largest success as targets for installation of genetic circuits for production of high-value molecules. For this purpose, plant cells and tissues have been used as bioreactors for production of a wide range of valuable compounds, including vaccines (Chan et al. 2016), enzymes with industrial applications (Daniell et al. 2019), pharmaceutical drugs for utilization in humans (Kwon et al. 2018) and vitamins (Yabuta et al. 2013). As alternative biotechnological applications, plastid engineering has also been used to improve plant traits in agriculture. For example, attempts to enhance plant photosynthetic efficiency have been performed by installing transgenes encoding Rubisco (ribulose-1,5-bisphosphate carboxylase/oxygenase) enzyme variants with improved kinetic properties for carbon fixation (Lin et al. 2014). Plastid genetic engineering has also been used to install valuable transgenes coding for insecticide molecules, such as the *Bacillus thuringiensis* toxin Bt and long dsRNAs, to produce plants with an improved resistance to insect pathogens (Kota et al. 1999; Zhang et al. 2015). However, despite the large success of utilization of plastids as bioreactors, at this time no transplastomic plants with improved traits have been approved for commercial use in agriculture.

Over the next decade, it is anticipated that the rapid expansion of cutting-edge techniques for improving the current toolbox for plastid genetic engineering will provide a valuable contribution to develop a new concept of plant biofactories. Furthermore, plastid genetic engineering will couple with other disciplines, from conventional breeding to novel synthetic biology tools, for defining novel concepts for sustainable agriculture to overcome future global challenges of climate change and food security. This book chapter will provide a comprehensive overview of plastid genetic engineering, describing both conventional plastid engineering techniques and current research progress in developing new tools. Bottlenecks and future challenges will also be discussed. Figure 6.1 provides an overview of the whole process of plastid genetic engineering and production of transplastomic plants in tissue culture.

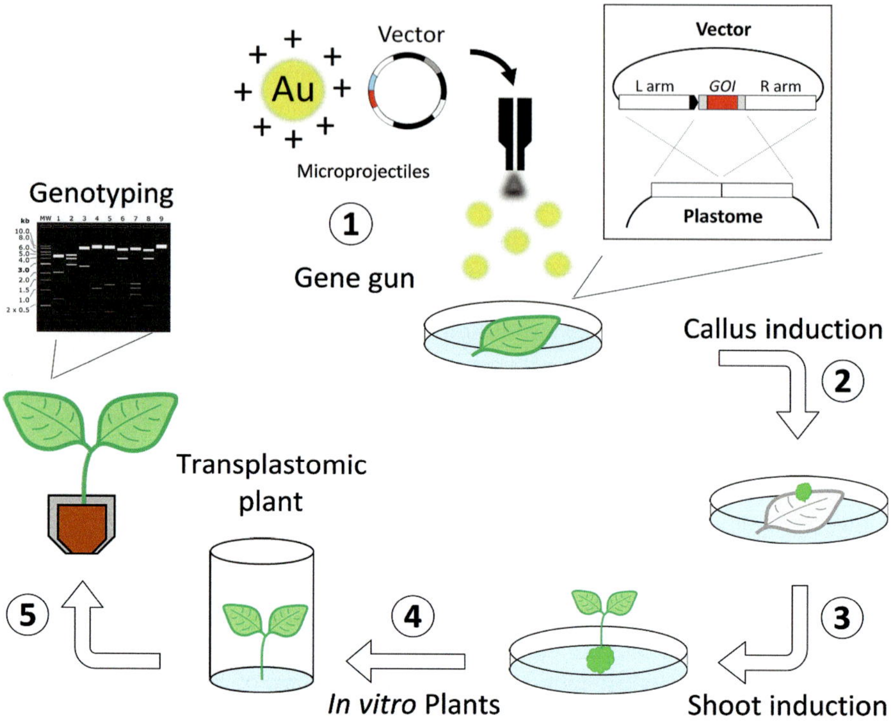

Fig. 6.1 Production of transplastomic plants in tissue culture. (1) Chloroplast transformation vectors are complexed on the surface of gold (Au) nanoparticles (microprojectiles) using electrostatic interactions. Microprojectiles (Au/vector) are delivered into leaf disks using biolistic bombardment (Gene-Gun apparatus). The transformation vector induces transgene (GOI) integration in the plastome by homologous recombination between homologous regions (arms) of the vector and plastome (see also Fig. 6.2). (2) Embryogenic green calli are produced in tissue culture from transformed leaf tissue using callus induction media with selection. (3) Transplastomic shoots are produced from green calli incubated in shooting induction media with selection. (4) Individual shoots are transferred in tissue culture vessels containing media for rooting with selection. (5) Transplastomic plants are genotyped and characterized for the functionality of the novel genetic circuit installed

6.2 Plastid Genetic Engineering as Valuable Alternative Strategy to Express Transgenes in Plant Cells, "Advantages and Potential Bottlenecks"

Since the first transplastomic plant was produced, plastid genetic engineering methods have been considered valuable alternative strategies for expressing transgenes in plant cells. In fact, plastid genetic engineering has several positive features compared to more widely used techniques for transformation of the nuclear genome (Daniell et al. 2002; Bock 2015). However, in association with these advantages, a few drawbacks must also be considered. In the following paragraphs the most important advantages and drawbacks associated with plastid genetic engineering will be discussed.

6.2.1 Positive Features of Plastid Genetic Engineering

One of the most important characteristics of plastid engineering is the ability to use homologous recombination (HR) to induce site-specific transgene integration into the plastome. Therefore, unlike binary transformation vectors used for nuclear genome engineering, the utilization of HR vectors avoids both potential transgene positional effects and multiple construct integration events that are typical in transformation of the nuclear genome (Verma and Daniell 2007; Lutz et al. 2007). While different transgenic lines for the nuclear genome are genetically different, transplastomic lines produced from different transformation events are genetically identical for the number of copies and position of transgene integrated. In this way, few transplastomic plants need to be characterized, the opposite of nuclear engineering, where dozens or even hundreds of independent lines must be screened to characterize the real biological effect of the transgene integrated.

Another positive feature of plastid genetic engineering is the ability to achieve very high levels of foreign protein accumulation. For plastid genetic engineering, foreign protein levels of up to the 70% of the total leaf soluble proteins were reported in the literature (Oey et al. 2009), largely exceeding the highest level of foreign protein accumulation reported for nuclear engineering. This characteristic of plastid genetic engineering is supported by the fact that multiple chloroplasts are contained in mesophyll cells, and each of these organelles, in turn, contains multiple copies of the plastome (Birky 1995), vastly increasing the number of transgenes available compared to nuclear genome engineering. For example, in the plant model *Arabidopsis thaliana*, a mature mesophyll cell contains 80 to 120 chloroplasts, each of these containing ~7 to 130 copies of plastome, for a total of thousands of transgene copies per cell (Crumpton-Taylor et al. 2012; Greiner et al. 2020). Plastid genetic engineering largely exceeds the transgene copy number achievable in nuclear engineered plants which would likely have 1–4 transgene copies, depending on number of integration events and on the ploidy of the plant. Another characteristic that improves transgene expression in plastids is the lack of typical mechanisms of gene silencing affecting expression of nuclear genes. In fact, expression of nuclear genes is frequently subjected to RNA interference (RNAi) or posttranscriptional gene silencing (PTGS), a mechanism used by plant cells to modulate expression of endogenous genes and as a defense mechanism against invading pathogenic sequences (Ruhlman et al. 2010; Verma et al. 2008). The components of gene silencing machinery are not localized in plastids, allowing higher levels of transgene expression and consequent improved accumulation of foreign protein.

Compared to their cyanobacterial ancestor, plant plastids maintain several features of the prokaryotic transcription and translation machineries. Similar to prokaryotic cells, one of the positive characteristics of plastids is the ability to coordinate the expression of multiple gene clusters into operons from the same polycistronic promoter, facilitating complex metabolic engineering. In contrast, the expression of multiple transgene pathways from the nuclear genome must be coordinated from individual expression cassettes. The control of optimal transgene expression levels of multi-gene clusters using individual monocistronic cassettes is extremely difficult and less flexible than organizing transgenes into operons (Howe et al. 2008a; Lu et al. 2013). An important advance in plastid genetic engineering was the characterization of intercistronic expression elements (IEEs), that are described as endogenous intergenic sequences involved in processing of polycistronic transcripts into monocistronic expression units in plastids. The utilization of IEEs along with other regulatory elements (promoters and untranslated regions (UTRs)) has enabled synthetic biologists to design complex synthetic operons for multi-gene expression in plastids (Lin et al. 2014; Zhou et al. 2007).

Another advantage of plastid genetic engineering is associated with the characteristic maternal mechanism of inheritance of plastids. This particular mode of inheritance represents a natural mechanism of bioconfinement, largely reducing the frequency of transgene escape through pollen (Daniell 2007; Ruf et al. 2007). In contrast, higher risk of gene flow to native plant species is associated with transgenes integrated in the nuclear genome which follows a classical Mendelian mechanism of inheritance. Furthermore, transgene expression in plastid genetic engineering is performed using endogenous plastomic regulatory elements (promoters and UTRs) without the utilization of pathogens and part of their genomes for both transgene delivery and expression in plant cells. Practically, all these positive features contribute to improve the environmental biosafety of plastid genetic engineering compared to conventional methods for transgene insertion in the nuclear genome that relay on the utilization of agrobacterium-mediated transgene transformation. These characteristics allow transgenic plants for the plastid genome to be exempted from regulation by governments (Daniell et al. 2019).

6.2.2 Potential Bottlenecks to Consider While Performing Plastid Genetic Engineering

The high level of foreign protein production achievable in plastid genetic engineering is an advantage for biotechnological applications that use plastids as bioreactors. On the other hand, overwhelming the endogenous metabolic system with mass reallocation of energy to sustain high foreign protein production is often associated with phenotypic penalties, including reduced plant growth and dwarf phenotypes (Oey et al. 2009; Schmidt et al. 2021; Zhou et al. 2008). While these phenotypic penalties are less evident in plants grown heterotrophically in bioreactors where the main goal is to produce high amounts of foreign protein per unit of biomass, the installation of valuable transgenes for improving plant traits requires transgenic plants with normal phenotypes able to grow efficiently in the field.

Depending on the application, transgene overexpression and high protein production is not always desirable. In fact, the efficient installation of complex enzymatic pathways requires optimal coordination of expression levels of multiple transgenes (Pfotenhauer et al. 2022). It is crucial to design expression cassettes using combinations of regulatory elements (promoters and UTRs) able to lead optimal levels of transgene expression and protein production for correct functionality of the new metabolic pathway. For this purpose, the development of large libraries of plastid-specific regulatory elements suitable for modular cloning (e.g., by Golden Gate) has accelerated the plant synthetic biology toolbox for fast end flexible assembly of transgene expression cassettes for plastid genetic engineering (Occhialini et al. 2019).

The utilization of optimal integration sites is another important aspect to consider for performing efficient plastid genetic engineering and for avoiding potential phenotypic penalties. Similar to prokaryotic genomes, the plastome has a compact structure with apparently fewer locations for foreign gene integration without interfering with endogenous genetic elements compared to large plant nuclear genomes. Therefore, to perform efficient plastid genetic engineering, the optimal integration site must be evaluated considering both the length of the DNA construct and its content of regulatory elements that may interfere with genetic elements in proximity to the integration site. Several integration sites either in the single copy regions (LSC and SSC) or inverted repeats (IRs) of the plastome have been tested for transgene integration, while only a few of them have been developed as routine vectors for plastid genetic engineering. The characteristics of different integration sites available for diverse biotech applications will be discussed in detail in the following paragraphs.

6.3 Delivery of Transformation Vectors into Plastids

Chloroplasts leaves of several species have been used as the main target in plastid genetic engineering. As alternative targets, plastids contained in dark-grown cell suspension cultures or embryogenic *calli* induced from root tissue have also been used (Langbecker et al. 2004; Ruf et al. 2019; Staub 2021). Plastids contained in dark-grown embryogenic cultures have been used as alternative starting material for plant species difficult to regenerate from leaf tissue and for those plant species, like *A. thaliana*, with mutation problems in plants regenerated in tissue culture from leaf tissue (Ruf et al. 2019). The high proliferation rates of dark-grown cell suspension cultures have also been used to shorten the time required to produce transplastomic plants *in vitro*. For instance, in *Nicotiana tabacum* (tobacco), the utilization of embryogenic cell suspensions has enabled the shortening of the production of transplastomic plants in tissue culture by several weeks. Despite these considerations, chloroplasts are the most widely used target in plastid genetic engineering due to several positive characteristics. Chloroplasts have approximately 50-fold larger size compared to plastids contained in dark-grown tissue (Langbecker et al. 2004), representing a better target for the commercially available size of nanoparticles used for transformation. Higher numbers of chloroplasts are also contained per leaf mesophyll cell, and each of these organelles contain a higher plastome copy number compared to undeveloped plastids of dark grown embryogenic cultures.

The biolistic particle delivery system still represents the most widely used method for delivery of transformation vectors into plastids of different plant species. However, as alternatives, protocols that rely on the use of different types of nanomaterials and polymers have also been described. Advantages and drawbacks of these delivery systems will be discussed in the following paragraphs.

6.3.1 Delivery of Transformation Vectors into Plastids Using the Biolistic Particle Delivery System

The invention of the biolistic system (Gene-Gun apparatus) for nanoparticle-mediated delivery of transformation vectors in plastids, along with the development of tissue culture methods for regeneration of transplastomic plants from transformed tissues have enabled plastid genetic engineering for over 30 years (Svab et al. 1990; Daniell 1993).

For this purpose, nanoparticles are complexed with transformation vectors and then physically "shot" into plastids contained in plant tissues using the Gene-Gun apparatus. While tungsten nanoparticles have been widely used in the past, the most recent protocols of plastid transformation relied on the utilization of gold particles. For sample preparation, transformation vectors are bound on the surface of nanoparticles using electrostatic interactions between the negatively charged phosphate groups of DNA and positively charged polyamines on the surface of nanoparticles. Thereafter, DNA-nanoparticle complexes are precipitated, homogeneously dispersed, and dried on the surface of specific macrocarriers, before being delivered into plant tissue (Verma et al. 2008; Occhialini et al. 2023). Particle delivery is performed in a vacuum chamber using pressurizing helium gas produced by the Gene Gun apparatus. Different transformation parameters including, the distance of sample from the gun, shooting pressure, vacuum level along with the type and size of nanoparticle must be considered for optimal DNA delivery into different types of tissues. Gold nanoparticles of 600 nm in size are used for transformation of leaf chloroplasts while smaller particles (>600 nm) are more suitable for smaller targets like plastids contained in embryogenic cell cultures (Langbecker et al. 2004; Staub 2021).

6.3.2 Alternative Plastid Transformation Methods

While the Gene-Gun is the most used method for plastid transformation in plant cells, its utilization is limited to laboratories equipped with this expensive equipment. The need for more simple and accessible methods of plastid transformation has led scientists working in plastid genetic engineering to develop alternative strategies. For this purpose, the well-known polymer poly-ethylene glycol (PEG), frequently used for nuclear genome transformation in protoplasts, has also been used for DNA delivery into plastids (Golds et al. 1993; O'Neill et al. 1993). The use of cellulolytic enzymes to digest the external cell wall of plant cells has enabled the production of cells in suspension, known as protoplasts, that are more suitable for transformation with exogenous DNA. In fact, PEG is a polyether that in combination with divalent cations (calcium or magnesium) can alter the dynamics of plasma membrane, enabling protoplasts' permeability to naked DNA (Kumar and Ling 2021). The utilization of PEG-mediated plastid transformation has the main advantage of allowing laboratories with limited resources to perform plastid genetic engineering without the need of expensive equipment. On another hand, this method suffers from important limitations such as the requirement of tissue culture methods for plant regeneration from transformed protoplasts along with genetic and epigenetic alterations frequently observed in plants regenerated from protoplasts. Despite these limitations, for a few plant species like *N. tabacum* and *Physcomitrella patens*, PEG-mediated transformation has been used as valuable alternative approach to the biolistic method (Díaz and Koop 2014; Sugita 2021).

Nanomaterials with specific physical and chemical properties have been used to deliver many types of biomolecules into eukaryotic cells (Wang et al. 2017; Deng et al. 2014). For this purpose, the properties of single-walled carbon nanotubes (SWNTs) to bind DNA and pass through the cell wall/membrane have been exploited to deliver transformation vectors into plant cells and organelles. Both the nanoparticle size and its surface charge affect the ability of plant cells to uptake nanoparticles, a process described by the lipid exchange envelope penetration model (LEEP) (Wong et al. 2016). In this group of nanomaterials, the chitosan-complexed single-walled carbon nanotube carriers (CS-SWNTs) are characterized by suitable high surface charge for chloroplast localization. CS-SWNT-based methods of leaf infiltration have been used to deliver DNA into leaf mesophyll chloroplasts of a few plant species including tobacco (*N. tabacum*) and spinach (*Spinacia oleracea*) (Kwak et al. 2019). While these proof-of-concept studies have demonstrated DNA delivery into plastids, CS-SWNT-based methods still require optimization before their adoption for routine use. In fact, many aspects of CS-SWNT methods still need to be optimized, like their reproducibility, transformation efficiency and potential cytotoxic effects of these nanomaterials in plant cells.

6.4 The Plastid Genome "the Plastome"

The study of the structure, evolution, and peculiar gene functions of the genome contained in plastids (plastome) is fundamental for a proper understanding of plastid genetic engineering. Thousands of plastomes have been characterized across the plant kingdom (from unicellular organisms to vascular plants) and these plastome sequences are available to the scientific community in public databases. In the last decade, large genomic studies on plastomes have been accelerated by both the fast improvement of sequencing technologies and the consequent decrease of sequencing cost of whole genomes.

The plastome has evolved as genetic material contained in plastids. Compared to the prokaryotic progenitor, the plastome has been subjected to a gradual size reduction and structural remodeling. In fact, during the evolution of terrestrial plants most genes from the cyanobacterial progenitor migrated to the nuclear genome through a gradual process of plastid-to-nucleus gene flow. Comparative studies between the *A. thaliana* genomic sequence and the genomes of other photosynthetic microorganisms have suggested that in this model plant >4500 genes encoding proteins have a cyanobacterial origin (Martin et al. 2002). This is also supported by studies showing that the whole plastid proteome comprises ~3000 proteins and the majority are encoded by the nuclear genome and then the translated preproteins are imported into plastid using chloroplast targeting peptides (cTPs) (Bruce 2000; Zoschke and Bock 2018; Kleffmann et al. 2004).

In modern plants the plastome consists of a relatively small genome with a wide size variation (45.7 to 217 kb) in different plant species (Schelkunov et al. 2015; Mower and Vickrey 2018). In most vascular plants the plastome is a circular molecule of DNA with a size of ~150 kb, subdivided in three structural regions, that includes two identical inverted repeat sequences (IRs) of ~20–25 kb each, separated by a single copy (SSC) and a large copy (LSC) of unique sequences (Daniell et al. 2016). While initial investigations have described the plastome as circular DNA, more recent investigations have also proposed the presence of complex genomic rearrangement and linear multibranched plastomic forms (Mower and Vickrey 2018; Morley et al. 2019). The plastome contains only ~120–130 genes because of structural remodeling and size reduction of the genome contained in the prokaryotic cyanobacterial progenitor (Timmis et al. 2004). Plastomic genes are subdivided into two groups, genes encoding non-coding RNA involved in protein translation and protein-encoding genes essential in plastid-specific metabolic functions. The first group of non-coding RNA genes are involved in the production of ribosomal RNAs (5S, 16S–23S-4.5S rRNA) of the 70S ribosomal complex, and genes for production of transfer RNAs (tRNAs). For instance, a set of 37 genes transcribing tRNA are present in the plastome of *A. thaliana*. In the second group, most gene-encoding proteins are involved in photosynthesis. This group of plastomic genes includes essential components of the electron transport chain (ETC) and ATP synthase, along with the large subunit (LSU) of the enzyme Rubisco involved in atmospheric carbon fixation. Other protein-encoding genes are involved in gene transcription and translation, including subunits of the plastid encoded RNA polymerase (PEP) core complex and structural proteins of the prokaryotic 70S ribosome, respectively. A small group of plastid genes also encode proteins involved in other metabolic pathways (e.g., fatty acid synthesis), and for proteins with unknown physiological functions.

6.5 Genetics of Transplastomic Plant Production

The plastome is contained in plastid microdomains anchored to the external membrane of thylakoids known as nucleoids. It is likely that these microdomains contain 10–20 plastome copies along with RNA molecules and protein components with either enzymatic or structural functions. In nucleoids, structural DNA binding proteins are involved in plastome organization in microdomains, where many protein components also function as scaffolds for co-localizing enzymes involved in essential reactions in plastids (Krupinska et al. 2013; Thomas and Rose 1983). Many plastid-specific pathways take place in nucleoids, supporting a central role of these microdomains to maintain correct plastid function. These pathways include DNA replication and repair, homologous recombination, and plastome segregation during plastid division, along with essential processes involved in gene transcription, RNA editing, and subsequent protein translation (Krupinska et al. 2013; Nishimura 2023). Like in prokaryotic nucleoids, the ability to localize multiple enzymes in close proximity without membrane physical separation is an advantage to facilitate the coordination of enzymatic pathways. For instance, it was demonstrated that the *A. thaliana*, the NusG protein binds both PEP-associated protein 9 (PAP9) and structural protein components of the 70S ribosome, supporting that RNA transcription and protein translation are coupled processes (Xiong et al. 2022).

Most protocols of biolistic delivery of plastid transformation vectors use chloroplasts from mesophyll cells as targets. It is likely that after delivery, the site-specific plastome integration of the transformation vector takes place into nucleoids. Each mesophyll cell contains >1000 plastome copies and reaching homoplasmy (all plastomes with the transgene integrated) in the transplastomic plant would require full vector integration in all copies of the plastid genome. The ability to reach homoplasmy is facilitated by tissue culture in which transformed leaf cells are dedifferentiated into totipotent cells that differentiate again into mature cells forming stable transplastomic plants. Compared to mature leaf mesophyll cells, meristematic cells contain few proplastid precursors (from 10 to 14), and each of these organelles contains only one or few copies of the plastome (Crumpton-Taylor et al. 2012; Greiner et al. 2020; Bendich 1987). Therefore, after transformation of chloroplasts and their dedifferentiation into proplastids, transgenic plastomes are concentrated into a few copies, before increasing again in copy number in re-differentiated mature cells of newly regenerated transplastomic plants. The plastome localization in nucleoids also facilitates homologous recombination events between transformed and unmodified plastomes increasing the number of genomes with the transgene integrated (Thomas and Rose 1983).

The use of selectable marker genes for production of transgenic plants in selective tissue culture media facilitates the gradual removal of unmodified plastomes with consequent increase of plastome copy number with transgene integrated until reaching full homoplasmy. Homoplasmy can be obtained after 1–3 vegetative generations in tissue culture. Time to reach homoplasmy depends on different variables, such as the type of plastome integration site and whether the procedures of tissue culture for plant regeneration require direct organogenesis (direct plant production) or somatic embryogenesis with an intermediate callus phase (Verma et al. 2008; Valkov et al. 2011).

6.6 Design of Transformation Vectors for Plastid Genetic Engineering

Conventional vectors for plastid genetic engineering rely on homologous recombination (HR) to induce site-specific transgene integration in the plastome (Fig. 6.2). For this, HR vectors are designed with two endogenous homologous sequences (left and right arms) cloned at both 5′ and 3′ ends of the transgene expression cassette to

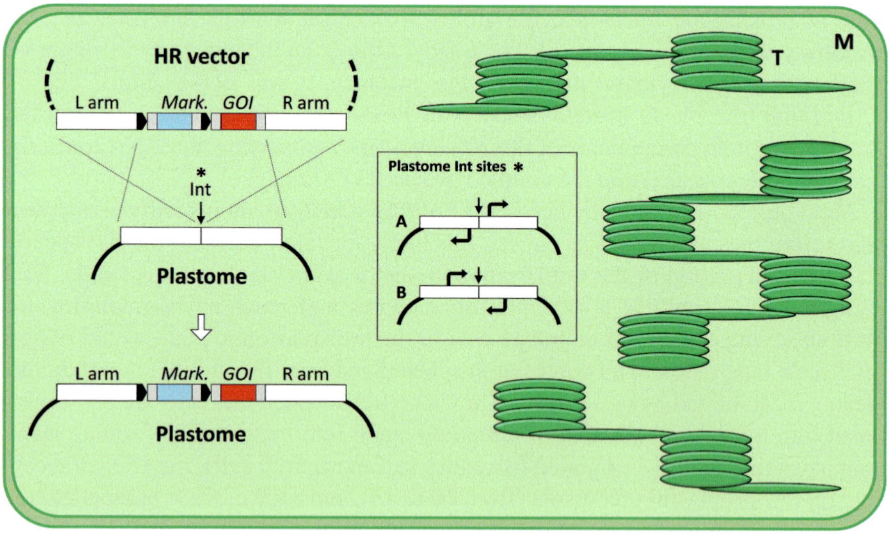

Fig. 6.2 Homologous recombination (HR) vector rearrangement in the plastome. HR vectors are designed with two 1–1.5 kb-long endogenous sequences, left (L) and right (R) arms, homologous to a specific integration site (int*) of the plastome. The selectable marker (mark) and transgene (GOI) expression cassettes cloned in between arms are integrated into the plastome by two homologous recombination events in between arms and cognate regions of the plastome. For ease representation, HR vector and plastome are represented in linear form. A representation of a transcriptionally silent (**a**) and active (**b**) integration site in indicated in the insert. T and M, indicate thylakoids and chloroplasts membranes, respectively

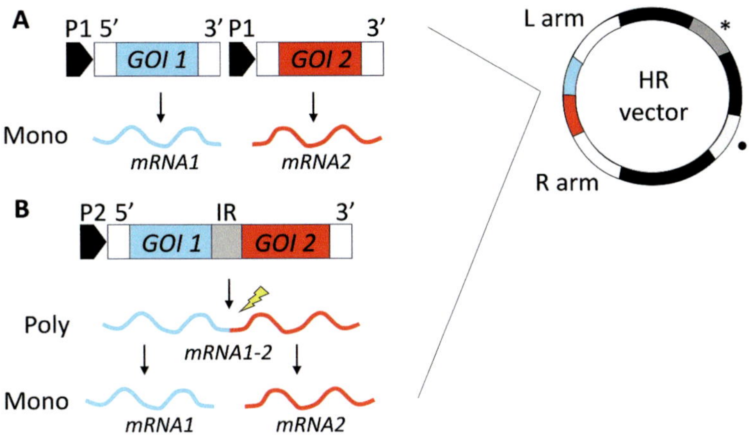

Fig. 6.3 Schematic representation of a homologous recombination (HR) vector. The HR vector contains transgene cassettes cloned in between two 1–1.5 kb homologous arms is indicated. Arms and expression cassettes are inserted in a pUC backbone plasmid for replication and maintenance in *Escherichia coli*. The bacteria *ori* and the selectable marker cassette are indicated with (*) and (•), respectively. The design of monocistronic (**a**) and polycistronic (**b**) cassettes (operons) for expression of two genes of interests (GOI1 and 2) are shown. For monocistronic cassettes, each GOI is expressed as a single transcriptional unit (mRNA1 and 2), while in an operon, genes are expressed into polycistronic transcripts (mRNA1–2) that are processed into monocistronic units (mRNA1 and 2). Monocistronic (P1) or polycistronic (P2) promoters along with 5′ and 3′ UTR (5′ and 3′) are shown in the figure. The design of polycistronic cassettes requires the insertion of intergenic regions (IR) containing intercistronic expression elements (IEE) for correct processing (arrow, yellow) of the polycistronic mRNAs into monocistronic units

integrate in the plastome. Therefore, transgene integration into a plastomic target site of interest is induced by simply designing suitable homologous arms. Generally, 1–2 kb-long arms of conventional HR vectors enable efficient transgene integration (Verma and Daniell 2007; Lutz et al. 2007). However, longer homologous sequences of ~3–5 kb can also be used to improve transformation efficiency, as it was demonstrated in *Solanum tuberosum* (potato) (Occhialini et al. 2020).

Optimal levels of transgene expression in plastids are achieved by using endogenous regulatory elements of the plastid genome (promoters and UTRs). Libraries of regulatory elements compatible for modular cloning are available in public databases, representing a valuable genetic resource for the scientific community working in plastid genetic engineering (Occhialini et al. 2019). Transgene expression cassettes along with homologous arms are cloned into a conventional pUC-based backbone plasmid for *Escherichia coli*, as these vectors are delivered in plastids by using biolistic without the use of pathogenic organisms.

The core components for designing conventional HR vectors will be discussed in the following paragraphs (Fig. 6.3). Novel genetic platforms for transformation of higher plant plastids using episomal non-integrating vectors will also be discussed as alternative strategies for expressing transgenes.

6.6.1 Design of Homologous Arms

The optimal design of homologous arms is fundamental to inducing efficient transgene integration in the plastome. Like the genome of prokaryotes, the plastid genome has a small size and compact structure, reducing the number of potential sites, "available space", for transgene integration. Foreign sequences must be inserted into suitable regions of the plastome, without altering both the sequence and the correct functionality of endogenous genetic elements (genes and regulatory elements). Therefore, the evaluation of optimal integration sites is fundamental to avoid any detrimental effects on plant viability and to preserve normal plastid function in the resulting transplastomic plant.

Several integration sites have been tested for transgene integration in the three structural regions of the plastome (LSC, SSC and IRs) (Table 6.1). Homologous arms designed based on the IR regions enable the integration of two transgene

Table 6.1 Plastomic regions used for transgene integration

Integration site			
Arm 1	Arm 2	Plastome region	References
trnH	*psbA*	Large single copy (LSC)	Carrer and Maliga (1995)
trnG	*trnfM*	Large single copy (LSC)	Bock and Maliga (1995)
ycf3	*trnS*	Large single copy (LSC)	Huang et al. (2002), Klaus et al. (2003)
rbcL	*accD*	Large single copy (LSC)	Svab and Maliga (1993)
petA	*psbJ*	Large single copy (LSC)	Huang et al. (2002), Klaus et al. (2003), Bock et al. (1994)
5′rps12	*clpP*	Large single copy (LSC)	Kuroda and Maliga (2003), Shikanai et al. (2001)
petD	*rpoA*	Large single copy (LSC)	Klaus et al. (2003), Suzuki and Maliga (2000)
ndhB	*rps7*	Inverted repeat (IR)	Hou et al. (2003)
3′rps12	*trnV*	Inverted repeat (IR)	Staub and Maliga (1992), Zoubenko et al. (1994)
trnV	*rrn16*	Inverted repeat (IR)	Staub and Maliga (1992, 1993)
rrn16	*trnI*	Inverted repeat (IR)	Svab et al. (1990), Staub and Maliga (1992)
trnI	*trnA*	Inverted repeat (IR)	Staub and Maliga (1992), Daniell et al. (1998)
trnR	*trnN*	Inverted repeat (IR)	Zou et al. (2003), Huang et al. (2002)
rpl32	*trnL*	Small single copy (SSC)	Eibl et al. (1999), Koop et al. (1996)
ndhG	*ndhI*	Small single copy (SSC)	Occhialini et al. (2020)

Integration sites are subdivided by their location in the three regions of the plastome, the large single copy region (LSC), the small single copy region (SSC), and the inverted repeats (IRs). Per each integration site the homologous arm1 and 2 are indicated with the genes located in the integration site copies per plastome, while only one transgene copy per plastome is integrated by using arms designed on single copy regions (LSC and SSC). Practically, the higher number of functional transgene expression cassettes integrated using HR vectors for IR regions are preferred in biotech applications that require higher transgene expression levels. Several HR vectors with sequence-specific homologous arms for transformation of different plastome sites of model plants, like *Nicotiana benthamiana* and *A. thaliana* (Svab and Maliga 1993; Yu et al. 2017), along with many agricultural important plant species including potato (*Solanum tuberosum*), tomato (*Lycopersicon esculentum*), lettuce (*Lactuca sativa*), soybean (*Glycine max*), and cotton (*Gossypium hirsuturm*) have been developed (Yabuta et al. 2013; Zhang et al. 2015; Lu et al. 2013; Dufourmantel et al. 2004). Table 6.2 presents a list of plants species, including higher plants, bryophytes and algae used in plastid genetic engineering.

Table 6.2 Species of plants and other organisms that have been used in plastid genetic engineering

Classification	Family	Binomial scientific name	Common name	References
Vascular plants	*Amaranthaceae*	*Beta vulgaris*	Sugar beet	De Marchis et al. (2009)
	Apiaceae	*Daucus carota*	Carrot	Kumar et al. (2004a)
	Asteraceae	*Artemisia annua*	Artemisia	Kaushal et al. (2020)
		Lactuca sativa	Lettuce	Lelivelt et al. (2005)
	Brassicaceae	*Arabidopsis thaliana*	Arabidopsis	Yu et al. (2017)
		Brassica capitata	Cabbage	Liu et al. (2007)
		Brassica napus	Oilseed rape	Hou et al. (2003)
		Brassica oleracea	Cauliflower	Nugent et al. (2006)
		Lesquerella fendleri	Lesquerella	Skarjinskaia et al. (2003)
	Cucurbitaceae	*Momordica charantia*	Bitter melon	Narra et al. (2018)
	Fabaceae	*Glycine max*	Soybean	Dufourmantel et al. (2004)
		Medicago sativa	Alfalfa	Wei et al. (2011)
	Malvaceae	*Gossypium hirsutum*	Cotton	Kumar et al. (2004b)
	Plantaginaceae	*Scoparia dulcis*	Sweet broom	Muralikrishna et al. (2016)
	Pinaceae	*Oryza sativa*	Rice	Lee et al. (2006)
	Salicaceae	*Populus alba*	Poplar	Okumura et al. (2006)

(continued)

Table 6.2 (continued)

Classification	Family	Binomial scientific name	Common name	References
	Solanaceae	*Capsicum annuum*	Pepper	Kota et al. (2019)
		Nicotiana benthamiana	Benthi	Davarpanah et al. (2009)
		Nicotiana sylvestris	Woodland tobacco	Maliga and Svab (2011)
		Nicotiana tabacum	Tobacco	Svab and Maliga (1993)
		Petunia hybrida	Petunia	Zubkot et al. (2004)
		Solanum lycopersicum	Tomato	Ruf et al. (2001)
		Solanum melongena	Eggplant	Singh et al. (2010)
		Solanum tuberosum	Potato	Valkov et al. (2011)
Bryophytes	*Funariaceae*	*Physcomitrella patens*	Physcomitrella	Sugiura and Sugita (2004)
	Marchantiales	*Marchantia polymorpha*	Umbrella liverwort	Chiyoda et al. (2007)
Green algae	*Chlamydomonadaceae*	*Chlamydomonas reinhardtii*	Chlamydomonas	Goldschmidt-Clermont (1991)
	Chlorodendraceae	*Platymonas (Tetraselmis) subcordiformis*	–	Cui et al. (2014)
	Dunaliellaceae	*Dunaliella tertiolecta*	–	Georgianna et al. (2013)
Microalgae	*Haematococcaceae*	*Haematococcus pluvialis*	–	Gutiérrez et al. (2012)
	Isochrysidaceae	*Tisochrysis lutea*	–	Bo et al. (2020)
	Monodopsidaceae	*Nannochloropsis oceanica*	–	Gan et al. (2018)
Red algae	*Bangiaceae*	*Pyropia yezoensis*	–	Kong et al. (2017)
	Cyanidiaceae	*Cyanidioschizon merolae*	–	Zienkiewicz et al. (2017)
	Porphyridiaceae	*Porphyridium sp. UTEX 637*	–	Lapidot et al. (2002)
Diatoms	*Phaeodactylaceae*	*Phaeodactylum tricornutum*	–	Xie et al. (2014)
Protists	*Euglenaceae*	*Euglena gracilis*	–	Doetsch et al. (2001)

Plants are subdivided into vascular plants, bryophyte, and algae (green, red and microalgae). Diatom and protist are also included in the table. The classification includes the family, the binomial scientific name, and the common name

The design of optimal HR vectors requires consideration of several intrinsic characteristics of different integration sites. Integration sites are subdivided into transcriptionally active or silent, enabling transgene integration in regions of the plastome with high transcriptional activity or transcriptionally silent regions, respectively. In transcriptionally silent regions (e.g., *trnV-3′rps12* located in the IR region) (Staub and Maliga 1992), regulatory elements of genes of the integration site have an opposite orientation compared to regulatory elements of the transgene cassette integrated. This particular organization reduces interfering effects of genetic elements of the integration site on genetic elements controlling transgene expression, and vice versa. Therefore, HR vectors for integration into silent regions are particularly indicated to preserve the stoichiometry of gene expression in multi-transgene constructs, or for comparing the native activity of regulatory elements without the interference of genetic elements of the integration site.

As an opposite strategy, HR vectors for transgene integration into transcriptionally active sites enable higher levels of transgene expression due to the additive effect of regulatory elements of the integration sites that have the same orientation of regulatory elements of the transgene cassette. The most used transcriptionally active site in plastid engineering integrates transgenes into an endogenous operon located in the IR *trnI/trnA* region of the plastome (Daniell et al. 1998). While silent integration sites are used to preserve optimal expression levels, transcriptionally active integration sites are used to induce high foreign protein production for utilization of plant tissues as bioreactors (Ruhlman et al. 2010). It should be noted that the induction of high level of transgene expression and foreign protein production requires high amounts of energy, potentially overwhelming the endogenous metabolic system. This mass reallocation of energy is associated with reduced plant growth and phenotypic alterations, making transcriptionally active sites suboptimal for installation of valuable transgenes for improving crop traits.

6.6.2 Genetic Modules for Transgene Expression in Plastids

The design of expression cassettes for transgene expression in plastids requires a full understanding of mechanisms that regulate gene transcription and protein translation in this class of organelles. As described in the previous paragraphs, one of the characteristics that make plastids similar to prokaryotic expression systems is their ability to express genes in both monocistronic and polycistronic expression units (Fig. 6.3). While monocistronic cassettes are designed to express single gene transcripts, polycistronic cassettes are suitable for expression of multiple-gene clusters (operons) from the same transcript, making plastids optimal compartments to perform metabolic engineering (Verma and Daniell 2007; Lutz et al. 2007).

The regulatory elements necessary for expression of a gene of interest (GOI) in plastids are specific for these compartments, and they are not interchangeable with their counterparts used for gene expression from the nuclear genome. The regulatory elements toolbox for plastid genetic engineering comprises several regulatory

Table 6.3 Genetic regulatory elements of the plastome that are commonly used in plastid genetic engineering

Regulatory element	Gene name	Function	Reference
Promoter PEP	*atpH*	ATP synthase CF0 subunit III	Miyagi et al. (1998)
	psbA	Photosystem II reaction center protein D1	Staub and Maliga (1994a), Hayashi et al. (2003)
	psbB	Photosystem II CP47 chlorophyll apoprotein	Tanaka et al. (1987)
	psbD	Photosystem II reaction center protein D2	Allison and Maliga (1995)
	rbcL	Ribulose-1,5-bisphosphate carboxylase/oxygenase large subunit	Shiina et al. (1998)
Promoter NEP	*accD*	acetyl-CoA carboxylase, carboxyl transferase subunit beta	Hirata et al. (2004)
	clpP	CLP protease ATP binding subunit	Sriraman et al. (1998)
	rpoB	RNA polymerase β subunit	Liere and Maliga (1999)
Promoter PEP/NEP	*atpB*	ATP synthase CF1 subunit beta	Xie and Allison (2002)
	atpI	ATP synthase CF0 subunit IV	Miyagi et al. (1998)
	rrn S16	16S ribosomal RNA	Allison et al. (1996), Suzuki et al. (2003)
5′UTR	*accD*	acetyl-CoA carboxylase, carboxyl transferase subunit beta	Caroca et al. (2013)
	atpB	ATP synthase CF1 subunit beta	Hirose and Sugiura (2004b), Yukawa et al. (2007)
	atpH	ATP synthase CF0 subunit III	Yukawa et al. (2007)
	atpI	ATP synthase CF0 subunit IV	Baecker et al. (2009)
	clpP	CLP protease ATP binding subunit	Kuroda and Maliga (2002)
	ndhD	NADH-plastoquinone oxidoreductase subunit 4	Hirose and Sugiura (1997)
	psbA	Photosystem II reaction center protein D1	Staub and Maliga (1994a), Zou et al. (2003)
	psbB	Photosystem II CP47 chlorophyll apoprotein	Yukawa et al. (2007)
	psbC	Photosystem II CP43 chlorophyll apoprotein	Kuroda et al. (2007)
	psbH	Photosystem II 10 kDa phosphoprotein	Hammani et al. (2012)
	psbN	Photosystem II protein N	Kuroda and Sugiura (2014)
	rbcL	Ribulose-1,5-bisphosphate carboxylase/oxygenase large subunit	Yukawa et al. (2007), Kuroda and Maliga (2001b)
	rps14	Ribosomal protein S14	Hirose et al. (1998)
	rps2	Ribosomal protein S2	Plader and Sugiura (2003)

(continued)

Table 6.3 (continued)

Regulatory element	Gene name	Function	Reference
	rrn 16S	16S ribosomal RNA	Allison et al. (1996)
3′UTR	*petD*	Cytochrome b6/f complex subunit 4	Tangphatsornruang et al. (2011)
	psaC	Photosystem I subunit VII (iron-sulfur center)	Hirose and Sugiura (1997)
	psbA	Photosystem II reaction center protein D1	Eibl et al. (1999)
	rbcL	Ribulose-1,5-bisphosphate carboxylase/oxygenase large subunit	Eibl et al. (1999)
	rpoA	RNA polymerase α subunit (PEP)	Tangphatsornruang et al. (2011)
UTR (IR)	IEE (*psbT*/ *psbH*)	Intercistronic expression element	Zhou et al. (2007)

Regulatory elements are subdivided into promoters (PEP, NEP, and PEP/NEP) and untranslated regions (5′ and 3′UTR, along with IR). Per each regulatory element, the corresponding gene and its function is also indicated

elements, including promoter sequences and UTRs located at both 5′ and 3′ ends of the gene sequence (Table 6.3). Regulatory elements are organized into the general gene expression cassette arrangement: *promoter::5′UTR::GOI::3′UTR* (Verma and Daniell 2007; Lutz et al. 2007). The design of optimal cassettes for transgene expression from either mono- or polycistronic transcripts requires the use of suitable regulatory elements specific for their requisite cassette arrangements. Other important aspects to consider for optimal transgene cassette design are the complex mechanisms of regulation of gene expression in plastids. In fact, gene expression in plastids is regulated by complex mechanisms of crosstalk between different cellular compartments that involve nuclear-encoded protein factors imported into plastids from the cytosol. A number of these protein factors are RNA-binding proteins interacting with specific *cis* regulatory elements essential to modulate the expression of downstream genes under different physiological and environmental conditions (Schindel et al. 2018; Barkan and Small 2014).

The development of large libraries of plastid-specific genetic modules compatible with Golden Gate cloning has provided a valuable contribution to the synthetic biology toolbox for plastid genetic engineering (Occhialini et al. 2019). Thousands of genetic regulatory elements with different activities are available in public databases, representing a valuable resource for designing optimal transgene expression cassette for plastid genetic engineering.

6.6.2.1 Promoters

The optimal design of a transgene expression cassette requires the understanding of complex mechanisms of gene transcription in plastids (Zhang et al. 2023). Promoters are defined as upstream sequences that contain *cis* regulatory elements for binding the RNA polymerase core complex necessary to initiate transcription of the downstream gene. Transcription starts from a transcription start site (TSS, +1) that identifies the boundary between upstream promoter and 5′ untranslated region (5′UTR). Two different types of RNA polymerase are involved in transcription of different types of genes characterized by different promoter regions (Palomar et al. 2022; Silhavy and Maliga 1998). The first type of RNA polymerase is a multi-subunit enzyme denominated plastid-encoded RNA polymerase (PEP). The core complex of PEP is composed of α, β, β′ and β″subunits encoded from the plastid genes, *rpoA*, *rpoB*, *rpoC1*, and *rpoC2*, respectively. The *rpoA* gene encoding for the α subunit is located in an operon along with other genes for ribosomal protein expression, while *rpoB*, *rpoC1*, and *rpoC2* genes encoding for β subunits are organized into the *rpoBC* operon (do Prado et al. 2024; Vergara-Cruces et al. 2024). The second type of RNA polymerase is defined as nuclear-encoded RNA polymerase (NEP), which is a single-subunit enzyme similar to RNA polymerases present in bacteriophages. Nuclear genes encoding NEP enzymes are classified into different types depending on the final organelle of enzyme accumulation and activity. Three genes *rpoTp*, *rpoTm* and *rpoTmp* are present in the genome of plants, encoding NEP enzymes targeted to plastids, mitochondria or to both of these endosymbiotic organelles (Tarasenko et al. 2016; Hajdukiewicz et al. 1997).

While the plant synthetic biology toolbox still requires substantial improvement, a number of plastid endogenous promoters have been characterized for their activity, representing a valuable resource in plastid genetic engineering. Plastid promoters are classified into PEP, NEP and PEP/NEP promoters depending on their specificity for different types of RNA polymerases involved in gene transcription in plastids. Generally, PEP promoters have a conserved organization of *cis* regulatory elements, while the organization of NEP promoters is more variable. Class I genes are transcribed from PEP-specific promoters, and they mainly include genes encoding proteins involved in photosynthesis, such as the *psbA* and *rbcL* genes. The rest of the genes involved in other plastid-specific pathways are classified into class II (e.g., *clpP* and the *rrn* operon) and III (e.g., *accD* and the *rpoB* operon), and they are expressed from promoter regions specific for both PEP and NEP or only NEP, respectively (Hajdukiewicz et al. 1997; Allison et al. 1996).

PEP promoters are easy to recognize for their similar structure to *E. coli* σ70 promoters. PEP promoters contain −10 (TATAAT) and −35 (TTGACT) consensus regions that are recognized by the cognate σ factor to initiate gene transcription. In plant cells, σ factors are subdivided into six subgroups for recognition of different PEP promoters in plastids. After σ factor promoter recognition, α and β subunits associate, forming the PEP core complex to initiate gene transcription (Isono et al. 1997; Lerbs-Mache 2011). While PEP promoters are characterized by the presence of conserved regions, the composition of *cis* regulatory elements of NEP promoters is variable. NEP promoters are subdivided into three groups, Ia, Ib and II. A particular YRTa domain is present in both Ia and Ib promoters, whereas in promoter Ib, the YRTa domain is also associated with an upstream GAA box. On the contrary, no conserved sequences have been characterized in promoters of class II (Liere and Maliga 1999; Kapoor and Sugiura 1999).

Several promoter regions (PEP, NEP, and PEP/NEP promoters) with different activities have been characterized, representing a valuable resource of genetic modules for transgene expression in plastid (Occhialini et al. 2019; Schindel et al. 2018). Specifically, PEP promoters control the expression of genes highly expressed in chloroplasts, such as many genes encoding for photosynthetic proteins accumulated at high level in mature chloroplasts of green tissues. The most commonly used PEP promoters for transgene overexpression in chloroplasts include the promoters of the *psbA* gene (encoding the protein D1 of photosystem II), the *rbcL* gene (encoding the Rubisco large subunit, LSU), and the *rrn* gene encoding the 16S ribosomal RNA (rRNA) (Ruhlman et al. 2010; Suzuki et al. 2003). Furthermore, expression of photosynthetic genes is also regulated by light irradiation, making the corresponding regulatory elements an optimal choice for biotech applications in chloroplasts.

While many PEP promoters are strong, NEP promoters are less active in mature chloroplasts of green tissues. Most of the genes expressed in non-green plastids, such as *clpP* (encoding an ATP-dependent Clp protease subunit) and *accD* (encoding for an acetyl-CoA carboxylase β subunit) genes, are controlled by NEP promoters, representing an optimal choice for transgene expression in amyloplasts of tubers and other dark-grown tissues (Valkov et al. 2011; Zhang et al. 2012).

6.6.2.2 Untranslated Regions at 5′ End (5′UTRs) of the Gene

The design of transgene expression cassettes requires suitable 5′UTRs to provide transcript stability in plastids and for correct initiation of protein translation. Regions forming stem-loop structures and other *cis*-regulatory elements are essential for correct function of 5′UTRs of different types of plastid promoters (PEP, NEP, and PEP/NEP promoters). While in prokaryotes the structure of 5′UTRs is well conserved, the structure of 5′UTRs in plastids is more variable and less well characterized (Schindel et al. 2018; Zhang et al. 2023).

In the 5′UTR region associated with several PEP promoters of genes encoding proteins is present an essential region for ribosome binding and initiation of protein translation (AGGAGG), known as ribosome-binding site (RBS) or Shine–Dalgarno (SD) sequence. In prokaryotes, RBSs are located 8 nt upstream of the start codon, whereas in 5′UTRs of plastid genes both the location and number of RBSs is variable (Kim and Mullet 1994; Hirose and Sugiura 1996, 2004a). The 5′UTR regions of the *psbA* and *rbcL* genes are well characterized, and they are used in combination with different promoter modules for transgene expression in chloroplasts. For correct design of optimal 5′UTR modules, all essential regions for transgene expression must be considered. The 5′UTR of *psbA* contains three RBSs, but only RBS1 and 2 are essential to provide optimal translation efficiency (Hirose and Sugiura

1996). The other two domains of the *psbA* 5′UTR required for optimal gene expression are an AU-box *cis* element important for light-induced translation (Staub and Maliga 1994a; Klaff and Gruissem 1995), and a 5′ stem-loop structure critical for RNA stability (Zou et al. 2003). Another example of a well-characterized genetic module used in plastid genetic engineering is the 5′UTR of the *rbcL* gene. The *rbcL* 5′UTR has a simple structure composed of two stem-loop regions essential for RNA stability, along with a single RBS site (Shiina et al. 1998). A commonly used regulatory element used for transgene overexpression in plastids consists of the fusion of the strong promoter of the *rrn* gene with the RBS of the *rbcL* 5′UTR (Kwon et al. 2013).

In synthetic biology, the rational design of synthetic modules by combining regulatory elements from different genetic sources enables the production of modules with a predictable activity. The functional characterization of 5′UTR *cis*-regulatory elements is fundamental for designing optimal synthetic modules to modulate transgene expression in plastids (Drechsel and Bock 2011). A commonly used synthetic 5′UTR for transgene overexpression in plastid has been designed by combining *cis*-regulatory elements from the *E. coli* phage T7 gene (*T7G10*) and a downstream box (DB) region including an RBS (Kuroda and Maliga 2001a; Mertens et al. 1996). While the genetic engineering toolbox for plastids includes a number of regulatory elements for constitutive expression, this system still needs to be improved with genetic modules for inducible expression. One of the few inducible systems for transgene expression in chloroplasts has been designed as synthetic 5′UTRs that function as riboswitches with high binding affinity for small molecules (Emadpour et al. 2015). The binding with the signal molecule induces a conformational change of the 5′UTR riboswitch enabling modulation of gene expression in response to specific chemical signals.

The genetic engineering toolbox for plastids also includes orthogonal systems for modulation of transgene expression. Orthogonal activator systems have been designed combining the maize pentatricopeptide repeat protein 10 (PPR10) and its *cis* regulatory element binding site located in the *atpH* 5′UTR (Zhang et al. 2023; Rojas et al. 2019; Yu et al. 2019). PPR10 is an RNA binding protein encoded by the nuclear genome that protects gene transcripts carrying the *atpH* 5′UTR from 5′–3′ exoribonucleolytic degradation, improving RNA stability and protein translation. Genetic circuits composed of a chloroplast targeted PPR10 and a transgene expression cassette controlled by the *atpH* 5′UTR binding site have been designed to induce high-fold changes in expression of target transgenes. For instance, a tissue specific PPR10/*atpH* 5′UTR orthogonal system has been used to drastically improve foreign protein production in amyloplasts, enabling biotech applications in potato tuber biofortification (Yu et al. 2019).

6.6.2.3 Untranslated Region at 3′ End (3′UTRs) and Intergenic Regions (IR) Necessary to Design Synthetic Operons

Untranslated regions at 3′ends of transcripts (3′UTRs) are essential regulatory elements for correct transgene expression in plastids. These regions contain essential *cis* regulatory elements to induce termination of gene transcription, and similar to 5′UTRs they also protect transcripts from 3′–5′ exonuclease RNA degradation. In plastids a canonical transcriptional termination signal is composed of a region forming a stem-loop structure (GC-rich region) and a downstream region rich in AT (Barkan 2011; Rott et al. 1998; Hayes et al. 1996). Compared to other types of regulatory elements (promoters and 5′UTRs), less research has been performed in designing efficient 3′UTRs for transgene expression, providing only a few 3′UTR modules available in the toolbox for plastid genetic engineering.

Transcriptional termination signals of *rbcL*, *psbA*, *petD*, and *rpoA* genes are commonly used 3′UTR modules to design transgene expression cassettes. The heterologous 3′UTR from the *E. coli rrnB* gene has also been used to provide efficient transcriptional termination in plastids, enabling increased transcript accumulation compared to other endogenous 3′UTRs (Tangphatsornruang et al. 2011). An important advance in plastid genetic engineering was the characterization of a 50 bp-long sequence forming stem loops (intercistronic expression element, IEE) located in the intergenic region of the operon between the 3′UTR of the *psbT* gene and the 5′UTR of the *psbH* gene. Such intergenic IEE sequence was characterized as the site of processing of the polycistronic transcript into monocistronic expression units by RNA endonucleases (Zhou et al. 2007). In plastid genetic engineering, IEEs are used as intergenic regions (IRs) between transgenes to provide correct processing of polycistronic transcripts produced from synthetic operons (Fig. 6.3), enabling the installation of complex metabolic pathways (Zhou et al. 2007; Macedo-Osorio et al. 2018).

6.6.3 *Selectable Marker Genes for Plastid Genetic Engineering*

The development of efficient selection systems for production of transplastomic plants in tissue culture has enabled plastid genetic engineering in several plant species. In particular, efficient selection systems are critical in plastid genetic engineering for elimination of all untransformed plastome copies until reaching homoplasmy for transgene integration. Several selectable marker genes have been tested in plastid genetic engineering (Table 6.4), while the most used selectable markers in higher plants belong to the group of positive dominant markers specific for prokaryotic systems (Day and Goldschmidt-Clermont 2011; Miki and McHugh 2004). Generally, positive marker genes encode detoxifying enzymes able to inactivate specific toxic molecules, such as antibiotics. Therefore, selectable marker expression cassettes integrated into the plastome enable transplastomic tissue to grow in tissue culture media in the presence of the cognate molecule, while wild-type tissue

Table 6.4 Selectable marker genes used in plastid genetic engineering

Functional group	Gene	Gene function	Mechanism	Species	References
Confer resistance to antibiotics	*aadA*	Aminoglycoside (3″) (Svab et al. 1990) adenylyltransferase	Confer resistance to spectinomycin and streptomycin	*N. tabacum*	Zoubenko et al. (1994)
	aac6-aph2	Aminoglycoside acetyltransferase (6′)-Ie/aminoglycoside phosphotransferase (2″)-Ia	Confer resistance to tobramycin	*N. tabacum*	Tabatabaei et al. (2017)
	aphA-6	Aminoglycoside 3′-phosphotransferase	Confer resistance to kanamycin and amikacin	*N. tabacum*	Huang et al. (2002)
	ble	Bleomycin resistance protein	Confer resistance to zeomycin	*Nannochloropsis oceanica*	Gan et al. (2018)
	cat	Chloramphenicol acetyltransferase (CAT)	Confer resistance to chloramphenicol	*N. tabacum*	Li et al. (2011)
	ereB	Erythromycin esterase	Confer resistance to erythromycin	*Dunaliella tertiolecta*	Georgianna et al. (2013)
	nptII	Neomycin phosphotransferase II	Confer resistance to kanamycin	*N. tabacum*	Carrer et al. (1993)
	rrnS	16S rRNA mutant	Confer resistance to spectinomycin and streptomycin	*N. tabacum*	Svab et al. (1990)
	rrnL	23S rRNA mutant	Confer resistance to erythromycin	*Chlamydomonas reinhardtii*	Newman et al. (1990)
Confer resistance to herbicides	*ahaS* (W492S)	Acetohydroxyacid synthase mutant	Confer resistance to Sulfometuron methyl	*Porphyridium* sp.	Lapidot et al. (2002)
	bar	Bialaphos resistance protein (BAR)	Confer resistance to phosphinothricin	*N. tabacum*	Iamtham and Day (2000)
	CP4 EPSPS	EPSPS (5-enolpyruvylshikimate 3-phosphate synthase)	Confer resistance to glyphosate	*N. tabacum*	Ye et al. (2003)
	HPPD	4-hydroxyphenylpyruvate dioxygenase	Improve tolerance to diketonitrile	*N. tabacum*	Dufourmantel et al. (2007)

	psbA	Resistant mutant of Photosystem II protein D1	Confer resistance to 3-(3,4-dichlorophenyl)-1,1-dimethylurea (DCMU) and metribuzin	*C. reinhardtii*	Przibilla et al. (1991)
Confer resistance to toxins	*hemL*	Glutamate 1-semialdehyde aminotransferase mutant	Confer resistance to gabaculine	*N. tabacum*	Bellucci et al. (2015)
Recovery metabolism capacity	*ARG9*	Plastid resident N-acetyl ornithine aminotransferase	Restoration of arginine prototrophy in an *arg9* mutant	*C. reinhardtii*	Remacle et al. (2009)
	ASA2	Feedback-insensitive anthranilate synthase (AS) alpha-subunit (ASA2)	Insensitivity to 4MI or 7MT	*N. tabacum*	Barone et al. (2009)
	BADH	Betaine aldehyde dehydrogenase	Confer resistance to betaine aldehyde	*N. tabacum*	Daniell et al. (2001)
	codA	Cytosine deaminase (CD)	Negative selection in media containing 5-fluorocytosine	*N. tabacum*	Serino and Maliga (1997)
	dao	D-amino acid oxidase	Confer tolerance to D-alanine and sensitivity to D-valine	*N. tabacum*	Gisby et al. (2012)
	dsdA	Codon optimized D-serine ammonia lyase	Confer resistance to D-serine	*N. tabacum*	Li et al. (2016)
	ptxD	Synthetic phosphite oxidoreductase	Ability to use phosphite as a source of phosphorus	*C. reinhardtii*	Changko et al. (2020)
Recovery photosynthetic abilities	*atpB*	CF1 ATPase beta-subunit	Recovery photosynthetic capacity in a *ΔatpB* mutant strain	*C. reinhardtii*	Boynton et al. (1988)
	psaA-3	Photosystem I P700 chlorophyll a apoprotein A1	Recovery photosynthetic capacity in recipient strain *psaA-3* mutant	*C. reinhardtii*	Redding et al. (1998)
	psaB	Photosystem I P700 chlorophyll a apoprotein A2	Recovery photosynthetic capacity in a *ΔpsaB* mutant strain	*C. reinhardtii*	Redding et al. (1998)
	petB	Cytochrome b6	Recovery photosynthetic capacity in a *ΔpetB* mutant strain	*C. reinhardtii*	Cheng et al. (2005)

(continued)

Table 6.4 (continued)

Functional group	Gene	Gene function	Mechanism	Species	References
	petA	Cytochrome f	Used to enhance selection by recovery photosynthetic capacity in a Δ*petA* mutant	*N. tabacum*	Klaus et al. (2003)
	rpoA	plastid-encoded RNA polymerase subunit alpha	Used to enhance selection by recovery photosynthetic capacity in a Δ*rpoA* mutant	*N. tabacum*	Klaus et al. (2003)
	rbcL	Rubisco large subunit (LSU)	Used to enhance selection by recovery photosynthetic capacity in a Δ*rbcL* mutant	*N. tabacum*	Kode et al. (2006)
	trnW (UCA)	*trnW* carrying a modified anticodon	Recovery photosynthetic capacity by translational read-through of *psaA-3* opal mutant	*C. reinhardtii*	Young and Purton (2016)
	tscA	Trans-splicing RNA co-factor	Recovery photosynthetic capacity in a Δ*tscA* mutant strain	*C. reinhardtii*	Goldschmidt-Clermont (1991)
	ycf3	Photosystem I assembly protein Ycf3	Used to enhance selection by recovery photosynthetic capacity in a Δ*ycf3* mutant	*N. tabacum*	Klaus et al. (2003)

Selectable markers are classified in genes that confer resistance to antibiotics, toxins, and herbicides, along with genes that restore endogenous metabolic functions and photosynthesis. The gene function, mechanism of selection and the plant species where the marker gene was first tested are indicated is completely removed. Many toxic molecules that specifically inactivate prokaryotic cells have also been tested for their ability to inactivate plastids, while only a few of them have been developed as routine selection systems (Day and Goldschmidt-Clermont 2011; Miki and McHugh 2004).

In the group of positive markers, the most efficient system developed for plastid genetic engineering of higher plants relies on the bacterial *aadA* gene that confers resistance to the aminoglycoside antibiotics spectinomycin and streptomycin (Svab and Maliga 1993; Rascón-Cruz et al. 2021; Murphy 1985). The *aadA* gene from *Staphylococcus aureus* transposon Tn554 encodes an aminoglycoside-3″-adenylyltransferase that catalyzes an adenylation reaction using spectinomycin as target substrate. The product of this chemical reaction is the 3″-adenylylstreptomycin, a non-toxic version of spectinomycin unable to bind the 70S ribosome of plastids (Miki and McHugh 2004; Murphy 1985). Despite the *aadA* gene/spectinomycin selection system enabling efficient plastid genetic engineering in a number of higher plant species (Valkov et al. 2011; Dufourmantel et al. 2004; Ruf et al. 2001), many agricultural important plant species (e.g., monocotyledons) and model plants (e.g., *A. thaliana*) are naturally tolerant to high spectinomycin concentrations. An elegant genetic engineering approach to improve spectinomycin selection in *A. thaliana* was the inactivation of the nuclear *ACC2* gene, encoding for a plastid acetyl-coenzyme A carboxylase. In fact, the CRISPR-Cas9 knock out of the *ACC2* gene produces an *acc2* genotype with much higher susceptibility to spectinomycin, improving the efficiency of plastid transformation by ~100-fold compared to the wild-type genotype (Ruf et al. 2019; Yu et al. 2017). A similar strategy will be a valuable genetic tool to improve plastid transformation efficiency of other recalcitrant plants species, such as monocotyledonous plants [e.g., wheat (*Triticum aestivum*), rice (*Oryza sativa*) and corn (*Zea mays*)].

While the *aadA* gene is still the most efficient and widely used selectable marker for plastid genetic engineering in higher plants, research has been performed to expand the current toolbox for plastid genetic engineering with alternative selection systems. In particular, these alternative systems have been tested in the plant model *N. tabacum* (tobacco), and only a few of them, such as the *nptII* (*neo*) gene, have been developed for routine plastid transformation (Verma and Daniell 2007; Lutz et al. 2007; Day and Goldschmidt-Clermont 2011). The bacterial *nptII* (*neo*) gene from the Tn7 transposon encodes a neomycin phosphotransferase enzyme able to detoxify a number of aminoglycoside antibiotic inhibitors of protein synthesis (kanamycin, neomycin, and geneticin G418) (Nap et al. 1992). This enzyme catalyzes the phosphorylation of aminoglycoside antibiotics, producing non-toxic molecules unable to bind the ribosome and inhibit protein synthesis. An alternative kanamycin resistance gene used to produce transplastomic plants in tobacco is the bacteria *aphA-6* gene encoding the aminoglycoside (3′) transferase (APH(3′)-VI) enzyme (Huang et al. 2002). Another example of a positive selection system tested for plastid genetic engineering of tobacco is the bacterial *cat* gene for detoxification of chloramphenicol, another inhibitor of protein synthesis (Li et al. 2011). This gene encodes a chloramphenicol acetyl transferase (CAT), that catalyzes the acetylation of chloramphenicol, producing a non-toxic product unable to bind the ribosome.

6.6.4 Reporter Genes Used in Plastid Genetic Engineering

Reporter genes are valuable synthetic biology tools for evaluating the activity of genetic modules (e.g., promoters and UTRs) and complex genetic circuits (e.g., orthogonal systems) as well as for subcellular localization of specific protein targets. While many reporter genes are available for eukaryotic systems, the toolbox for plastid genetic engineering comprises only a few reporter genes that have been tested in plastid transformation. The *uidA* gene from the bacterium *E. coli* encoding a β-glucuronidase (GUS) enzyme was one of the first reporter genes utilized in plastid engineering (Staub and Maliga 1993). GUS catalyzes an enzymatic reaction that leads to the conversion of a colorless substrate (X-gluc) into a blue molecule product easily visualized by histochemical staining of plant tissues. Other GUS-specific substrates, such as p-nitrophenyl β-D-glucuronide and the 4-methylumbelliferyl-β-D-glucuronide (MUG), are commonly used for spectrophotometric and fluorometric quantitative assays using plant cell extracts, respectively.

A wide range of fluorescent reporters isolated from either natural organisms or from synthetic design are available in scientific databases (FPbase) (Lambert 2019). The *GREEN FLUORESCENT PROTEIN* (*GFP*) gene from the jellyfish *Aequorea victoria* and synthetic versions are the most common fluorescent reporter genes for plastid genetic engineering (Lambert 2019). The conventional *GFP* gene encodes a protein that produces green-fluorescent signal when excited with a light source in the blue/ultraviolet range of spectrum. The GFP signal can be visualized directly in living cells, representing an advantage compared to other reporter systems that require histochemical cell fixation or homogenization. Moreover, as an alternative genetic tool for plastid genetic engineering, the *GFP* gene was also fused to the *aadA* marker gene for producing a fusion protein with both functions of visual and chemical selection of transformed tissues (Khan and Maliga 1999).

6.7 Episomal Strategies for Plastid Genetic Engineering

Methods that relied on homologous recombination (HR) vectors were the most used strategies for plastid genetic engineering for over 30 years (Svab et al. 1990; Svab and Maliga 1993). Innovation in plastid genetic engineering requires the introduction of novel strategies for improving flexibility and efficiency for installation of complex genetic circuits in plastids. For this purpose, strategies that rely on non-integrating episomal plasmids have been described as alternative approaches for expressing transgenes in plastids (Occhialini et al. 2022; Jakubiec et al. 2021).

The use of episomal platforms for transformation of higher plant plastids has been inspired by the particular plastome organization of a group of photosynthetic microorganisms known as dinoflagellates. Compared to the canonical plastome structure in higher plants, in dinoflagellates plastid genes are subdivided into minicircles, i.e., multiple replicating units containing an origin of replication (*ori*) (Howe et al. 2008b; Koumandou and Howe 2007). The description of the unique minicircle organization of plastomic DNA in dinoflagellates together with the characterization of higher plant *ori* involved in plastome replication have inspired the scientific community to design novel episomal platforms for plastid genetic engineering (Occhialini et al. 2022; Min et al. 2015a, b; Staub and Maliga 1994b). An example of an episomal platform developed for genetic engineering of higher plant plastids is the mini-synplastome. With a similar structure to minicircles, the mini-synplastome is a plasmid with a plastomic endogenous sequence of ~7.9 kb containing plastid-specific *ori* (*ori* A2-A1 and *Nicotiana* plastid extrachromosomal element, NICE1) located at a site opposite to the transgene expression cassettes (Occhialini et al. 2022; Staub and Maliga 1994b, 1995). Mini-synplastomes designed with this ~7.9 kb functional region are maintained in episomal form in higher plant chloroplasts, supporting transgene expression without perturbing the endogenous plastome (Fig. 6.4).

As an alternative strategy, the replication machinery of the beet curly top geminivirus (BCTV) was also used to enable episomal replication in plastids (Jakubiec et al. 2021). For use in plastid genetic engineering, this heterologous system requires two components, a viral Replicase (Rep) helper protein that initiates DNA replication and Rep-binding DNA sequences, known as viral origin of replications (VORs), flanking the transgene expression cassette. Rep specifically interacts with VOR, inducing the formation of a DNA nick that is required to initiate DNA replication in plastids. After Rep-mediated initiation, DNA replication relies entirely on the endogenous replication system of plastids, enabling the formation of episomes (minichromosomes) containing the transgene expression cassette and VOR sequence (Lozano-Durán 2016; Laufs et al. 1995; Jeske et al. 2001) (Fig. 6.4). The viral Rep protein required for this episomal platform can be expressed directly from the plastid genome, or as an alternative strategy, from the nuclear genome as a chloroplast-targeted protein (Jakubiec et al. 2021).

6.8 Marker-Free Strategies for Plastid Genetic Engineering

Given that the real ecological risks associated with potential transgene flow in the environment have been discussed for many years (Miki and McHugh 2004; Breyer et al. 2014; Philips et al. 2022; Yau and Stewart Jr. 2013), the complete removal of

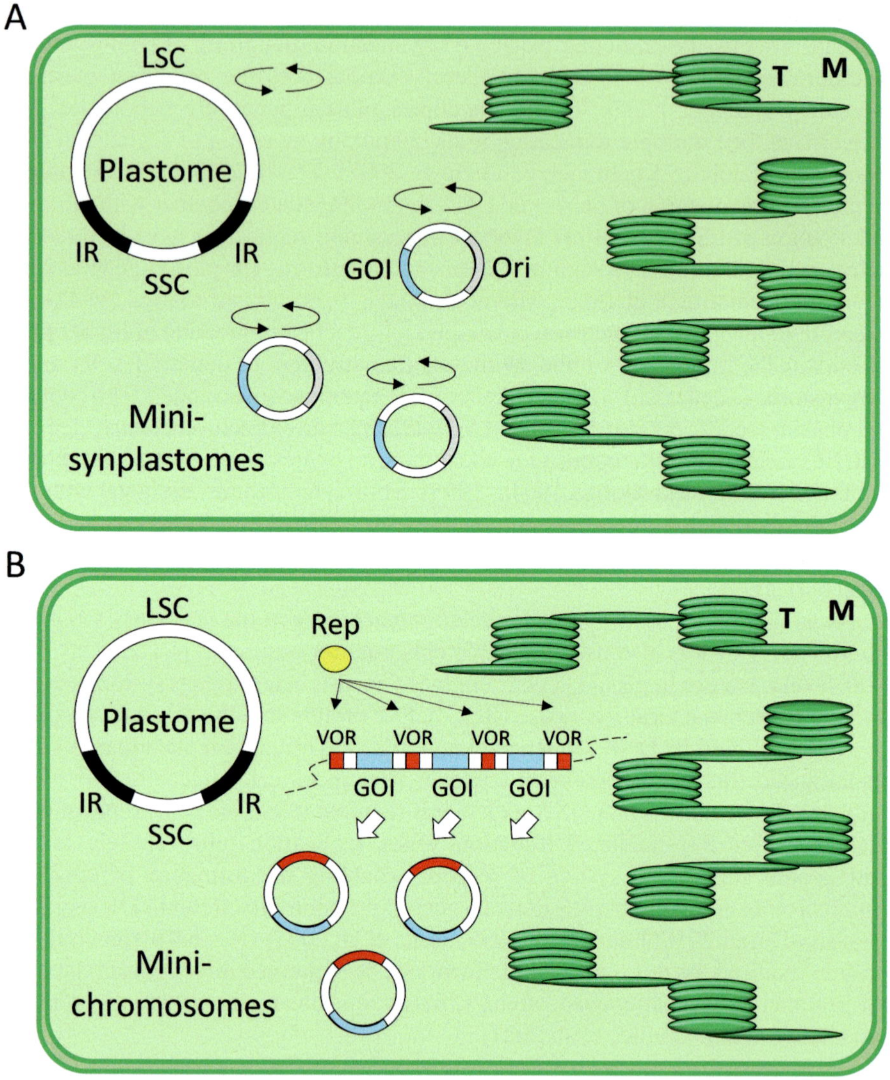

Fig. 6.4 Episomal strategies for transgene expression in plastids. (**a**) The mini-synplastome is a plasmid designed with a transgene expression cassette (GOI, blue) and a plastomic homologous region of ~7.6 kb containing sequences involved in replication (*ori* A2-A1 and NICE1, gray). Mini-synplastomes can be maintained as episomes in chloroplast stroma (black arrows). (**b**) Components of the replication system of the beet curly top geminivirus (BCTV) are used to produce minichromosomes in the chloroplast stroma. In this system the replication initiator protein (Rep, yellow) initiates replication by binding viral origin of replications (VORs, red) flanking transgene expression cassettes (GOI, blue). After Rep-mediated initiation, the endogenous replication system of chloroplasts produces episomes known as minichromosomes containing a transgene expression cassette and a VOR sequence. In both figures, T and M, indicate thylakoids and chloroplasts membranes, respectively

selectable marker genes from transgenic tissues is certainly advantageous to accelerating the path for commercialization of novel transgenic plant varieties. Furthermore, the removal of the marker gene cassette is also an advantage to prevent unnecessary energy costs to endogenous plant metabolism, and to prevent any potential toxic or allergenic effects due to overaccumulation of foreign proteins in food crops.

Recombinase enzymes such as the Cre/LoxP system from P1 phage were used as the first genetic tools to induce excision of marker genes integrated into the plastome (Corneille et al. 2001; Hajdukiewicz et al. 2001) (Fig. 6.5). For this purpose, the marker gene expression cassette is inserted in between two 34 bp-long recognition sites (LoxP) with the same orientation specific for the Cre enzyme. This enzyme is a tyrosine recombinase, able to catalyze a site-specific recombination event in between LoxP recognition sites that leads to excision of the internal sequence containing the marker gene when LoxP sites have the same orientation. The Cre enzyme can be either expressed from the plastid genome, or from the nuclear genome and then targeted to plastids by adding a chloroplast targeting peptide (cTP). Several potential drawbacks are associated with the use of recombinases, making these systems laborious to use. In fact, recombinases leave genetic scars after marker gene removal, and the possibility of unexpected recombination events affecting correct

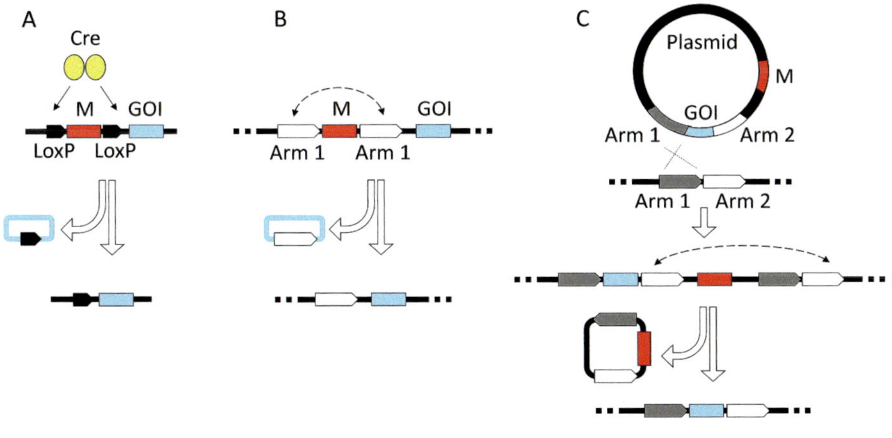

Fig. 6.5 Techniques for marker gene removal in plastids. (**a**) In the Cre/LoxP system, a marker gene expression cassette (M, red) is flanked by LoxP recognition sites with the same orientation. The Cre recombinase (Cre, yellow) catalyzes a site-specific recombination event that leads to the excision of the marker gene cassette between LoxP sites, producing marker-free plants with the transgene of interest integrated (GOI, blue). (**b**) For homologous recombination excision, the marker gene cassette (M, red) is cloned between two direct repeat sequences (Arm-1, white). Homologous recombination between repeats leads to marker gene loop out, producing marker-free plants with the transgene of interest integrated (GOI, blue). (**c**) For inducing marker excision using transient cointegration, a homologous recombination vector is designed with a marker gene cassette (M, red) inserted in the backbone. Vector integration using one arm (Arm-1), followed by a second homologous recombination between Arm-2, leads to the production of marker-free plants with the transgene of interest integrated (GOI, blue)

genome integrity cannot be excluded. Strategies that express Cre from the nuclear genome also require an extra step to remove both the *Cre* gene and its selectable marker gene cassette from the nuclear genome.

The plastid's ability to perform efficient homologous recombination is used to excise marker genes integrated into the plastid genome (Fig. 6.5). For this purpose, the design of marker-free vectors includes two direct DNA repeat sequences flanking the transgene expression cassette. Homologous recombination between direct repeats induces loop-out of the internal marker gene cassette, producing marker-free plants. While marker gene excision can be induced using short direct repeats of ~170 bp, an increased excision frequency was demonstrated by using longer sequences of >400 bp. Such homologous recombination strategies have been used to produce marker-free plants for several plant species, including soybean, lettuce, and tobacco (Daniell et al. 2019, 2020; Iamtham and Day 2000; Dufourmantel et al. 2007; Kumari et al. 2019; Park et al. 2020).

Another homologous recombination method to produce transplastomic marker-free plants relies on transient cointegration vectors (Fig. 6.5). For this purpose, transient cointegration vectors are designed with the marker gene cassette located in the backbone, outside the two arms for transgene integration. Integration of these vectors produce an unstable integration event with large direct repeats containing the marker gene cassette. A second homologous recombination event between direct repeats induces marker gene excision, leaving the transgene cassette integrated in the plastome (Klaus et al. 2004). While homologous recombination methods are the most-used platforms to induce marker gene excision in plastids, they require optimal design of homologous arms to avoid any unexpected secondary recombination events compromising plastome integrity.

6.9 Conclusions and Future Challenges

Although in the last decade the plastid genetic engineering toolbox has been rapidly improved, there is still a lot of unexplored potential for utilization of these plant organelles in metabolic engineering. In fact, many bottlenecks associated with plastid engineering platforms still need to be overcome in order to achieve similar efficiency and flexibility of metabolic engineering performed in prokaryotic systems. The rapid expansion of libraries of plastid-specific genetic modules compatible with Golden Gate cloning has accelerated the design-build-test cycle for plastid genetic engineering, enabling the assembly of numerous DNA constructs in a short time. However, in synthetic biology, the major bottleneck that still prevents large screening approaches of "thousands" of genetic circuits in a short time is the lack of an efficient transient expression system for transgene expression from the plastid genome. Currently, the functional validation of plastid transformation vectors is performed in transplastomic plants, enabling testing only a relatively small number of genetic circuits at the same time. To accelerate this design-build-test cycle, specific cell-free systems have been developed for higher plant plastids, representing a valuable innovation for testing DNA constructs *in vitro* (Böhm et al. 2024; Clark et al. 2024). As an alternative strategy, the use of lower plants (e.g., *Physcomitrella patents*) or other unicellular photosynthetic organisms (e.g., cyanobacteria and algae) with fast growth and shorter life cycles could also represent valuable models for functional evaluation of novel genetic pathways.

It is anticipated that the development of efficient transient expression systems will accelerate the synthetic biology toolbox for plastid engineering in the next decade. The design of novel genetic modules and orthogonal systems will enable the installation of complex logic gates in plastids, allowing the precise modulation of transgene expression. In fact, while transgene overexpression is a routine approach to using plastids as bioreactors, the installation of complex metabolic pathways requires complex genetic circuits to control the expression of multiple genes. Novel episomal platforms for engineering higher plant plastids could potentially improve many aspects of plastid metabolic engineering (Occhialini et al. 2022; Jakubiec et al. 2021; Occhialini and Lenaghan 2023). A potential advantage of using episomal systems includes the ability to mix-and-match genetic modules in multiple episomes, enabling more complex one-step metabolic engineering. Furthermore, in episomes transgene expression is performed without perturbing endogenous genetic elements of integration sites, contributing to maintaining the functional integrity of the plastome.

An important future challenge in genetic engineering of higher plant plastids is to increase the number target plant species to include many agricultural important monocots like wheat, rice, and corn. Plastid transformation in monocots has many challenges, including their natural resistance to high concentrations of spectinomycin that has prevented efficient recovery of transplastomic plants using the *aadA* selectable marker. Therefore, the plastid genetic engineering toolbox for monocots requires the introduction of novel and more efficient selection systems. The installation of valuable genetic circuits in monocot plastids will be a game changer, expanding the number of biotech applications for improving crop production and, in the context of climate change, for producing plants more resilient to environmental biotic and abiotic factors.

Acknowledgement This book chapter is supported by Biotechnology Risk Assessment Research Grants Program grant award no. 2022-33522-38289 from the USDA National Institute of Food and Agriculture. The author also acknowledges USDA Hatch grant.

References

Allison LA, Maliga P (1995) Light-responsive and transcription-enhancing elements regulate the plastid *psbD* core promoter. EMBO J 14(15):3721–3730. https://doi.org/10.1002/j.1460-2075.1995.tb00042.x

Allison LA, Simon LD, Maliga P (1996) Deletion of *rpoB* reveals a second distinct transcription system in plastids of higher plants. EMBO J 15(11):2802–2809

Baecker JJ, Sneddon JC, Hollingsworth MJ (2009) Efficient translation in chloroplasts requires element(s) upstream of the putative ribosome binding site from *atpI*. Am J Bot 96(3):627–636. https://doi.org/10.3732/ajb.0800259

Barkan A (2011) Expression of plastid genes: organelle-specific elaborations on a prokaryotic scaffold. Plant Physiol 155(4):1520–1532. https://doi.org/10.1104/pp.110.171231

Barkan A, Small I (2014) Pentatricopeptide repeat proteins in plants. Annu Rev Plant Biol 65:415–442. https://doi.org/10.1146/annurev-arplant-050213-040159

Barone P, Zhang XH, Widholm JM (2009) Tobacco plastid transformation using the feedback-insensitive anthranilate synthase [alpha]-subunit of tobacco (ASA2) as a new selectable marker. J Exp Bot 60(11):3195–3202. https://doi.org/10.1093/jxb/erp160

Bellucci M, De Marchis F, Ferradini N, Pompa A, Veronesi F, Rosellini D (2015) A mutant Synechococcus gene encoding glutamate 1-semialdehyde aminotransferase confers gabaculine resistance when expressed in tobacco plastids. Plant Cell Rep 34(12):2127–2136. https://doi.org/10.1007/s00299-015-1856-z

Bendich AJ (1987) Why do chloroplasts and mitochondria contain so many copies of their genome? BioEssays 6(6):279–282. https://doi.org/10.1002/bies.950060608

Birky CW Jr (1995) Uniparental inheritance of mitochondrial and chloroplast genes: mechanisms and evolution. Proc Natl Acad Sci USA 92(25):11331–11338. https://doi.org/10.1073/pnas.92.25.11331

Bo Y, Wang K, Wu Y, Cao H, Cui Y, Wang L (2020) Establishment of a chloroplast transformation system in *Tisochrysis lutea*. J Appl Phycol 32(5):2959–2965. https://doi.org/10.1007/s10811-020-02159-4

Bock R (2015) Engineering plastid genomes: methods, tools, and applications in basic research and biotechnology. Annu Rev Plant Biol 66:211–241. https://doi.org/10.1146/annurev-arplant-050213-040212

Bock R, Maliga P (1995) *In vivo* testing of a tobacco plastid DNA segment for guide RNA function in *psbL* editing. Mol Gen Genet 247(4):439–443. https://doi.org/10.1007/bf00293145

Bock R, Kössel H, Maliga P (1994) Introduction of a heterologous editing site into the tobacco plastid genome: the lack of RNA editing leads to a mutant phenotype. EMBO J 13(19):4623–4628. https://doi.org/10.1002/j.1460-2075.1994.tb06784.x

Böhm CV, Inckemann R, Burgis M, Baumann J, Brinkmann CK, Lipińska KE, et al (2024) Chloroplast cell-free systems from different plant species as a rapid prototyping platform. *bioRxiv*, 2024.2002.2019.580994. https://doi.org/10.1101/2024.02.19.580994

Boynton JE, Gillham NW, Harris EH, Hosler JP, Johnson AM, Jones AR et al (1988) Chloroplast transformation in Chlamydomonas with high velocity microprojectiles. Science 240(4858):1534–1538. https://doi.org/10.1126/science.2897716

Breyer D, Kopertekh L, Reheul D (2014) Alternatives to antibiotic resistance marker genes for *in vitro* selection of genetically modified plants – Scientific Developments, Current Use, Operational Access and Biosafety Considerations. Crit Rev Plant Sci 33(4):286–330. https://doi.org/10.1080/07352689.2013.870422

Bruce BD (2000) Chloroplast transit peptides: structure, function and evolution. Trends Cell Biol 10(10):440–447. https://doi.org/10.1016/s0962-8924(00)01833-x

Caroca R, Howell KA, Hasse C, Ruf S, Bock R (2013) Design of chimeric expression elements that confer high-level gene activity in chromoplasts. Plant J 73(3):368–379. https://doi.org/10.1111/tpj.12031

Carrer H, Maliga P (1995) Targeted insertion of foreign genes into the tobacco plastid genome without physical linkage to the selectable marker gene. Bio/Technology 13(8):791–794. https://doi.org/10.1038/nbt0895-791

Carrer H, Hockenberry TN, Svab Z, Maliga P (1993) Kanamycin resistance as a selectable marker for plastid transformation in tobacco. Mol Gen Genet 241(1–2):49–56. https://doi.org/10.1007/bf00280200

Chan HT, Xiao Y, Weldon WC, Oberste SM, Chumakov K, Daniell H (2016) Cold chain and virus-free chloroplast-made booster vaccine to confer immunity against different poliovirus serotypes. Plant Biotechnol J 14(11):2190–2200. https://doi.org/10.1111/pbi.12575

Changko S, Rajakumar PD, Young REB, Purton S (2020) The phosphite oxidoreductase gene, *ptxD* as a bio-contained chloroplast marker and crop-protection tool for algal biotechnology using Chlamydomonas. Appl Microbiol Biotechnol 104(2):675–686. https://doi.org/10.1007/s00253-019-10258-7

Cheng Q, Day A, Dowson-Day M, Shen GF, Dixon R (2005) The *Klebsiella pneumoniae* nitrogenase Fe protein gene (*nifH*) functionally substitutes for the *chlL* gene in *Chlamydomonas reinhardtii*. Biochem Biophys Res Commun 329(3):966–975. https://doi.org/10.1016/j.bbrc.2005.02.064

Chiyoda S, Linley PJ, Yamato KT, Fukuzawa H, Yokota A, Kohchi T (2007) Simple and efficient plastid transformation system for the liverwort *Marchantia polymorpha* L. suspension-culture cells. Transgenic Res 16(1):41–49. https://doi.org/10.1007/s11248-006-9027-1

Choi H, Yi T, Ha SH (2021) Diversity of plastid types and their interconversions. Front Plant Sci 12:692024. https://doi.org/10.3389/fpls.2021.692024

Clark L, Voigt CA, Jewett MC (2024) A chloroplast cell-free system for measuring ribosome binding site strengths. *bioRxiv*. 2024.2002.2017.580832. https://doi.org/10.1101/2024.02.17.580832

Corneille S, Lutz K, Svab Z, Maliga P (2001) Efficient elimination of selectable marker genes from the plastid genome by the CRE-lox site-specific recombination system. Plant J 27(2):171–178. https://doi.org/10.1046/j.1365-313x.2001.01068.x

Crumpton-Taylor M, Grandison S, Png KM, Bushby AJ, Smith AM (2012) Control of starch granule numbers in Arabidopsis chloroplasts. Plant Physiol 158(2):905–916. https://doi.org/10.1104/pp.111.186957

Cui Y, Qin S, Jiang P (2014) Chloroplast transformation of *Platymonas* (*Tetraselmis*) *subcordiformis* with the *bar* gene as selectable marker. PLoS One 9(6):e98607. https://doi.org/10.1371/journal.pone.0098607

Daniell H (1993) Foreign gene expression in chloroplasts of higher plants mediated by tungsten particle bombardment. Methods Enzymol 217:536–556. https://doi.org/10.1016/0076-6879(93)17088-m

Daniell H (2007) Transgene containment by maternal inheritance: effective or elusive? Proc Natl Acad Sci USA 104(17):6879–6880. https://doi.org/10.1073/pnas.0702219104

Daniell H, McFadden BA (1987) Uptake and expression of bacterial and cyanobacterial genes by isolated cucumber etioplasts. Proc Natl Acad Sci USA 84(18):6349–6353. https://doi.org/10.1073/pnas.84.18.6349

Daniell H, Datta R, Varma S, Gray S, Lee SB (1998) Containment of herbicide resistance through genetic engineering of the chloroplast genome. Nat Biotechnol 16(4):345–348. https://doi.org/10.1038/nbt0498-345

Daniell H, Wiebe PO, Millan AF (2001) Antibiotic-free chloroplast genetic engineering – an environmentally friendly approach. Trends Plant Sci 6(6):237–239. https://doi.org/10.1016/s1360-1385(01)01949-5

Daniell H, Khan MS, Allison L (2002) Milestones in chloroplast genetic engineering: an environmentally friendly era in biotechnology. Trends Plant Sci 7(2):84–91. https://doi.org/10.1016/s1360-1385(01)02193-8

Daniell H, Lin CS, Yu M, Chang WJ (2016) Chloroplast genomes: diversity, evolution, and applications in genetic engineering. Genome Biol 17(1):134. https://doi.org/10.1186/s13059-016-1004-2

Daniell H, Ribeiro T, Lin S, Saha P, McMichael C, Chowdhary R et al (2019) Validation of leaf and microbial pectinases: commercial launching of a new platform technology. Plant Biotechnol J 17(6):1154–1166. https://doi.org/10.1111/pbi.13119

Daniell H, Mangu V, Yakubov B, Park J, Habibi P, Shi Y et al (2020) Investigational new drug enabling angiotensin oral-delivery studies to attenuate pulmonary hypertension. Biomaterials 233:119750. https://doi.org/10.1016/j.biomaterials.2019.119750

Davarpanah SJ, Jung SH, Kim YJ, Park Y-I, Min SR, Liu JR et al (2009) Stable plastid transformation in *Nicotiana benthamiana*. J Plant Biol 52(3):244–250. https://doi.org/10.1007/s12374-009-9023-0

Day A, Goldschmidt-Clermont M (2011) The chloroplast transformation toolbox: selectable markers and marker removal. Plant Biotechnol J 9(5):540–553. https://doi.org/10.1111/j.1467-7652.2011.00604.x

De Marchis F, Wang Y, Stevanato P, Arcioni S, Bellucci M (2009) Genetic transformation of the sugar beet plastome. Transgenic Res 18(1):17–30. https://doi.org/10.1007/s11248-008-9193-4

Deng X, Cao M, Zhang J, Hu K, Yin Z, Zhou Z et al (2014) Hyaluronic acid-chitosan nanoparticles for co-delivery of MiR-34a and doxorubicin in therapy against triple negative breast cancer. Biomaterials 35(14):4333–4344. https://doi.org/10.1016/j.biomaterials.2014.02.006

Díaz AH, Koop HU (2014) *Nicotiana tabacum*: PEG-mediated plastid transformation. Methods Mol Biol 1132:165–175. https://doi.org/10.1007/978-1-62703-995-6_9

Doetsch NA, Favreau MR, Kuscuoglu N, Thompson MD, Hallick RB (2001) Chloroplast transformation in *Euglena gracilis*: splicing of a group III twintron transcribed from a transgenic *psbK* operon. Curr Genet 39(1):49–60. https://doi.org/10.1007/s002940000174

Drechsel O, Bock R (2011) Selection of Shine-Dalgarno sequences in plastids. Nucleic Acids Res 39(4):1427–1438. https://doi.org/10.1093/nar/gkq978

Dufourmantel N, Pelissier B, Garçon F, Peltier G, Ferullo JM, Tissot G (2004) Generation of fertile transplastomic soybean. Plant Mol Biol 55(4):479–489. https://doi.org/10.1007/s11103-004-0192-4

Dufourmantel N, Dubald M, Matringe M, Canard H, Garcon F, Job C et al (2007) Generation and characterization of soybean and marker-free tobacco plastid transformants over-expressing a bacterial 4-hydroxyphenylpyruvate dioxygenase which provides strong herbicide tolerance. Plant Biotechnol J 5(1):118–133. https://doi.org/10.1111/j.1467-7652.2006.00226.x

Eibl C, Zou Z, Beck a, Kim M, Mullet J, Koop HU (1999) *In vivo* analysis of plastid *psbA*, *rbcL* and *rpl32* UTR elements by chloroplast transformation: tobacco plastid gene expression is controlled by modulation of transcript levels and translation efficiency. Plant J 19(3):333–345. https://doi.org/10.1046/j.1365-313x.1999.00543.x

Emadpour M, Karcher D, Bock R (2015) Boosting riboswitch efficiency by RNA amplification. Nucleic Acids Res 43(10):e66. https://doi.org/10.1093/nar/gkv165

Gan Q, Jiang J, Han X, Wang S, Lu Y (2018) Engineering the chloroplast genome of oleaginous marine microalga *Nannochloropsis oceanica*. Front Plant Sci 9:439. https://doi.org/10.3389/fpls.2018.00439

Georgianna DR, Hannon MJ, Marcuschi M, Wu S, Botsch K, Lewis AJ et al (2013) Production of recombinant enzymes in the marine alga *Dunaliella tertiolecta*. Algal Res 2(1):2–9. https://doi.org/10.1016/j.algal.2012.10.004

Gisby MF, Mudd EA, Day A (2012) Growth of transplastomic cells expressing D-amino acid oxidase in chloroplasts is tolerant to D-alanine and inhibited by D-valine. Plant Physiol 160(4):2219–2226. https://doi.org/10.1104/pp.112.204107

Golds T, Maliga P, Koop HU (1993) Stable plastid transformation in PEG-treated protoplasts of *Nicotiana tabacum*. Bio/Technology 11(1):95–97. https://doi.org/10.1038/nbt0193-95

Goldschmidt-Clermont M (1991) Transgenic expression of aminoglycoside adenine transferase in the chloroplast: a selectable marker of site-directed transformation of chlamydomonas. Nucleic Acids Res 19(15):4083–4089. https://doi.org/10.1093/nar/19.15.4083

Gould SB, Waller RF, McFadden GI (2008) Plastid evolution. Annu Rev Plant Biol 59:491–517. https://doi.org/10.1146/annurev.arplant.59.032607.092915

Greiner S, Golczyk H, Malinova I, Pellizzer T, Bock R, Börner T et al (2020) Chloroplast nucleoids are highly dynamic in ploidy, number, and structure during angiosperm leaf development. Plant J 102(4):730–746. https://doi.org/10.1111/tpj.14658

Gutiérrez CL, Gimpel J, Escobar C, Marshall SH, Henríquez V (2012) Chloroplast genetic tool for the green microalgae *Haematococcus pluvialis* (*Chlorophyceae*, *Volvocales*). J Phycol 48(4):976–983. https://doi.org/10.1111/j.1529-8817.2012.01178.x

Hajdukiewicz PT, Allison LA, Maliga P (1997) The two RNA polymerases encoded by the nuclear and the plastid compartments transcribe distinct groups of genes in tobacco plastids. EMBO J 16(13):4041–4048. https://doi.org/10.1093/emboj/16.13.4041

Hajdukiewicz PT, Gilbertson L, Staub JM (2001) Multiple pathways for Cre/lox-mediated recombination in plastids. Plant J 27(2):161–170. https://doi.org/10.1046/j.1365-313x.2001.01067.x

Hammani K, Cook WB, Barkan A (2012) RNA binding and RNA remodeling activities of the half-a-tetratricopeptide (HAT) protein HCF107 underlie its effects on gene expression. Proc Natl Acad Sci USA 109(15):5651–5656. https://doi.org/10.1073/pnas.1200318109

Hayashi K, Shiina T, Ishii N, Iwai K, Ishizaki Y, Morikawa K et al (2003) A role of the −35 element in the initiation of transcription at *psbA* promoter in tobacco plastids. Plant Cell Physiol 44(3):334–341. https://doi.org/10.1093/pcp/pcg041

Hayes R, Kudla J, Schuster G, Gabay L, Maliga P, Gruissem W (1996) Chloroplast mRNA 3′-end processing by a high molecular weight protein complex is regulated by nuclear encoded RNA binding proteins. EMBO J 15(5):1132–1141

Hirata N, Yonekura D, Yanagisawa S, Iba K (2004) Possible involvement of the 5′-flanking region and the 5'UTR of plastid *accD* gene in NEP-dependent transcription. Plant Cell Physiol 45(2):176–186. https://doi.org/10.1093/pcp/pch021

Hirose T, Sugiura M (1996) *Cis*-acting elements and *trans*-acting factors for accurate translation of chloroplast *psbA* mRNAs: development of an *in vitro* translation system from tobacco chloroplasts. EMBO J 15(7):1687–1695

Hirose T, Sugiura M (1997) Both RNA editing and RNA cleavage are required for translation of tobacco chloroplast *ndhD* mRNA: a possible regulatory mechanism for the expression of a chloroplast operon consisting of functionally unrelated genes. EMBO J 16(22):6804–6811. https://doi.org/10.1093/emboj/16.22.6804

Hirose T, Sugiura M (2004a) Functional Shine-Dalgarno-like sequences for translational initiation of chloroplast mRNAs. Plant Cell Physiol 45(1):114–117. https://doi.org/10.1093/pcp/pch002

Hirose T, Sugiura M (2004b) Multiple elements required for translation of plastid *atpB* mRNA lacking the Shine-Dalgarno sequence. Nucleic Acids Res 32(11):3503–3510. https://doi.org/10.1093/nar/gkh682

Hirose T, Kusumegi T, Sugiura M (1998) Translation of tobacco chloroplast *rps14* mRNA depends on a Shine-Dalgarno-like sequence in the 5′-untranslated region but not on internal RNA editing in the coding region. FEBS Lett 430(3):257–260. https://doi.org/10.1016/s0014-5793(98)00673-5

Hou BK, Zhou YH, Wan LH, Zhang ZL, Shen GF, Chen ZH et al (2003) Chloroplast transformation in oilseed rape. Transgenic Res 12(1):111–114. https://doi.org/10.1023/a:1022180315462

Howe CJ, Barbrook AC, Nisbet RE, Lockhart PJ, Larkum AW (2008a) The origin of plastids. Philos Trans R Soc Lond Ser B Biol Sci 363(1504):2675–2685. https://doi.org/10.1098/rstb.2008.0050

Howe CJ, Nisbet RE, Barbrook AC (2008b) The remarkable chloroplast genome of dinoflagellates. J Exp Bot 59(5):1035–1045. https://doi.org/10.1093/jxb/erm292

Huang FC, Klaus SM, Herz S, Zou Z, Koop HU, Golds TJ (2002) Efficient plastid transformation in tobacco using the *aphA-6* gene and kanamycin selection. Mol Gen Genomics 268(1):19–27. https://doi.org/10.1007/s00438-002-0738-6

Iamtham S, Day A (2000) Removal of antibiotic resistance genes from transgenic tobacco plastids. Nat Biotechnol 18(11):1172–1176. https://doi.org/10.1038/81161

Isono K, Shimizu M, Yoshimoto K, Niwa Y, Satoh K, Yokota A et al (1997) Leaf-specifically expressed genes for polypeptides destined for chloroplasts with domains of sigma 70 fac-

tors of bacterial RNA polymerases in *Arabidopsis thaliana*. Proc Natl Acad Sci USA 94(26):14948–14953. https://doi.org/10.1073/pnas.94.26.14948
Jakubiec A, Sarokina A, Choinard S, Vlad F, Malcuit I, Sorokin AP (2021) Replicating minichromosomes as a new tool for plastid genome engineering. Nat Plants 7(7):932–941. https://doi.org/10.1038/s41477-021-00940-y
Jarvis P, López-Juez E (2013) Biogenesis and homeostasis of chloroplasts and other plastids. Nat Rev Mol Cell Biol 14(12):787–802. https://doi.org/10.1038/nrm3702
Jeske H, Lütgemeier M, Preiss W (2001) DNA forms indicate rolling circle and recombination-dependent replication of Abutilon mosaic virus. EMBO J 20(21):6158–6167. https://doi.org/10.1093/emboj/20.21.6158
Jin S, Daniell H (2015) The engineered chloroplast genome just got smarter. Trends Plant Sci 20(10):622–640. https://doi.org/10.1016/j.tplants.2015.07.004
Kapoor S, Sugiura M (1999) Identification of two essential sequence elements in the nonconsensus type II *PatpB-290* plastid promoter by using plastid transcription extracts from cultured tobacco BY-2 cells. Plant Cell 11(9):1799–1810. https://doi.org/10.1105/tpc.11.9.1799
Kaushal C, Abdin MZ, Kumar S (2020) Chloroplast genome transformation of medicinal plant *Artemisia annua*. Plant Biotechnol J 18(11):2155–2157. https://doi.org/10.1111/pbi.13379
Khan MS, Maliga P (1999) Fluorescent antibiotic resistance marker for tracking plastid transformation in higher plants. Nat Biotechnol 17(9):910–915. https://doi.org/10.1038/12907
Kim J, Mullet JE (1994) Ribosome-binding sites on chloroplast *rbcL* and *psbA* mRNAs and light-induced initiation of D1 translation. Plant Mol Biol 25(3):437–448. https://doi.org/10.1007/bf00043872
Klaff P, Gruissem W (1995) A 43 kD light-regulated chloroplast RNA-binding protein interacts with the *psbA* 5′ non-translated leader RNA. Photosynth Res 46(1–2):235–248. https://doi.org/10.1007/bf00020436
Klaus SM, Huang FC, Eibl C, Koop HU, Golds TJ (2003) Rapid and proven production of transplastomic tobacco plants by restoration of pigmentation and photosynthesis. Plant J 35(6):811–821. https://doi.org/10.1046/j.1365-313x.2003.01838.x
Klaus SM, Huang FC, Golds TJ, Koop HU (2004) Generation of marker-free plastid transformants using a transiently cointegrated selection gene. Nat Biotechnol 22(2):225–229. https://doi.org/10.1038/nbt933
Kleffmann T, Russenberger D, von Zychlinski A, Christopher W, Sjölander K, Gruissem W et al (2004) The *Arabidopsis thaliana* chloroplast proteome reveals pathway abundance and novel protein functions. Curr Biol 14(5):354–362. https://doi.org/10.1016/j.cub.2004.02.039
Kode V, Mudd EA, Iamtham S, Day A (2006) Isolation of precise plastid deletion mutants by homology-based excision: a resource for site-directed mutagenesis, multi-gene changes and high-throughput plastid transformation. Plant J 46(5):901–909. https://doi.org/10.1111/j.1365-313X.2006.02736.x
Kong F, Zhao H, Liu W, Li N, Mao Y (2017) Construction of plastid expression vector and development of genetic transformation system for the seaweed *Pyropia yezoensis*. Mar Biotechnol (NY) 19(2):147–156. https://doi.org/10.1007/s10126-017-9736-x
Koop HU, Steinmüller K, Wagner H, Rössler C, Eibl C, Sacher L (1996) Integration of foreign sequences into the tobacco plastome via polyethylene glycol-mediated protoplast transformation. Planta 199(2):193–201. https://doi.org/10.1007/bf00196559
Kota M, Daniell H, Varma S, Garczynski SF, Gould F, Moar WJ (1999) Overexpression of the *Bacillus thuringiensis* (Bt) Cry2Aa2 protein in chloroplasts confers resistance to plants against susceptible and Bt-resistant insects. Proc Natl Acad Sci USA 96(5):1840–1845. https://doi.org/10.1073/pnas.96.5.1840
Kota S, Lakkam R, Kasula K, Narra M, Qiang H, Rao Allini V et al (2019) Construction of a species-specific vector for improved plastid transformation efficiency in *Capsicum annuum* L. 3 Biotech 9(6):226. https://doi.org/10.1007/s13205-019-1747-z
Koumandou VL, Howe CJ (2007) The copy number of chloroplast gene minicircles changes dramatically with growth phase in the dinoflagellate *Amphidinium operculatum*. Protist 158(1):89–103. https://doi.org/10.1016/j.protis.2006.08.003

Krupinska K, Melonek J, Krause K (2013) New insights into plastid nucleoid structure and functionality. Planta 237(3):653–664. https://doi.org/10.1007/s00425-012-1817-5

Kumar AU, Ling APK (2021) Gene introduction approaches in chloroplast transformation and its applications. J Genet Eng Biotechnol 19(1):148. https://doi.org/10.1186/s43141-021-00255-7

Kumar S, Dhingra A, Daniell H (2004a) Plastid-expressed betaine aldehyde dehydrogenase gene in carrot cultured cells, roots, and leaves confers enhanced salt tolerance. Plant Physiol 136(1):2843–2854. https://doi.org/10.1104/pp.104.045187

Kumar S, Dhingra A, Daniell H (2004b) Stable transformation of the cotton plastid genome and maternal inheritance of transgenes. Plant Mol Biol 56(2):203–216. https://doi.org/10.1007/s11103-004-2907-y

Kumari U, Singh R, Ray T, Rana S, Saha P, Malhotra K et al (2019) Validation of leaf enzymes in the detergent and textile industries: launching of a new platform technology. Plant Biotechnol J 17(6):1167–1182. https://doi.org/10.1111/pbi.13122

Kuroda H, Maliga P (2001a) Complementarity of the 16S rRNA penultimate stem with sequences downstream of the AUG destabilizes the plastid mRNAs. Nucleic Acids Res 29(4):970–975. https://doi.org/10.1093/nar/29.4.970

Kuroda H, Maliga P (2001b) Sequences downstream of the translation initiation codon are important determinants of translation efficiency in chloroplasts. Plant Physiol 125(1):430–436. https://doi.org/10.1104/pp.125.1.430

Kuroda H, Maliga P (2002) Overexpression of the *clpP* 5′-untranslated region in a chimeric context causes a mutant phenotype, suggesting competition for a *clpP*-specific RNA maturation factor in tobacco chloroplasts. Plant Physiol 129(4):1600–1606. https://doi.org/10.1104/pp.004986

Kuroda H, Maliga P (2003) The plastid *clpP1* protease gene is essential for plant development. Nature 425(6953):86–89. https://doi.org/10.1038/nature01909

Kuroda H, Sugiura M (2014) Processing of the 5'-UTR and existence of protein factors that regulate translation of tobacco chloroplast *psbN* mRNA. Plant Mol Biol 86(6):585–593. https://doi.org/10.1007/s11103-014-0248-z

Kuroda H, Suzuki H, Kusumegi T, Hirose T, Yukawa Y, Sugiura M (2007) Translation of *psbC* mRNAs starts from the downstream GUG, not the upstream AUG, and requires the extended Shine-Dalgarno sequence in tobacco chloroplasts. Plant Cell Physiol 48(9):1374–1378. https://doi.org/10.1093/pcp/pcm097

Kwak SY, Lew TTS, Sweeney CJ, Koman VB, Wong MH, Bohmert-Tatarev K et al (2019) Chloroplast-selective gene delivery and expression in planta using chitosan-complexed single-walled carbon nanotube carriers. Nat Nanotechnol 14(5):447–455. https://doi.org/10.1038/s41565-019-0375-4

Kwon KC, Verma D, Jin S, Singh ND, Daniell H (2013) Release of proteins from intact chloroplasts induced by reactive oxygen species during biotic and abiotic stress. PLoS One 8(6):e67106. https://doi.org/10.1371/journal.pone.0067106

Kwon KC, Sherman A, Chang WJ, Kamesh A, Biswas M, Herzog RW et al (2018) Expression and assembly of largest foreign protein in chloroplasts: oral delivery of human FVIII made in lettuce chloroplasts robustly suppresses inhibitor formation in haemophilia A mice. Plant Biotechnol J 16(6):1148–1160. https://doi.org/10.1111/pbi.12859

Lambert TJ (2019) FPbase: a community-editable fluorescent protein database. Nat Methods 16(4):277–278. https://doi.org/10.1038/s41592-019-0352-8

Langbecker CL, Ye GN, Broyles DL, Duggan LL, Xu CW, Hajdukiewicz PT et al (2004) High-frequency transformation of undeveloped plastids in tobacco suspension cells. Plant Physiol 135(1):39–46. https://doi.org/10.1104/pp.103.035410

Lapidot M, Raveh D, Sivan A, Arad SM, Shapira M (2002) Stable chloroplast transformation of the unicellular red alga *Porphyridium species*. Plant Physiol 129(1):7–12. https://doi.org/10.1104/pp.011023

Laufs J, Traut W, Heyraud F, Matzeit V, Rogers SG, Schell J et al (1995) *In vitro* cleavage and joining at the viral origin of replication by the replication initiator protein of tomato yellow leaf curl virus. Proc Natl Acad Sci USA 92(9):3879–3883. https://doi.org/10.1073/pnas.92.9.3879

Lee SM, Kang K, Chung H, Yoo SH, Xu XM, Lee SB et al (2006) Plastid transformation in the monocotyledonous cereal crop, rice (*Oryza sativa*) and transmission of transgenes to their progeny. Mol Cells 21(3):401–410

Lelivelt CLC, McCabe MS, Newell CA, deSnoo CB, van Dun KMP, Birch-Machin I et al (2005) Stable plastid transformation in lettuce (*Lactuca sativa* L.). Plant Mol Biol 58(6):763–774. https://doi.org/10.1007/s11103-005-7704-8

Lerbs-Mache S (2011) Function of plastid sigma factors in higher plants: regulation of gene expression or just preservation of constitutive transcription? Plant Mol Biol 76(3–5):235–249. https://doi.org/10.1007/s11103-010-9714-4

Li W, Ruf S, Bock R (2011) Chloramphenicol acetyltransferase as selectable marker for plastid transformation. Plant Mol Biol 76(3–5):443–451. https://doi.org/10.1007/s11103-010-9678-4

Li Y, Wang R, Hu Z, Li H, Lu S, Zhang J et al (2016) Expression of a codon-optimized *dsdA* gene in tobacco plastids and rice nucleus confers D-serine tolerance. Front Plant Sci 7:640. https://doi.org/10.3389/fpls.2016.00640

Liere K, Maliga P (1999) *In vitro* characterization of the tobacco *rpoB* promoter reveals a core sequence motif conserved between phage-type plastid and plant mitochondrial promoters. EMBO J 18(1):249–257. https://doi.org/10.1093/emboj/18.1.249

Lin MT, Occhialini A, Andralojc PJ, Parry MA, Hanson MR (2014) A faster Rubisco with potential to increase photosynthesis in crops. Nature 513(7519):547–550. https://doi.org/10.1038/nature13776

Liu CW, Lin CC, Chen JJ, Tseng MJ (2007) Stable chloroplast transformation in cabbage (*Brassica oleracea* L. var. capitata L.) by particle bombardment. Plant Cell Rep 26(10):1733–1744. https://doi.org/10.1007/s00299-007-0374-z

Lozano-Durán R (2016) Geminiviruses for biotechnology: the art of parasite taming. New Phytol 210(1):58–64. https://doi.org/10.1111/nph.13564

Lu Y, Rijzaani H, Karcher D, Ruf S, Bock R (2013) Efficient metabolic pathway engineering in transgenic tobacco and tomato plastids with synthetic multigene operons. Proc Natl Acad Sci USA 110(8):E623–E632. https://doi.org/10.1073/pnas.1216898110

Lutz KA, Azhagiri AK, Tungsuchat-Huang T, Maliga P (2007) A guide to choosing vectors for transformation of the plastid genome of higher plants. Plant Physiol 145(4):1201–1210. https://doi.org/10.1104/pp.107.106963

Macedo-Osorio KS, Pérez-España VH, Garibay-Orijel C, Guzmán-Zapata D, Durán-Figueroa NV, Badillo-Corona JA (2018) Intercistronic expression elements (IEE) from the chloroplast of *Chlamydomonas reinhardtii* can be used for the expression of foreign genes in synthetic operons. Plant Mol Biol 98(4–5):303–317. https://doi.org/10.1007/s11103-018-0776-z

Maliga P, Svab Z (2011) Engineering the plastid genome of *Nicotiana sylvestris*, a diploid model species for plastid genetics. Methods Mol Biol 701:37–50. https://doi.org/10.1007/978-1-61737-957-4_2

Martin W, Rujan T, Richly E, Hansen A, Cornelsen S, Lins T et al (2002) Evolutionary analysis of Arabidopsis, cyanobacterial, and chloroplast genomes reveals plastid phylogeny and thousands of cyanobacterial genes in the nucleus. Proc Natl Acad Sci USA 99(19):12246–12251. https://doi.org/10.1073/pnas.182432999

McFadden GI (2014) Origin and evolution of plastids and photosynthesis in eukaryotes. Cold Spring Harb Perspect Biol 6(4):a016105. https://doi.org/10.1101/cshperspect.a016105

Mertens N, Remaut E, Fiers W (1996) Increased stability of phage T7g10 mRNA is mediated by either a 5′- or a 3′-terminal stem-loop structure. Biol Chem 377(12):811–817. https://doi.org/10.1515/bchm3.1996.377.12.811

Miki B, McHugh S (2004) Selectable marker genes in transgenic plants: applications, alternatives and biosafety. J Biotechnol 107(3):193–232. https://doi.org/10.1016/j.jbiotec.2003.10.011

Min SR, Davarpanah SJ, Jung SH, Park Y-i, Liu JR, Jeong W-J (2015a) An episomal vector system for plastid transformation in higher plants. Plant Biotechnol Reports 9(6):443–449. https://doi.org/10.1007/s11816-015-0381-4

Min SR, Jung SH, Liu JR, Jeong W-J (2015b) The fate of extrachromosomal DNAs in the progeny of plastid-transformed tobacco plants. Plant Biotechnol Reports 9(6):431–442. https://doi.org/10.1007/s11816-015-0380-5

Miyagi T, Kapoor S, Sugita M, Sugiura M (1998) Transcript analysis of the tobacco plastid operon *rps2/atpI/H/F/A* reveals the existence of a non-consensus type II (NCII) promoter upstream of the *atpI* coding sequence. Mol Gen Genet 257(3):299–307. https://doi.org/10.1007/s004380050651
Morley SA, Ahmad N, Nielsen BL (2019) Plant organelle genome replication. Plants (Basel) 8(10). https://doi.org/10.3390/plants8100358
Mower JP, Vickrey TL (2018) Chapter nine – structural diversity among plastid genomes of land plants. In: Chaw SM, Jansen RK (eds) Advances in botanical research, vol 85. Academic, pp 263–292
Muralikrishna N, Srinivas K, Kumar KB, Sadanandam A (2016) Stable plastid transformation in *Scoparia dulcis* L. Physiol Mol Biol Plants 22(4):575–581. https://doi.org/10.1007/s12298-016-0386-7
Murphy E (1985) Nucleotide sequence of a spectinomycin adenyltransferase AAD(9) determinant from *Staphylococcus aureus* and its relationship to AAD(3″) (9). Mol Gen Genet 200(1):33–39. https://doi.org/10.1007/bf00383309
Nap JP, Bijvoet J, Stiekema WJ (1992) Biosafety of kanamycin-resistant transgenic plants. Transgenic Res 1(6):239–249. https://doi.org/10.1007/bf02525165
Narra M, Kota S, Velivela Y, Ellendula R, Allini VR, Abbagani S (2018) Construction of chloroplast transformation vector and its functional evaluation in *Momordica charantia* L. 3 Biotech 8(3):140. https://doi.org/10.1007/s13205-018-1160-z
Newman SM, Boynton JE, Gillham NW, Randolph-Anderson BL, Johnson AM, Harris EH (1990) Transformation of chloroplast ribosomal RNA genes in Chlamydomonas: molecular and genetic characterization of integration events. Genetics 126(4):875–888. https://doi.org/10.1093/genetics/126.4.875
Nishimura Y (2023) Plastid nucleoids: insights into their shape and dynamics. Plant Cell Physiol. https://doi.org/10.1093/pcp/pcad090
Nugent GD, Coyne S, Nguyen TT, Kavanagh TA, Dix PJ (2006) Nuclear and plastid transformation of *Brassica oleracea* var. botrytis (cauliflower) using PEG-mediated uptake of DNA into protoplasts. Plant Sci 170(1):135–142. https://doi.org/10.1016/j.plantsci.2005.08.020
O'Neill C, Horváth GV, Horváth E, Dix PJ, Medgyesy P (1993) Chloroplast transformation in plants: polyethylene glycol (PEG) treatment of protoplasts is an alternative to biolistic delivery systems. Plant J 3(5):729–738
Occhialini A, Lenaghan SC (2023) Plastid engineering using episomal DNA. Plant Cell Rep 42(7):1125–1132. https://doi.org/10.1007/s00299-023-03020-x
Occhialini A, Piatek AA, Pfotenhauer AC, Frazier TP, Stewart CN Jr, Lenaghan SC (2019) MoChlo: a versatile, modular cloning toolbox for chloroplast biotechnology. Plant Physiol 179(3):943–957. https://doi.org/10.1104/pp.18.01220
Occhialini A, Pfotenhauer AC, Frazier TP, Li L, Harbison SA, Lail AJ et al (2020) Generation, analysis, and transformation of macro-chloroplast potato (*Solanum tuberosum*) lines for chloroplast biotechnology. Sci Rep 10(1):21144. https://doi.org/10.1038/s41598-020-78237-x
Occhialini A, Pfotenhauer AC, Li L, Harbison SA, Lail AJ, Burris JN et al (2022) Mini-synplastomes for plastid genetic engineering. Plant Biotechnol J 20(2):360–373. https://doi.org/10.1111/pbi.13717
Occhialini A, Pfotenhauer AC, Daniell H, Stewart CN Jr, Lenaghan SC (2023) Genetic engineering of potato (*Solanum tuberosum*) chloroplasts using the small synthetic plastome "mini-synplastome". In: Yang B, Harwood W, Que Q (eds) Plant Genome Engineering: Methods and Protocols. Springer US, New York, NY, pp 73–92
Oey M, Lohse M, Kreikemeyer B, Bock R (2009) Exhaustion of the chloroplast protein synthesis capacity by massive expression of a highly stable protein antibiotic. Plant J 57(3):436–445. https://doi.org/10.1111/j.1365-313X.2008.03702.x
Okumura S, Sawada M, Park YW, Hayashi T, Shimamura M, Takase H et al (2006) Transformation of poplar (*Populus alba*) plastids and expression of foreign proteins in tree chloroplasts. Transgenic Res 15(5):637–646. https://doi.org/10.1007/s11248-006-9009-3
Palomar VM, Jaksich S, Fujii S, Kuciński J, Wierzbicki AT (2022) High-resolution map of plastid-encoded RNA polymerase binding patterns demonstrates a major role of transcription in chloroplast gene expression. Plant J 111(4):1139–1151. https://doi.org/10.1111/tpj.15882

Park J, Yan G, Kwon KC, Liu M, Gonnella PA, Yang S et al (2020) Oral delivery of novel human IGF-1 bioencapsulated in lettuce cells promotes musculoskeletal cell proliferation, differentiation and diabetic fracture healing. Biomaterials 233:119591. https://doi.org/10.1016/j.biomaterials.2019.119591

Pfotenhauer AC, Occhialini A, Nguyen MA, Scott H, Dice LT, Harbison SA et al (2022) Building the plant SynBio toolbox through combinatorial analysis of DNA regulatory elements. ACS Synth Biol 11(8):2741–2755. https://doi.org/10.1021/acssynbio.2c00147

Philips JG, Martin-Avila E, Robold AV (2022) Horizontal gene transfer from genetically modified plants – regulatory considerations. Front Bioeng Biotechnol 10:971402. https://doi.org/10.3389/fbioe.2022.971402

Plader W, Sugiura M (2003) The Shine-Dalgarno-like sequence is a negative regulatory element for translation of tobacco chloroplast *rps2* mRNA: an additional mechanism for translational control in chloroplasts. Plant J 34(3):377–382. https://doi.org/10.1046/j.1365-313x.2003.01732.x

do Prado PFV, Ahrens FM, Liebers M, Ditz N, Braun HP, Pfannschmidt T et al (2024) Structure of the multi-subunit chloroplast RNA polymerase. Mol Cell 84(5):910–925.e915. https://doi.org/10.1016/j.molcel.2024.02.003

Przibilla E, Heiss S, Johanningmeier U, Trebst A (1991) Site-specific mutagenesis of the D1 subunit of photosystem II in wild-type Chlamydomonas. Plant Cell 3(2):169–174. https://doi.org/10.1105/tpc.3.2.169

Rascón-Cruz Q, González-Barriga CD, Iglesias-Figueroa BF, Trejo-Muñoz JC, Siqueiros-Cendón T, Sinagawa-García SR et al (2021) Plastid transformation: advances and challenges for its implementation in agricultural crops. Electron J Biotechnol 51:95–109. https://doi.org/10.1016/j.ejbt.2021.03.005

Redding K, MacMillan F, Leibl W, Brettel K, Hanley J, Rutherford AW et al (1998) A systematic survey of conserved histidines in the core subunits of Photosystem I by site-directed mutagenesis reveals the likely axial ligands of P700. EMBO J 17(1):50–60. https://doi.org/10.1093/emboj/17.1.50

Remacle C, Cline S, Boutaffala L, Gabilly S, Larosa V, Barbieri MR et al (2009) The ARG9 gene encodes the plastid-resident N-acetyl ornithine aminotransferase in the green alga *Chlamydomonas reinhardtii*. Eukaryot Cell 8(9):1460–1463. https://doi.org/10.1128/ec.00108-09

Rojas M, Yu Q, Williams-Carrier R, Maliga P, Barkan A (2019) Engineered PPR proteins as inducible switches to activate the expression of chloroplast transgenes. Nat Plants 5(5):505–511. https://doi.org/10.1038/s41477-019-0412-1

Rott R, Liveanu V, Drager RG, Stern DB, Schuster G (1998) The sequence and structure of the 3′-untranslated regions of chloroplast transcripts are important determinants of mRNA accumulation and stability. Plant Mol Biol 36(2):307–314. https://doi.org/10.1023/a:1005943701253

Ruf S, Hermann M, Berger IJ, Carrer H, Bock R (2001) Stable genetic transformation of tomato plastids and expression of a foreign protein in fruit. Nat Biotechnol 19(9):870–875. https://doi.org/10.1038/nbt0901-870

Ruf S, Karcher D, Bock R (2007) Determining the transgene containment level provided by chloroplast transformation. Proc Natl Acad Sci USA 104(17):6998–7002. https://doi.org/10.1073/pnas.0700008104

Ruf S, Forner J, Hasse C, Kroop X, Seeger S, Schollbach L et al (2019) High-efficiency generation of fertile transplastomic Arabidopsis plants. Nat Plants 5(3):282–289. https://doi.org/10.1038/s41477-019-0359-2

Ruhlman T, Verma D, Samson N, Daniell H (2010) The role of heterologous chloroplast sequence elements in transgene integration and expression. Plant Physiol 152(4):2088–2104. https://doi.org/10.1104/pp.109.152017

Schelkunov MI, Shtratnikova VY, Nuraliev MS, Selosse MA, Penin AA, Logacheva MD (2015) Exploring the limits for reduction of plastid genomes: a case study of the *Mycoheterotrophic orchids*, *Epipogium aphyllum* and *Epipogium roseum*. Genome Biol Evol 7(4):1179–1191. https://doi.org/10.1093/gbe/evv019

Schindel HS, Piatek AA, Stewart CN Jr, Lenaghan SC (2018) The plastid genome as a chassis for synthetic biology-enabled metabolic engineering: players in gene expression. Plant Cell Rep 37(10):1419–1429. https://doi.org/10.1007/s00299-018-2323-4

Schmidt JA, Richter LV, Condoluci LA, Ahner BA (2021) Mitigation of deleterious phenotypes in chloroplast-engineered plants accumulating high levels of foreign proteins. Biotechnol Biofuels 14(1):42. https://doi.org/10.1186/s13068-021-01893-2

Serino G, Maliga P (1997) A negative selection scheme based on the expression of cytosine deaminase in plastids. Plant J 12(3):697–701. https://doi.org/10.1046/j.1365-313x.1997.00697.x

Shiina T, Allison L, Maliga P (1998) *rbcL* transcript levels in tobacco plastids are independent of light: reduced dark transcription rate is compensated by increased mRNA stability. Plant Cell 10(10):1713–1722. https://doi.org/10.1105/tpc.10.10.1713

Shikanai T, Shimizu K, Ueda K, Nishimura Y, Kuroiwa T, Hashimoto T (2001) The chloroplast *clpP* gene, encoding a proteolytic subunit of ATP-dependent protease, is indispensable for chloroplast development in tobacco. Plant Cell Physiol 42(3):264–273. https://doi.org/10.1093/pcp/pce031

Sierra J, Escobar-Tovar L, Leon P (2023) Plastids: diving into their diversity, their functions, and their role in plant development. J Exp Bot 74(8):2508–2526. https://doi.org/10.1093/jxb/erad044

Silhavy D, Maliga P (1998) Mapping of promoters for the nucleus-encoded plastid RNA polymerase (NEP) in the iojap maize mutant. Curr Genet 33(5):340–344. https://doi.org/10.1007/s002940050345

Singh AK, Verma SS, Bansal KC (2010) Plastid transformation in eggplant (*Solanum melongena* L.). Transgenic Res 19(1):113–119. https://doi.org/10.1007/s11248-009-9290-z

Skarjinskaia M, Svab Z, Maliga P (2003) Plastid transformation in *Lesquerella fendleri*, an oilseed Brassicacea. Transgenic Res 12(1):115–122. https://doi.org/10.1023/a:1022110402302

Soll J, Schleiff E (2004) Protein import into chloroplasts. Nat Rev Mol Cell Biol 5(3):198–208. https://doi.org/10.1038/nrm1333

Sriraman P, Silhavy D, Maliga P (1998) The phage-type *PclpP-53* plastid promoter comprises sequences downstream of the transcription initiation site. Nucleic Acids Res 26(21):4874–4879. https://doi.org/10.1093/nar/26.21.4874

Staub JM (2021) Transformation of the plastid genome in tobacco suspension cell cultures. Methods Mol Biol 2317:167–175. https://doi.org/10.1007/978-1-0716-1472-3_8

Staub JM, Maliga P (1992) Long regions of homologous DNA are incorporated into the tobacco plastid genome by transformation. Plant Cell 4(1):39–45. https://doi.org/10.1105/tpc.4.1.39

Staub JM, Maliga P (1993) Accumulation of D1 polypeptide in tobacco plastids is regulated via the untranslated region of the *psbA* mRNA. EMBO J 12(2):601–606. https://doi.org/10.1002/j.1460-2075.1993.tb05692.x

Staub JM, Maliga P (1994a) Translation of *psbA* mRNA is regulated by light via the 5′-untranslated region in tobacco plastids. Plant J 6(4):547–553. https://doi.org/10.1046/j.1365-313x.1994.6040547.x

Staub JM, Maliga P (1994b) Extrachromosomal elements in tobacco plastids. Proc Natl Acad Sci USA 91(16):7468–7472. https://doi.org/10.1073/pnas.91.16.7468

Staub JM, Maliga P (1995) Marker rescue from the *Nicotiana tabacum* plastid genome using a plastid/*Escherichia coli* shuttle vector. Mol Gen Genet 249(1):37–42. https://doi.org/10.1007/bf00290233

Sugita M (2021) Plastid transformation in *Physcomitrium* (*Physcomitrella*) *patens*: an update. Methods Mol Biol 2317:321–331. https://doi.org/10.1007/978-1-0716-1472-3_19

Sugiura C, Sugita M (2004) Plastid transformation reveals that moss tRNA(Arg)-CCG is not essential for plastid function. Plant J 40(2):314–321. https://doi.org/10.1111/j.1365-313X.2004.02202.x

Suzuki JY, Maliga P (2000) Engineering of the *rpl23* gene cluster to replace the plastid RNA polymerase alpha subunit with the *Escherichia coli* homologue. Curr Genet 38(4):218–225. https://doi.org/10.1007/s002940000141

Suzuki JY, Sriraman P, Svab Z, Maliga P (2003) Unique architecture of the plastid ribosomal RNA operon promoter recognized by the multisubunit RNA polymerase in tobacco and other higher plants. Plant Cell 15(1):195–205. https://doi.org/10.1105/tpc.007914

Svab Z, Maliga P (1993) High-frequency plastid transformation in tobacco by selection for a chimeric *aadA* gene. Proc Natl Acad Sci USA 90(3):913–917. https://doi.org/10.1073/pnas.90.3.913

Svab Z, Hajdukiewicz P, Maliga P (1990) Stable transformation of plastids in higher plants. Proc Natl Acad Sci USA 87(21):8526–8530. https://doi.org/10.1073/pnas.87.21.8526

Tabatabaei I, Ruf S, Bock R (2017) A bifunctional aminoglycoside acetyltransferase/phosphotransferase conferring tobramycin resistance provides an efficient selectable marker for plastid transformation. Plant Mol Biol 93(3):269–281. https://doi.org/10.1007/s11103-016-0560-x

Tanaka M, Obokata J, Chunwongse J, Shinozaki K, Sugiura M (1987) Rapid splicing and stepwise processing of a transcript from the *psbB* operon in tobacco chloroplasts: determination of the intron sites in *petB* and *petD*. Mol Gen Genet 209(3):427–431. https://doi.org/10.1007/bf00331145

Tangphatsornruang S, Birch-Machin I, Newell CA, Gray JC (2011) The effect of different 3′ untranslated regions on the accumulation and stability of transcripts of a *GFP* transgene in chloroplasts of transplastomic tobacco. Plant Mol Biol 76(3–5):385–396. https://doi.org/10.1007/s11103-010-9689-1

Tarasenko VI, Katyshev AI, Yakovleva TV, Garnik EY, Chernikova VV, Konstantinov YM et al (2016) RPOTmp, an Arabidopsis RNA polymerase with dual targeting, plays an important role in mitochondria, but not in chloroplasts. J Exp Bot 67(19):5657–5669. https://doi.org/10.1093/jxb/erw327

Thomas MR, Rose RJ (1983) Plastid number and plastid structural changes associated with tobacco mesophyll protoplast culture and plant regeneration. Planta 158(4):329–338. https://doi.org/10.1007/bf00397335

Timmis JN, Ayliffe MA, Huang CY, Martin W (2004) Endosymbiotic gene transfer: organelle genomes forge eukaryotic chromosomes. Nat Rev Genet 5(2):123–135. https://doi.org/10.1038/nrg1271

Valkov VT, Gargano D, Manna C, Formisano G, Dix PJ, Gray JC et al (2011) High efficiency plastid transformation in potato and regulation of transgene expression in leaves and tubers by alternative 5′ and 3′ regulatory sequences. Transgenic Res 20(1):137–151. https://doi.org/10.1007/s11248-010-9402-9

Vergara-Cruces Á, Pramanick I, Pearce D, Vogirala VK, Byrne MJ, Low JKK et al (2024) Structure of the plant plastid-encoded RNA polymerase. Cell 187(5):1145–1159.e1121. https://doi.org/10.1016/j.cell.2024.01.036

Verma D, Daniell H (2007) Chloroplast vector systems for biotechnology applications. Plant Physiol 145(4):1129–1143. https://doi.org/10.1104/pp.107.106690

Verma D, Samson NP, Koya V, Daniell H (2008) A protocol for expression of foreign genes in chloroplasts. Nat Protoc 3(4):739–758. https://doi.org/10.1038/nprot.2007.522

Wang H, Mukherjee S, Yi J, Banerjee P, Chen Q, Zhou S (2017) Biocompatible chitosan-carbon dot hybrid nanogels for NIR-imaging-guided synergistic photothermal-chemo therapy. ACS Appl Mater Interfaces 9(22):18639–18649. https://doi.org/10.1021/acsami.7b06062

Wei Z, Liu Y, Lin C, Wang Y, Cai Q a, Dong Y et al (2011) Transformation of alfalfa chloroplasts and expression of green fluorescent protein in a forage crop. Biotechnol Lett 33(12):2487–2494. https://doi.org/10.1007/s10529-011-0709-2

Wong MH, Misra RP, Giraldo JP, Kwak SY, Son Y, Landry MP et al (2016) Lipid exchange envelope penetration (LEEP) of nanoparticles for plant engineering: a universal localization mechanism. Nano Lett 16(2):1161–1172. https://doi.org/10.1021/acs.nanolett.5b04467

Xie G, Allison LA (2002) Sequences upstream of the YRTA core region are essential for transcription of the tobacco *atpB* NEP promoter in chloroplasts *in vivo*. Curr Genet 41(3):176–182. https://doi.org/10.1007/s00294-002-0293-z

Xie WH, Zhu CC, Zhang NS, Li DW, Yang WD, Liu JS et al (2014) Construction of novel chloroplast expression vector and development of an efficient transformation system for the diatom

Phaeodactylum tricornutum. Mar Biotechnol (NY) 16(5):538–546. https://doi.org/10.1007/s10126-014-9570-3

Xiong HB, Pan HM, Long QY, Wang ZY, Qu WT, Mei T et al (2022) AtNusG, a chloroplast nucleoid protein of bacterial origin linking chloroplast transcriptional and translational machineries, is required for proper chloroplast gene expression in *Arabidopsis thaliana*. Nucleic Acids Res 50(12):6715–6734. https://doi.org/10.1093/nar/gkac501

Yabuta Y, Tanaka H, Yoshimura S, Suzuki A, Tamoi M, Maruta T et al (2013) Improvement of vitamin E quality and quantity in tobacco and lettuce by chloroplast genetic engineering. Transgenic Res 22(2):391–402. https://doi.org/10.1007/s11248-012-9656-5

Yau YY, Stewart CN Jr (2013) Less is more: strategies to remove marker genes from transgenic plants. BMC Biotechnol 13:36. https://doi.org/10.1186/1472-6750-13-36

Ye GN, Colburn SM, Xu CW, Hajdukiewicz PT, Staub JM (2003) Persistence of unselected transgenic DNA during a plastid transformation and segregation approach to herbicide resistance. Plant Physiol 133(1):402–410. https://doi.org/10.1104/pp.103.021949

Young RE, Purton S (2016) Codon reassignment to facilitate genetic engineering and biocontainment in the chloroplast of *Chlamydomonas reinhardtii*. Plant Biotechnol J 14(5):1251–1260. https://doi.org/10.1111/pbi.12490

Yu Q, Lutz KA, Maliga P (2017) Efficient plastid transformation in Arabidopsis. Plant Physiol 175(1):186–193. https://doi.org/10.1104/pp.17.00857

Yu Q, Barkan A, Maliga P (2019) Engineered RNA-binding protein for transgene activation in non-green plastids. Nat Plants 5(5):486–490. https://doi.org/10.1038/s41477-019-0413-0

Yukawa M, Kuroda H, Sugiura M (2007) A new *in vitro* translation system for non-radioactive assay from tobacco chloroplasts: effect of pre-mRNA processing on translation *in vitro*. Plant J 49(2):367–376. https://doi.org/10.1111/j.1365-313X.2006.02948.x

Zhang J, Ruf S, Hasse C, Childs L, Scharff LB, Bock R (2012) Identification of *cis*-elements conferring high levels of gene expression in non-green plastids. Plant J 72(1):115–128. https://doi.org/10.1111/j.1365-313X.2012.05065.x

Zhang J, Khan SA, Hasse C, Ruf S, Heckel DG, Bock R (2015) Pest control. Full crop protection from an insect pest by expression of long double-stranded RNAs in plastids. Science 347(6225):991–994. https://doi.org/10.1126/science.1261680

Zhang Y, Tian L, Lu C (2023) Chloroplast gene expression: recent advances and perspectives. Plant Commun 4(5):100611. https://doi.org/10.1016/j.xplc.2023.100611

Zhou F, Karcher D, Bock R (2007) Identification of a plastid intercistronic expression element (IEE) facilitating the expression of stable translatable monocistronic mRNAs from operons. Plant J 52(5):961–972. https://doi.org/10.1111/j.1365-313X.2007.03261.x

Zhou F, Badillo-Corona JA, Karcher D, Gonzalez-Rabade N, Piepenburg K, Borchers AM et al (2008) High-level expression of human immunodeficiency virus antigens from the tobacco and tomato plastid genomes. Plant Biotechnol J 6(9):897–913. https://doi.org/10.1111/j.1467-7652.2008.00356.x

Zienkiewicz M, Krupnik T, Drożak A, Golke A, Romanowska E (2017) Transformation of the *Cyanidioschyzon merolae* chloroplast genome: prospects for understanding chloroplast function in extreme environments. Plant Mol Biol 93(1–2):171–183. https://doi.org/10.1007/s11103-016-0554-8

Zoschke R, Bock R (2018) Chloroplast translation: structural and functional organization, operational control, and regulation. Plant Cell 30(4):745–770. https://doi.org/10.1105/tpc.18.00016

Zou Z, Eibl C, Koop HU (2003) The stem-loop region of the tobacco *psbA* 5′UTR is an important determinant of mRNA stability and translation efficiency. Mol Gen Genomics 269(3):340–349. https://doi.org/10.1007/s00438-003-0842-2

Zoubenko OV, Allison LA, Svab Z, Maliga P (1994) Efficient targeting of foreign genes into the tobacco plastid genome. Nucleic Acids Res 22(19):3819–3824. https://doi.org/10.1093/nar/22.19.3819

Zubkot MK, Zubkot EI, van Zuilen K, Meyer P, Day A (2004) Stable transformation of petunia plastids. Transgenic Res 13(6):523–530. https://doi.org/10.1007/s11248-004-2374-x